NATO ASI Series

Advanced Science Institutes Series

A series presenting the results of activities sponsored by the NATO Science Committee, which aims at the dissemination of advanced scientific and technological knowledge, with a view to strengthening links between scientific communities.

The Series is published by an international board of publishers in conjunction with the NATO Scientific Affairs Division

A **Life Sciences** B **Physics**	Plenum Publishing Corporation London and New York
C **Mathematical and Physical Sciences** D **Behavioural and Social Sciences** E **Applied Sciences**	Kluwer Academic Publishers Dordrecht, Boston and London
F **Computer and Systems Sciences** G **Ecological Sciences** H **Cell Biology** I **Global Environmental Change**	Springer-Verlag Berlin Heidelberg New York London Paris Tokyo Hong Kong Barcelona Budapest

PARTNERSHIP SUB-SERIES

1. **Disarmament Technologies**	Kluwer Academic Publishers
2. **Environment**	Springer-Verlag
3. **High Technology**	Kluwer Academic Publishers
4. **Science and Technology Policy**	Kluwer Academic Publishers
5. **Computer Networking**	Kluwer Academic Publishers

The Partnership Sub-Series incorporates activities undertaken in collaboration with NATO's Cooperation Partners, the countries of the CIS and Central and Eastern Europe, in Priority Areas of concern to those countries.

NATO-PCO DATABASE

The electronic index to the NATO ASI Series provides full bibliographical references (with keywords and/or abstracts) to about 50000 contributions from international scientists published in all sections of the NATO ASI Series. Access to the NATO-PCO DATABASE compiled by the NATO Publication Coordination Office is possible in two ways:

- via online FILE 128 (NATO-PCO DATABASE) hosted by ESRIN, Via Galileo Galilei, I-00044 Frascati, Italy.
- via CD-ROM "NATO Science & Technology Disk" with user-friendly retrieval software in English, French and German (© WTV GmbH and DATAWARE Technologies Inc. 1992).

The CD-ROM can be ordered through any member of the Board of Publishers or through NATO-PCO, Overijse, Belgium.

Series I: Global Environmental Change, Vol. 36

Springer-Verlag Berlin Heidelberg GmbH

Diachronic Climatic Impacts on Water Resources

with Emphasis on the Mediterranean Region

Edited by

Andreas N. Angelakis
National Foundation for Agricultural Research
Institute of Iraklio
71110 Iraklio, Greece

Arie S. Issar
Ben-Gurion University of the Negev
The Jacob Blaustein Institute for Desert Research
Desert Hydrology Unit
Sede Boker Campus, 84990 Israel

Springer
Published in cooperation with NATO Scientific Affairs Division

Initiated during the NATO Advanced Research Workshop "Diachronic Climatic Impacts on Water Resources", held in Iraklio, Crete, October 17–23, 1993

Cataloging-in-Publication Data applied for

Die Deutsche Bibliothek - CIP-Einheitsaufnahme

Diachronic climatic impacts on water resources : with emphasis on the mediterranean region ; [proceedings of the NATO Advanced Research Workshop on Diachronic Climatic Impacts on Water Resources, held in Iraklio, Crete, October 17 - 23, 1993] / ed. by Andreas N. Angelakis ; Arie S. Issar. Publ. in cooperation with NATO Scientific Affairs Division. - Berlin ; Heidelberg ; New York ; Barcelona ; Budapest ; Hong Kong ; London ; Milan ; Paris ; Santa Clara ; Singapore ; Tokyo : Springer, 1996
(NATO ASI series : Ser. I, Global environmental change ; Vol. 36)

NE: Angelakēs, Andreas N. [Hrsg.]; Advanced Research Workshop on Diachronic Climatic Impacts on Water Resources <1993, Hērakleion>; NATO: NATO ASI series / I

ISBN 978-3-642-64686-7 ISBN 978-3-642-61084-4 (eBook)
DOI 10.1007/978-3-642-61084-4

Softcover reprint of the hardcover 1st edition 1996

Typesetting: Camera ready by authors/editors
Printed on acid-free paper
SPIN: 10083571 31/3137 - 5 4 3 2 1 0

PROLEGOMENA

"Στη μέση του θαμπού γιαλού ένα νησί είναι, η Κρήτη,
όμορφη και πολύκαρπη και θαλασσοκλεισμένη,
πόχει κατοίκους άπειρους κι έχει χωριά εννενήντα"
(Οδύσσεια, 19, 172-174).

"There is an island called Crete in the midst of the wine-dark sea,
a fair land and rich, begird with water,
and there in are innumerable men, and ninety cities"
(Odyssey, 19, 172-174)

Nearly three thousand years ago, Homer related that Crete was heavily forested and blessed with fertile irrigated soil that supported a teeming population. Even earlier, in the Minoan period (*ca.* 3000-1100 B.C.), a sustained cultural growth occurred which may be related to the advanced water management techniques practised in Crete at that time.

Since the beginning of Man's existence on Earth, adequacy of available water was as important as any other factor for man's survival and prosperity. It's not much different today. Assuring an adequate supply of water to an ever growing and developing population, is a world-wide problem. Even areas thought to be blessed with an abundance of available-water are recognizing water as a natural resource superior to all others. In certain areas of the world, such as the Mediterranean region, water has long been the recipient of intense planning and development. The impact of climatic changes is potentially one of man's most pressing environmental problems, posing major scientific, environmental, economic, social and potential challenges. Such changes in global climate between the present time and the middle of the twenty first century are likely to be dominated by the influence of global warming due to increasing concentrations of CO_2 and other greenhouse gases in the atmosphere. The latter among other things may well be related to water resources and the marine environment.

Many important economic and social decisions made today, such as those relating to water resources management, coastal engineering projects, urban and energy planning, nature conservation etc., are based on the assumption that past climate data provide a reliable guide to the future. This is no longer a safe assumption; climatic change must be considered, particularly in view of the current population explosion, the increasing utilization of coastal areas for industry, agriculture, fisheries and tourism. To assess the environmental problems associated with the marine environment and adjacent coastal areas and to identify suitable policy option and response measures, the potential impact of expected climatic changes must be considered.

The comprehensive reviews included in this volume were authored by the main lecturers of a NATO-Advanced Research Workshop held in Iraklio, Crete from 17 to 23 of October 1993. *Diachronic Climatic Impact on Water Resources* is written for civil and agricultural engineers, hydrologists, geologists, environmental scientists, and research scientists. The book is also useful as a reference to consulting engineers, agriculturists, environmentalists as well as students at colleges and universities studying environmental and water sciences.

Publication of a book of this scope and magnitude could be accomplished only with the help of many people, and our appreciation is gratefully offered to the authors of the chapters. Their contribution to the quality of this book is evident. Similarly, the assistance and contribution of the review committee members is gratefully acknowledged.

We would like to express our gratitude to the Scientific Affairs Division of the North Atlantic Treaty Organization (NATO) and to its Director, Dr. da Cunha for partially financing the implementation of the NATO-AR Workshop. Special appreciation is also due to the reviewers who provided valuable assistance and authoritative guidance in the review of this book. These are: (1) Prof. O. Davis, University of Arizona, USA, (2) Prof. E. Diamadopoulos, Technical University of Crete, Greece, (3) Prof. W. Dragoni, Perugia University, Italy, (4) Prof. J. Ganoulis, Aristotle University of Thessaloniki, Greece, (5) Dr. S. Jelgersma, Geological Survey, Haarlem, The Netherlands, (6) Prof. D. Metaxas, University of Ioannina, Greece, (7) Prof. M. Mimikou, National Technical University of Athens, Greece, (8) Prof. I. Minguez, ETSIA, Ciudad Universitaria, Madrid, Spain, (9) Prof. D. Monopolis, Technical University of Crete, Greece, (10) Dr. T. Nelsen, NOAA-AOML-OCD, Miami, Florida, USA, (11) Dr. N. Petit-Maire, CNRS - Luminy, Marseille, France, (12) Prof. M. Salgot, University of Barcelona, Spain, (13) Prof. S. Spyridakis, University of California, Davis, USA, and (14) Prof. I. Tsanis, Mc Master University, Ontario, Canada.

The production of the final copy was done from submitted computer discs by Mrs. M. Kazanaki and A. Athanasaki at TYPOKRETA Inc.; their efforts are gratefully acknowledged. Thanks are also due to the personnel of the Iraklio Institute of the National Foundation for Agricultural Research and to that of the Municipal Enterprise for Water Supply and Sewerage of Iraklio for their continuous support and help during the preparation of this book. Finally, we are immeasurably indebted to our wives for the many hours they put into assisting us in editing this work.

Iraklio, Greece
July, 1995

Andreas N. Angelakis and
Arie S. Issar

CONTENTS

CHAPTER 1

CLIMATIC CHANGE AND CULTURAL CHANGE IN THE SAHARA OVER THE LAST 130 Ka[1]

NICOLE PETIT-MAIRE

International Union of Geological Sciences,
URA 164, CNRS, Case 907, Luminy,
13288 Marseille Cedex 9, FRANCE

1-1. INTRODUCTION

Man being very sensitive to environmental variations, the equilibrium of our societies is particularly fragile in those areas which suffer frequent and severe alterations of their biotope; the Saharo-Sahelian belts belong to these unstable zones due to the high variability of monsoonal precipitation (Fairbridge, 1986). In the last ten years, an increasing number of observations on lake sediments, paleosols and anthropological material has allowed the correlation of their climatic evolution with the global changes recorded in oceanic and ice cores (Lorius *et al.*, 1993 and Shackleton 1982), as well as with the dynamics of human populations and cultures.

1-2. THE CLIMATIC OPTIMUM AT 125,000 B.P. AND THE END OF THE ACHEULIAN

The rapid rise of sea level and temperatures which culminated at *ca.* 125 Ka[1] B.P. relates with a previous marked increase in insolation (Bard *et al.*, 1990 and Berger, 1992). In the central Sahara, this climatic optimum coincided with a major lacustrine development. In central Libya (27°30'N) where the present-day annual rainfall amounts to 30 mm, the Shati lake, fed by aquifer rise and local run off, extended during isotopic stage 5e over 2 000 km^2 with salinities varying from 3 to 10‰ and depths varying from 40 to 50 m (Gaven *et al.*, 1981; Petit-Maire *et al.*, 1980 and Petit-Maire, 1982).

Surface archaeological sites are found throughout the area (Aumassip and Petit-Maire, 1982): terminal Acheulian on the substructural flats above the valley; small (<8 cm) piriform bifaces, mixed with terminal Acheulian and Levallois flakes or blades along the upper shorelines of the lake; exclusively Levallois assemblages at the foot of the thick

[1] Ka, Thousand years read as

NATO ASI Series, Vol. I 36
Diachronic Climatic Impacts on Water Resources
Edited by Andreas N. Angelakis and Arie S. Issar

lumachelle mounds indicating a molluscan thanatocoenosis due to lake regression; Aterian and Neolithic unassociated with the shores. According to their location relative to the lake deposits, the Levallois sites may be situated in time as a little younger than the lake main regressive phase, which is consistent with the age suggested by McBurney for the earliest Levallois-Mousterian in Cyrenaica, *ca.* -80 Ka. We leave archaeologists to interprete these data. However, one can underline the correlation of an important humid phase in the Sahara during the Mid-Pleistocene with the end of the old Acheulian civilisation. The occurrence and long duration of a hospitable biotope probably allowed, together with sedentary life, invention, easier contacts and transmission of new techniques.

Since 1982, other evidence for this lake phase in the Sahara was found in Nubia (Wendorf *et al.*, 1987), in Tunisia (Ballais and Ben Ouezdou, 1987 and Gasse *et al.*, 1987) and in northern Mali (Petit-Maire *et al.*, 1994).

1-3. UPPER PLEISTOCENE CLIMATES AND THE ATERIAN

The progressive deterioration of climate registered by oceanic isotopic curves between stages 5 and 2 was not regular. In the Sahara, repeated fluctuations between environmental degradation and regeneration are recorded. They differ widely with latitude (Durand *et al.*, 1983 and Petit-Maire, 1989) and imply either heavy stress or new territorial possibilities for prehistoric groups, with probable response in terms of migrations and cultural change. The climatic and archaeological sequence of the Sahara between 100 Ka and 50 Ka is still little known. An arid phase occurred in southern Niger between 50 Ka and 40 Ka (Durand *et al.*, 1983). It was followed between 40 Ka and 20 Ka by one or two humid episodes that have been observed right across the continent, from the Red Sea to the Atlantic, and which are associated with the Aterian (Clark 1980 and Tillet, 1983). In the Fezzan, the large Shati lake became a saline sebkha, but episodes of fluvial activity or small lakes are U/Th and ^{14}C dated at 40 Ka and 26 Ka on freshwater molluscs (Petit-Maire, 1986). Aterian pedunculate points occur on the banks of the wadis or channels (Aumassip and Petit-Maire, 1982). In northern Mali, extensions of lacustrine limestones or large clayey pans, both with *Melania tuberculata*, are found in flat areas between 19°N and 22°30'N and a lake shore thanatocoenosis, indicating rapid evaporation, is dated 21 Ka (Petit-Maire, 1986). Aterian unifacial pedunculate points generally line the palaeobeaches (Chavaillon and Fabre, 1960).

The dates proposed for the Alterian in northern Niger by Tillet (1983; 40-20 Ka) and Durand *et al.* (1983; 26-20 Ka) are consistent with the environmental data indicating favorable climatic change. The Aterian could be defined as the culture born and developed during isotopic stage 3.

The end of the Aterian is also related to climatic change. In the late Pleistocene, a severe arid phase, culminating between 20 and 15 Ka correlates with the last glacial episode and a major lowering of sea level and temperatures. Rainfall may have been reduced to at least 20% of modern values, saharan dune fields extending South to 13°N (Talbot 1984). Even the mountainous areas had sparse vegetation. At 23°15'N, an Aterian site was found *in situ*, just beneath the lower layers of a 3 m thick palaeodune (Raimbault, 1994), demonstrating the causes of the end of the Aterian in the Sahara; it became impossible for plants or animals to survive in areas where, today, annual rainfall varies from < 5 to 20 mm. The Aterians had to migrate. But where to? Mountains, river valleys and lake shores are classical refuges, but competition must have been severe along the Niger, Senegal and Nile rivers which registered strong changes in their water balance. From the western and northern Sahara, man migrated to the coasts of the Atlantic and Mediterranean, where large areas of the continental plateau had emerged during the -130 m oceanic regression. Unfortunately, those sites are now under sea level. Evidence for such migrations may be found in physical anthropology. In the late Pleistocene of the Maghreb (Camps *et al.,* 1973 and Roche, 1963) and Upper Egypt, a new population appears: the north african Cromagnoids. They share several morphological traits with the scarce Aterian remains we dispose of (Ferembach, 1985). The Aterians possibly evolved quite rapidly, due either to environmental change inducing change in selective pressure, or to the crossbreeding and drift induced by migration.

However, the available archaeological evidence does not support such an hypothesis: in the Maghreb, the Iberomaurusian follows the Aterian after a hiatus of as yet unknown duration and, in the southern Sahara, the Ounanian appears after the aeolian gap noted previously. Roche (1963) denies any affinity between Aterian and Iberomaurusian, and Tillet (1983) sees no connection with the Ounanian. We propose that cultural change could have paralleled genetic change, for the same reasons.

1-4. THE HOLOCENE INTERGLACIAL AND NEOLITHISATION

Recent multidisciplinary studies have defined with great precision the Holocene environmental changes in the most arid areas of the Sahara. In brief, a widespread lacustrine episode took place from *ca.* 9.5 to *ca.* 4.5 Ka, between 24° and 22°N, from 9.5 to 4 Ka between 22° and 20°N, and from 9.5 to 3 Ka between 20° and 18°N (Petit-Maire, 1993 and Yan and Petit-Maire, 1994). Lake or swamp deposits are so extensive that the importance of surface fresh water implies precipitations varying from South to North from 500 to 200 mm, in an area that today receives from 50 to 5 mm. After 6.7 Ka, the climate progressively deteriorated and changes in lake balance became frequent and severe, leading to frequent chemical changes, saline deposits, or large mud cracks surface . Thus, together with a warmer global change, the paratropical hyperarid belts probably disappeared completely during the Holocene (Petit-Maire *et al.*, 1994).

This hypothesis is supported by anthropological data from northern Mali where the remains of 117 individuals were collected from burials dated *ca.* 7 to 4.5 Ka. They have the same Cromagnoid features (Dutour, 1989 and Petit-Maire and Dutour, 1987) as those listed *supra* in the Maghreb : they could, between 10 and 7 Ka, come from the North, across a continuous steppe with available fresh water.

The lag between the dates for lacustrine evolution and those for archaeological sites is striking. 1500 to 2000 years were necessary to provide a full biological response in a region which had been hyperarid for some 10,000 years. Increased rainfall induced growth of a Gramineae steppe to the North of the area and of a Sudanese savanna South of the 20th parallel. A biologic chain of Foraminifera, Ostracods, Molluscs, large fish (up to 1.60 m), crocodiles, turtles, hippopotamus, rhinoceros, phacochoerus, elephants and large antelopes *(Alcelaphus buselaphus, Limnotragus spekei, Hippotragus equinus,* etc.) was established up to 22°N (Petit-Maire and Riser, 1983 and Petit-Maire *et al.*, 1993).

Man took advantage of the optimum to develop a new culture, the Neolithic. The wide extension of a Gramineae steppe correlates with the ubiquity and density of mortars and grinders. The abundance of large fish in shallow lakes must be related to the development of bone hooks and harpoons (115 harpoons were found in a single 200 m^2 site, at 19°15'N; Decobert and Petit-Maire, 1985). The possibility of a sedentary life around clayey pans allowed the invention and expansion of a rich ceramics industry (Commelin 1984). The variety and quality of pottery decoration and of stone and bone adornments (Camps-Fabrer, 1983), suggest spare time for refinement, incompatible with hard struggle for life in a hostile environment.

After 4.5 Ka, the archaeological sites are found on the dried lake beds, probably around the remaining water holes or wells at the center of the depressions. The large mammalian fauna is gone by 5 Ka and freshwater molluscs could no longer survive, except in the Azawad (Petit-Maire and Riser, 1987). Like the Aterians some 15,000 years earlier, man had to migrate. The populations of the eastern Taoudenni basin moved to the Tilemsi valley (19°N) by 4 Ka where they mixed with negroid groups (Smith, 1980 and Petit-Maire, 1986). From inner Mauretania and the western Sahara, man migrated to the Atlantic coast where fresh water holes and sea food allowed them to live until 2.5 Ka (Petit-Maire, 1980). They also mixed there with southern populations (Petit-Maire and Dutour, 1987). At 3.5 Ka the Sahara basins were desert again and the interglacial humid phase came to an end.

1-5. THE PRESENT AND THE FUTURE

The climatic evolution of hyperarid Sahara correlates well with the astronomical, ice and oceanic isotopic curves for the last 130,000 years. Humid lacustrine or paludal phases follow the maximal insolation and minimal cryosphere peaks while arid phases correlate with the opposite ones. It was shown (Petit-Maire *et al.*, 1991 and 1994) that those changes also grossly fit the variations of the methane atmospheric ratio, due to the vastness of the concerned fluctuating areas. The origin and development of human cultures, promoted by favorable biotopes and sedentarisation, is also closely linked with these changes and climatic landmarks clearly stake out the onset and the end of prehistoric civilizations. The Lower Palaeolithic ended with the last interglacial (isotopic stage 5e); the Aterian developed during the swampy episodes marking stage 3 and ended with the occurrence of aridity at stage 2; the Neolithic began with the Holocene climatic optimum. Thus, palaeoclimates have always been one of the fundamental factors in our evolution.

The astronomical insolation curve (Berger , 1992) fits global climatic change and would be the best prediction for the long term future, if man did not interfere. The variations of summer insolation will bring the Earth to a new cold extreme about 60,000 years from now: a new glacial would thus set in and correlate once again with a major extension of the arid belts, as 20,000 years ago; the trend would be progressive and irregular in the future as it has been in the past. Man would have to migrate into his old refuges and to new ones arranged by our modern technology, until a new warmer and humid phase returns. However, the natural patterns are now being changed by the quick atmospheric rise of man-made radiative gases, highly enhancing the natural greenhouse effect. No real palaeoanalogue exists for such a situation, since astronomical forcing is not the cause of such expected global warming. Models vary in the anticipation of this future man-made climate, but agree that the changes will probably occur much faster and to a higher level than any of the recent Quaternary interglacial scenarios. However, the examples of the last climatic optima still provide, for the moment being, the best available realistic image of the possible "greenhouse future" of the Saharo-Sahelian belt, mainly an increase in the activity and range of the monsoon (Petit-Maire, 1990).

REFERENCES

Aumassip G. and Petit-Maire N. (1982). Les industries préhistoriques de la vallée du Shati. *In*: Le Shati, lac pléistocene du Fezzan (Libye) N. Petit-Maire (Ed.) : 86-88. CNRS, Paris / Marseille, France.

Ballais J.L. and Ben Ouezdou H. (1987). Formes et dépôts du Quaternaire continental de la bordure présaharienne du Maghreb oriental. Essai de synthese provisoire. Conférence PICG., 210: 18-19.

Bard E., Hamelin B., and Fairbanks R.G. (1990). U-Th ages obtained by mass spectrometry in corals from Barbados : sea level during the past 130,000 years. Nature, 346 : 456-458.

Berger A. (1992). Le Climat de la Terre. Un passé pour quel avenir ? De Böeck-Wesmael, pp. 475.

Camps G., Delibrias G., and Thommeret J. (1973). Chronologie des civilisations préhistoriques du Nord de l'Afrique d'après le radiocarbone. Libyca, 21: 65-69.

Camps-Fabrer H. (1983). Parures et engins de pêche. *In*: N. Petit-Maire and J. Riser (Eds). Sahara ou Sahel? Quaternaire récent du Bassin de Taoudenni, pp. 367-401. CNRS, Marseille, France.

Chavaillon N. and Fabre J. (1960). L'Atérien d' El Guettara. B.S.P.F., 57 : 346-354.

Clark J. D. (1980). Human populations and cultural adaptations in the Sahara and the Nile during prehistoric times. *In*: M.A.J. Williams and H. Faure (Eds.).The Sahara and the Nile: 527-582. Balkema, Rotterdam , NL.

Commelin D. (1984). La céramique néolithique dans le Bassin de Taoudenni (Sahara malien). Thèse, Université de Marseille, France.

Decobert M. and Petit-Maire N. (1985). An early Neolithic midden and necropolis in Malian Sahara. Nyame Akuma, 26: 26-32.

Durand A., Lang J., Morel A.,and Roset J.P. (1983). Evolution géomorphologique, stratigraphique et paléoclimatique au Pléistocène supérieur et à l'Holocène de l'Aîr oriental (Sahara méridional, Niger). Revue Géologie Dynamique et Géographie Physique, 24: 47-59.

Dutour O. (1989). Hommes fossiles du Sahara. Peuplement holocène du Mali septentrional. CNRS, Marseille: pp. 342.

Fairbridge Ph. (1986). Monsoons and Paleomonsoons. Episodes, 9 (3) : 143-149.

Gasse F., Fontes J.C., Plaziat J.C., Carbonel P., Kaczmarska I., de Deckker P., Soulié-Marshe I., Callot Y., and Dupeuble P. (1987). Biological remains, geochemistry and stable isotopes for the reconstruction of environmental and hydrological changes in the Holocene lakes from North Sahara. Palaeogeography, Palaeoclimatology, Palaeoecology, 60 : 1-46.

Gaven C., Hillaire-Marcel C., and Petit-Maire N. (1981). A Pleistocene lacustrine episode in southeastern Libya. Nature, 290: 131-133.

Lorius C., Jouzel J., Ritz C., Merlivat L., Barkov Y.S., Korotkevich Y.S., and Kotlyakov V.M. (1993). The ice core record : climate sensitivity and future greenhouse warming. Nature, 347 : 139-145.

Petit-Maire N. (1980). Holocene biogeographical variations along the northwestern African coast (28o-19oN). Paleoclimatic interpretations. *In*: Sarnthein M., Seibold E. and Rognon P. (Eds): Sahara and the Surrounding Seas. Palaeoecology of Africa, 12 : 365-377. Balkema, Rotterdam, NL.

Petit-Maire N., Delibrias G., and Gaven C. (1980b). Pleistocene lakes in the Shati area, Fezzan (27° 30' N). *In*: Sarnthein M., Seibold E. and Rognon P. (Eds). Sahara and the Surrounding Seas. Palaeoecology of Africa, 12 : 289-295. Balkema, Rotterdam, NL

Petit-Maire N. (Ed.) (1982). Le Shati, lac pléistocène du Fezzan. CNRS, Paris / Marseille, France, pp.182.

Petit-Maire N. and Riser J. (Eds.). (1983). Sahara ou Sahel? Quaternaire récent du bassin de Taoudenni (Mali). CNRS, Marseille, France.

Petit-Maire N. (1986). Paleoclimates in the Sahara of Mali: a multidisciplinary study. Episodes, 9: 7-16.

Petit-Maire N. and Dutour O. (1987). Holocene populations of the western and southern Sahara: Mechtoids and paleoclimates. *In*:. Close A. (Ed). Prehistory of Arid North Africa : 259-286. Southern Methodist University Press, Dallas, USA.

Petit-Maire N. and Riser J. (1988). Le Sahara à l'Holocène: Mali. - Paris (FR): CGMW, 1988. Map 1/1,000,000.

Petit-Maire N. (1989). Interglacial environments in presently hyperarid Sahara : paleoclimatic implications. *In*: Leinen M., Sarnthein M. (Eds.). Paleoclimatology and Paleometeorology and Paleometeorology: Modern and Past Patterns of Global Atmospheric Transport, pp. 637-661. Dordrecht (NL), Kluwer.

Petit-Maire N. (1990). Will Greenhouse green the Sahara ? Episodes, 13/(2): 103-107.

Petit-Maire N. (1991). Paléoenvironnements du Sahara. Lacs holocènes a Taoudenni (Mali). Editions du CNRS, Paris-Marseille, pp. 242.

Petit-Maire N., Fontugne M., and Rouland C. (1991). Atmospheric methane ratio and environmental changes in the Sahara and Sahel during the last 130 kyrs. Palaeogeography, Palaeoclimatology, Palaeoecology, 86: 197-204.

Petit-Maire N. (1992). Environnements et climats de la ceinture tropicale nord-africaine depuis 140 000 ans. *In*: Miskovsky J.C. (Dir. Publ.). Les Applications de la Géologie à la Connaissance de l'Environnement de l'Homme Fossile. Paris : Mémoire Société Géologique de France, 160 : 27-34.

Petit-Maire N., Page N., and Marchand J. (1994). The Sahara in the Holocene. Map 1/5,000,000. UNESCO and CGMW, Paris.

Petit-Maire N., Reyss J.L., and Fabre J. (1994). Un paléolac du dernier interglaciaire au Sahara Malien actuellement hyperide (23°N). Comptes Rendus de l'Académie des Sciences. Paris, 319, II: 805-809.

Raimbault M. (1994). Sahara malien: environnement, populations et industries préhistoriques. Thèse Doctorat d'Etat, Université de Provence (Aix-en-Provence), pp. 1095.

Roche J. (1963). L'épipaléolithique marocain. Didier, Paris, France.

Shackleton N.J. (1982). The deep sea sediment records of climate variability. Proc. Oceanogr., 2: 199-218.

Smith A.B. (1980). The neolithic traditions in the Sahara. *In*: Williams M.A.J. and Faure H. (Eds). The Sahara and the Nile: 451-466. Balkema, Rotterdam, NL.

Talbot M. (1984). Late pleistocene rainfall and dune building in the Sahel. Palaeoecology of Africa, 16: 203-214.

Tillet T. (1983). Le Paléolithique du bassin Tchadien Septentrional. CNRS, Marseille, France.

Wendorf F., Close A., and Schild R. (1987). Recent work on the Middle Paleolithic of the eastern Sahara. A.A.R., 5: 49-63.

Yan Z. and Petit-Maire N. (1994). The last 140 Ka in the Afro-Asian arid / semi-arid transitional zone. Palaeogeography, Palaeoclimatology, Palaeoecology, 110 :217-233.

CHAPTER 2

CLIMATIC CHANGES IN THE SPANISH CENTRAL ZONE DURING THE LAST 3000 B.P. BASED ON POLLINIC ANALYSIS

BLANCA RUIZ ZAPATA, MARIA JOSÉ GIL GARCIA AND MIRIAM DORADO VALIÑO

Departamento de Geología, Universidad de Alcalá de Henares,
Facultad de Ciencias, Carretera N-II, Km 33,600,
28871 - Alcalá de Henares, Madrid, SPAIN.

2-1. INTRODUCTION

Palinological data coming from different deposits of peat of the Spanish Central zone are presented in order to determine the climatic impact on the vegetation development, along 400 km with W-E direction, in the most western area of the mediterranean region. Given the geographic characteristics of the study area, the data coming from mountain areas (in the "Sistema Central Español") and "Meseta" ("Submeseta Sur") are compared, in order to understand the different replies of the vegetation according to the altitude, latitude and orientation, against local and global changes. Besides due to the studied period of time, special mention about the human impact manifested in the zone will be effected.

2-2. GEOGRAPHIC SITUATION

The Spanish Central Region is formed by two great physiographic unities:
a) The "Sistema Central" or mountain range which, with ENE-WSW direction, separates two unities in the "Meseta Central" and acts like watershed of Duero and Tajo basins. It is 620 km large, including the Portuguese mountain ranges and it can be divided into a series of independent mountain unities, formed by minor geographic groups, united under the name of "sierras" (Acaso, 1983), delimitated between them by passes or corridors which make the "Sistema Central" not to present an altimetric continuity. "Gredos", next to "Guadarrama", "Peña de Francia", "Gata" and "Serra da Estrela", are the main elements which form the "Sistema Central". These elements can be grouped in three great sectors (Fig.1); Eastern, which has a simple structure, because it is a main mountain range (Guadarrama and Somosierra) with some secondary branches. Central, formed by "Gredos" and minor groups and western, formed by "Pena de Francia", "Gata" and "Estrela", which present a great deal of "sierras", with E-W direction and, to

NATO ASI Series, Vol. I 36
Diachronic Climatic Impacts on Water Resources
Edited by Andreas N. Angelakis and Arie S. Issar

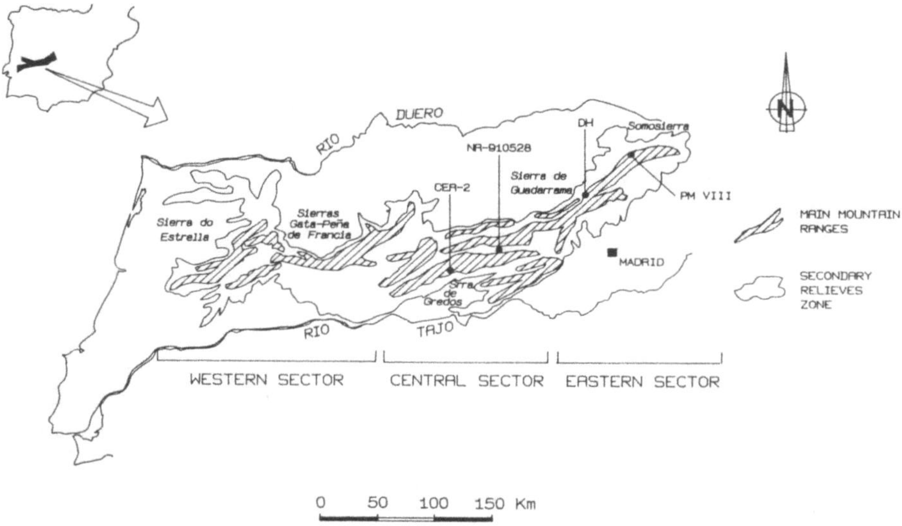

Figure 2-1. Geographic situation and deposits locations

certain extent, independent; all this provides the pattern in the form of a lying "H" presented by the whole range (Pedraza, 1978). As far as the altitude is concerned, the highest altitudes are in the central sector, with the 2592 m of "Pico Almanzor" (in "Sierra de Gredos"); in the eastern sector "Peñalara" with its 2430 m (in "Sierra de Guadarrama") and "Pico del Lobo" with 2273 m (in "Somosierra") are significant, nevertheless in the western sector the maximum altitude is 1993 m which corresponds to "Serra da Estrela" peak (Portugal). It is formed by plutonic rocks (granites) and metamorphic rocks (gneiss, quartzites, schists and shales) with Paleozoical and Precambrian age, affected by the Hercinian orogeny; later on, a strong erosion, previous to Mesozoic, demolished it, originating the "Meseta". The Alpine orogeny affected this area producing elevations in ENE-WSW direction and subdividing the "Meseta" in two parts tilting it to the west at the same time. During Quaternary period, the fluvial, periglacial and glacial erosions modeled the landscape. The known mediterranean character of the "Sistema Central", is not caused by the scarce pluviosity, which is about 1000 mm in many "sierras", but by the distribution of the precipitations (Fig. 2-2). These are concentrated in autumm and spring, and during summer, only when the local conditions are adequate, there are storms with a strong electric discharge. There are also numerous local variations, such as the glacial circus and valleys, where the medium values

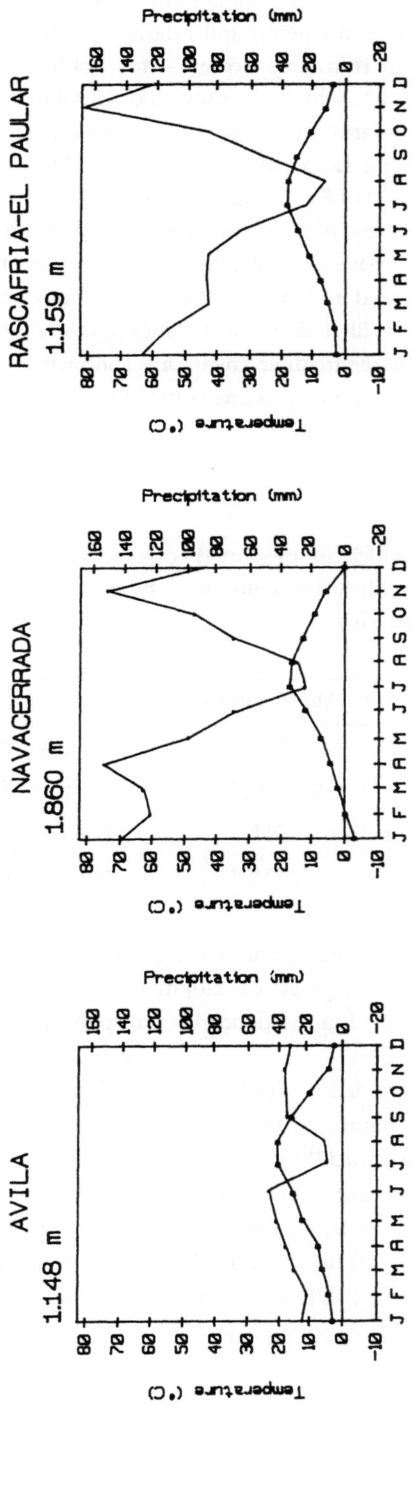

Figure 2-2. Ombroclimatic diagrams

of temperature are cooler, or the warmer temperature in the sierras with SW orientation; precipitations are less frequent in the sierras and valleys protected by mountain elevations, which originate shadows of rain (this explains, for example, the more dryness in "Villafranca" and "Paramera" which are more like Guadarrama's than the nearest Gredos's); however, in "Somosierra" and "Ayllon" the estival dryness is lower than in the rest of the sierras because of the longer distance from the influence of Azores anticyclon. For this tabla, in there area forest of *Betula* and *Fagus* no tipicad.

b) The "Meseta", is a more or less plane surface (plateau), with a medium altitude of 600 m. It is formed by mesozoic materials and after the Alpine orogeny, which elevates the "Sistema Central", is subdivided into two great basins ("Duero" to the North and "Tajo" to the South), which will be refilled along the Tertiary period. From the climatic point of view, it rather corresponds to the mediterranean one, continental, and only lower values of temperature and a higher pluviosity define some of the peculiar characteristics of the "Submeseta Norte".

2-3. METHODOLOGY

The Pollinic Diagrams which are presented correspond to some of the more significant deposits in order to establish the evolution in time and in space, in our study area. The deposit report is the following:

"Sierra"	Abbreviation	Place
1.- Sierra de Gredos	CER - 2	El Cervunal (Avila)
2.- Sierra. de Avila	NR-910528	Narrillos del Rebollar (Avila)
3.- Sierra. de Guadarrama	DH	Peñalara (Madrid)
4.- Somosierra	PM VIII	Pto. Morcuera (Madrid)

Of all four deposits, two of them (1 and 3) are glacio-lacustrine originated, formed as a consequence of the morainic deposits development which act as dikes that obstruct the drain (obstruction lagoon); 2 and 4 deposits correspond to a small depression with periglacial origin.

The lithology (peat more or less detritic) and the high mountain character of the deposits, make the use of manual drillings (DACHNOSVSKY type, modified HILLER and/or RUSA) necessary to extract the core.

About the chemical treatment of the samples, the procedure, based on the lithologic nature (chemical attack by acids, alkalis and, in some cases, by dense liquids which favour the pollen concentration) has been used. The statistics work has been done with the programs G1PAL (Goeury, 1987) and TILIA (Grimm, 1992).

2-4. RESULTS

As the aim of this work is to recognize the climatic changes occurred during the last 3000 B.P. through the pollinic analysis, we will comment on some of the more significant data of the presented profiles. It will also be possible to establish the temporal and spacial evolution of the most important taxa which have conformed the vegetal landscape of the study area.

"Macizo Central de Gredos": P.D: CER-2

It is an old intramorainic lake ("El Cervunal"), framed by the side moraines of "Gredos" and "El Pinar" glaciers (Fig. 2-3), with an altitude of 1840 m. The deposit has thickness higher than 7 m, but only the first three meters are peat and provide continuous search under pollinic point of view (Ruiz and Acaso, 1981 and 1984). Along these, it can be observed a lower level (I) determined by the highest A.P. values in all the profile. Other significant data are the *Betula* percentage against a group formed by *Pinus, Salix* and *Corylus* with low and diminishing representation. This makes the lower zone (Ia) to be a forestal mass with more variety than the upper zone (Ib) which has less variety and a larger development of *Ericaceae* and *Poaceae.*

Although ^{14}C data have not been provided for this deposit, because of the similitude with other neighbouring deposits as the case of "Lagoa Comprida" (Serra da Estrela. Van den Brink, 1985), which reveals ages higher than 8000 B.P. (Atlantic climatic period) and where *Betula* is the main landscape element, the same as "Baterna" deposit in "Valle de Ambles" (Dorado, 1993) which is dated about 6000 B.P., this level I could be placed in the Atlantic climatic period (when the environmental humidity conditions were larger).

The *Betula* descent, the *Pinus* development, together with a general loss of the damp non-arborean taxa and the *Quercus* increase, determine the level II, whose A.P. values fluctuate about 50%. It indicates a drastic change in the environmental conditions which could correspond to the Subboreal climatic period or, most probably, to the Subboreal-Subatlantic transition. The Subatlantic climatic period is clearly represented in the level III, where the most significant data are the decrease of the arborean cover and the development of *Quercus*. It is not possible to appreciate any changes occurred during the last 2500 B.P.

"Sierra de Avila": P.D: NR910528

It is a deposit with periglacial origin, formed by a more or less detritic peat, with a thickness of 330 cm and placed in "Narrillos del Rebollar" at 1560 m of altitude (Fig. 2-4). The vegetation development (Dorado, 1993) reflects the existence of a relatively close pine-wood (A.P. higher than 50%) as the main component, accompanied by *Quercus*-c and *Quercus*-p, whose relative low values are constant in the whole profile. The non-arborean vegetation is mainly formed by *Cistaceae, Ericaceae* and *Artemisia; Poaceae*

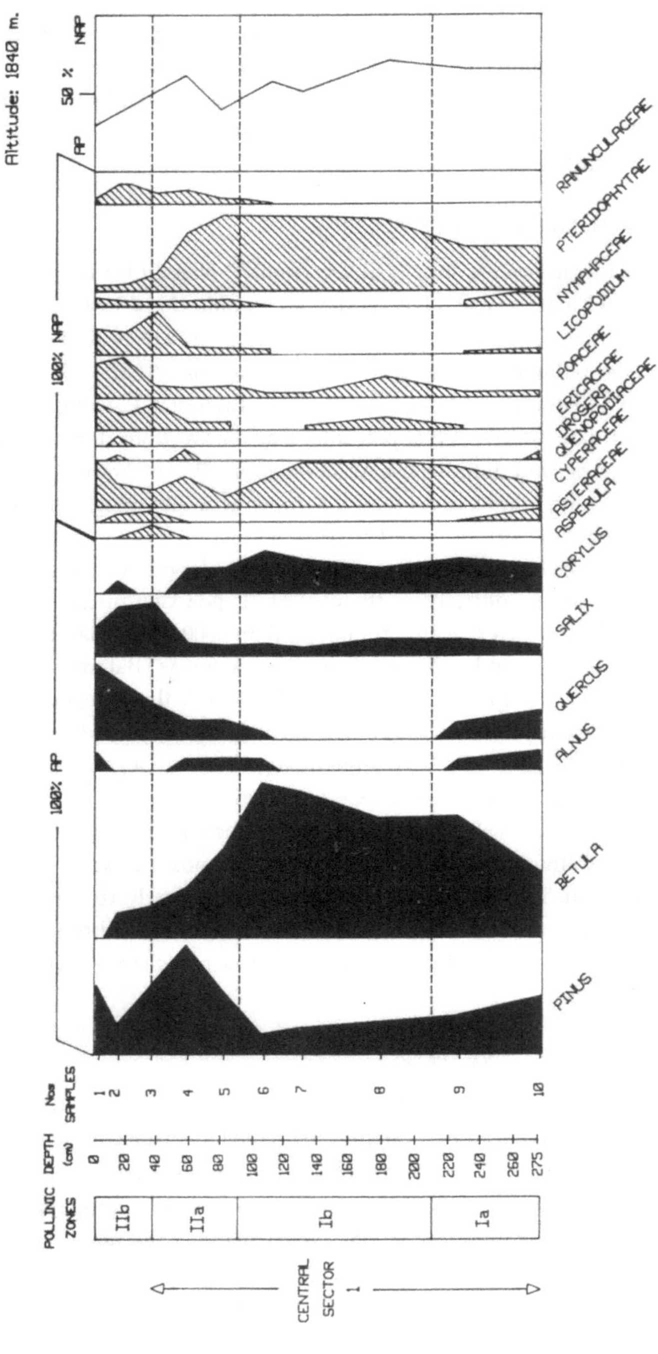

Figure 2-3. Pollinic diagram of El Cervunal (CER-2)

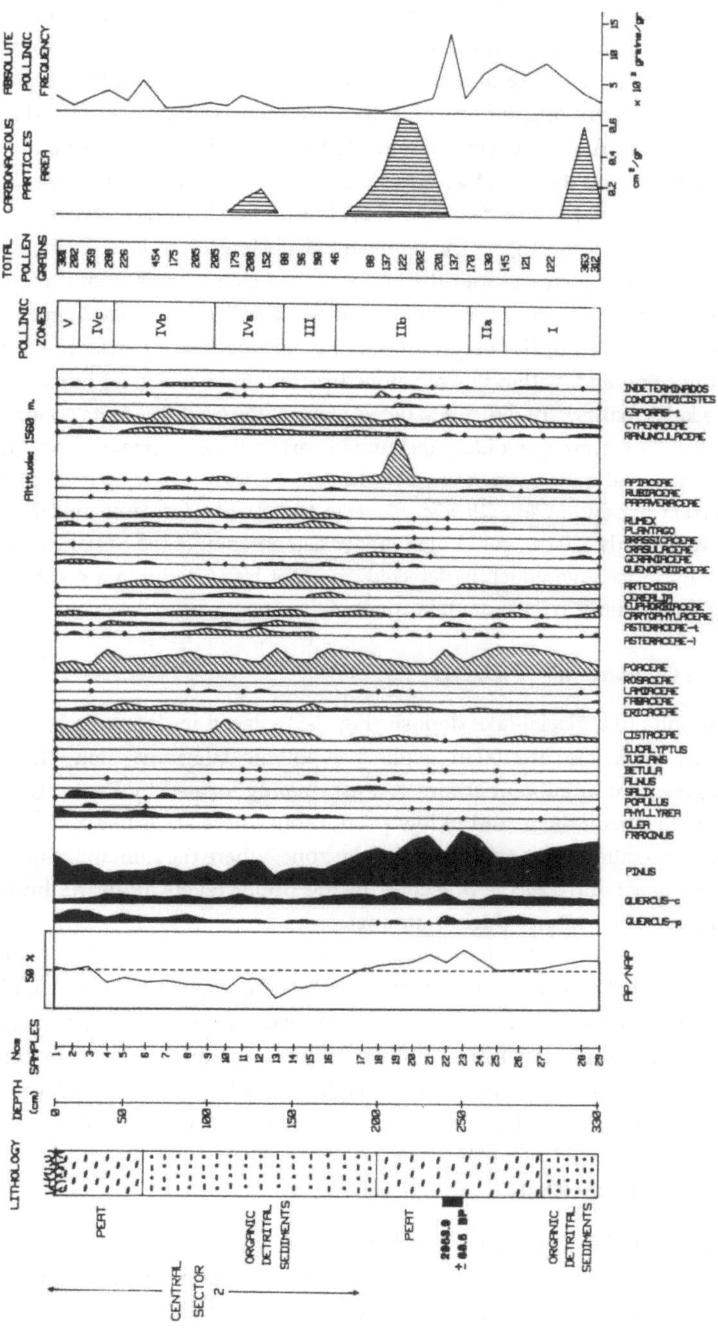

Figure 2-4. Pollinic diagram of Narrillos del Rebollar (NR-910528)

increases progressively. In level II there are the highest A.P. values of the profile, mainly caused by the *Pinus* development instead of *Quercus*-c and *Quercus*-p. From 2953±68 B.P., the A.P. start to decrease and the new herbaceous taxa like *Quenopodiaceae* and *Plantago* are developed; others have an important increase like *Poaceae* and *Artemisia*. The lower values of arborean cover are refflected in level III. It is important to point out the *Olea* appearance and the *Cistaceae, Artemisia, Plantago* and *Rumex* development.

A new A.P. recovering is observed in level IV, together with a larger development of taxa which indicate anthropical activity. The dominant characteristics in level V, the upper one, are the *Pinus* and Olea increases and the non-arborean components stabilization. So, this Pollinic Diagram shows the existence of a Pinus wood dominating the landscape during the Subboreal climatic period; the beginning of the A.P. decrease marks the Subboreal-Subatlantic transition and, within this mentioned period, there is a great human use of the territory. The latter makes very difficult to detect some of the climatic events occurred during the last 3000 B.P., however, by avoiding the human part on the environment, the transition to warmer and drier conditions is determined by the A.P. decrease; a light increase of *Quercus*-c, *Alnus* and *Betula* presence, *Olea, Cerealia, Plantago* and *Rumex* decreases and the *Artemisia* slight increase, towards 120 cm, could mark a low deterioration of the climatic conditions (it could be the Little Ice Age which is detected, more or less briefly, in a great part of the Pollinic Diagrams, mainly in the mountain zones of this area).

"Sierra de Guadarrama": P.D: DH

It corresponds to a glacial-lake deposit (Fig. 2-5), placed in the named "La Hoya de Dos Hermanas" (Peñalara) at 2100 m height. The pollinic data show, along the 110 cm of thickness, a general dominion of arborean mass, mainly formed by *Pinus, Quercus* and *Juniperus*. Together with them, and in low proportions, there are *Betula* and *Corylus*. The non-arborean vegetation is the typical one of the zone, where the human influence is rare and only the *Rumex* and *Poaceae* presence, in the upper levels, indicate limited cattle activity in the altitude pastures during summer.

The descent of A.P. curve, and of *Pinus, Quercus* and *Corylus*, about 90 cm of depth, which corresponds to the increase of *Juniperus, Betula* and damp needed taxa, could indicate larger precipitations (Little Ice Age). The remaining vegetation evolution is determined by the human history, although, given the deposit altitude, this does not clearly refflect the cultivation or other things which are detected in other deposits of the zone, at lower altitudes (Alia *et al.*, 1957; Jimenez *et al.*, 1985; Ruiz *et al.*, 1986 ,1988 and Vazquez, 1992).

"Somosierra": P.D:PMVIII

It is a fluvio-glacial deposit, with a thickness of 2 m, placed at 1740 m of altitude ("Pto. Morcuera. Madrid). Based on its components development, the Pollinic Diagram (Fig. 2-6) reflects (Gil, 1992) an inferior zone "a", dominated by arborean vegetation,

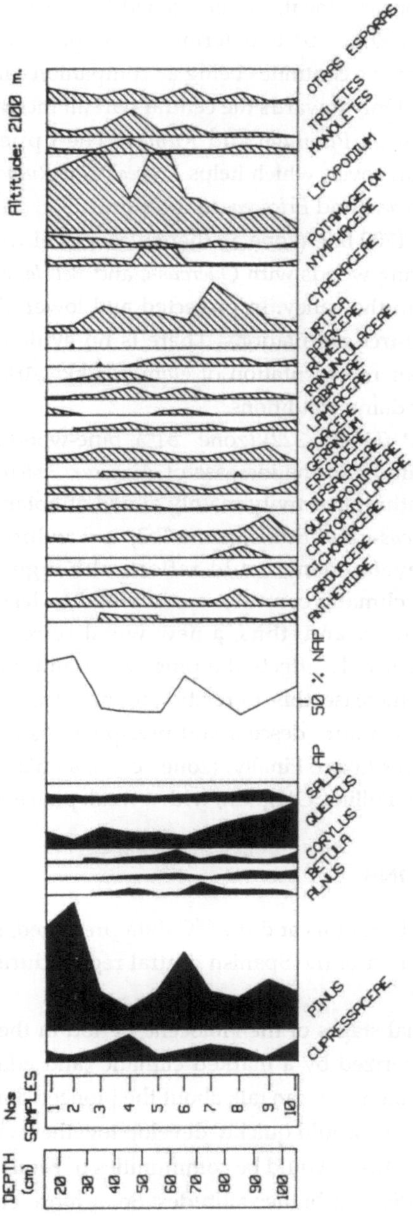

Figure 2-5. Pollinic diagram of La Hoya de Dos Hermanas (DH)

being *Pinus* the responsible for it, accompanied by *Quercus*-c, *Betula* and sporadically *Corylus* and *Juniperus*. Zone "b" is determined by the marked fluctuation of the A.P. curve and *Pinus*. *Quercus*-c continues being as companions and, in sporadic way, *Betula, Corylus* and *Juniperus*. Only towards the central part an increase of *Quercus*-c is observed together with *Artemisia, Plantago* and *Rumex*. The upper zone ("c"), presents an impoverished arborean cover, which helps *Poaceae* and *Rumex* to develop, and elements like *Artemisia, Cichorioideae* and *Ericaceae* to disappear.

Therefore, before 1700 B.P. (zone a), there was a local wood with relative density of *Pinus*, and neighbouring woods with *Quercus*-c and *Betula*, as well as *Alnus* and *Corylus* riverside formations in the valley. In protected and lower altitude zones (Lopez, 1982), there would be olive-tree cultivations. There is no evidence of human activity, that, together with the poor representation of elements like *Artemisia*, could determine the existence of relatively damp conditions.

From 1700±90 B.P. (UGRA-322) (zone "b1"), pine-wood and birch-tree retrocessions are observed coinciding with the increase of *Apiaceae, Asteroidaceae, Poaceae* and *Rumex*, that shows certain anthropic activity mainly aimed at obtaining pastures. A new wood recovery (*Pinus, Quercus*-c and *Betula*) (zone "b3) and anthropogenic taxa descent as well as the *Quercus*-p development, could reflect, although these phenomena are not synchronous, better climatic conditions combined to less human activity. This is a relatively short recovery and, thus, a new wood retrocession appears (b3) again, although in this case it only affects the pine-wood and oak-wood because the birch-wood suffers a slight increase; this, in relation to the retrocession of the drier taxa, could be indicative of a temperature descent and precipitations increase which would favour the development of this taxon. Finally, (zone "c"), a similar situation to the present one can be observed in the Pollinic Diagram, with a weak pine expansion.

2-5. CONCLUSIONS

With the Pollinic Diagrams and the ^{14}C data presented, some of the most important aspects on the vegetation of the Spanish central region during the Holocene period can be observed (Fig. 2-7).

Actually, the initial stages of the Holocene period in the study area are not known, although it is characterized by a marked climatic and edaphic unstability. However, based on the existing data, one can talk about the pioneer character of *Betula* and *Pinus* in the Atlantic period; these would quickly develop together with *Juniperus* and, only when the weather stabilized, there would be communities of *Betula, Quercus*-c and *Quercus*-p at lower altitudes and *Pinus* at higher altitudes. So, a more eurosiberian vegetation, with dominant *Betula* woods, are the general characteristics in the Atlantic climatic period in the "Sierras" of the central region (Arila-Dorado, 1993; Bejar-Atienza, 1991 and Gredos-Ruiz and Acaso, 1981), while in the lower and more plane zones (La Mancha) (Garcia *et*

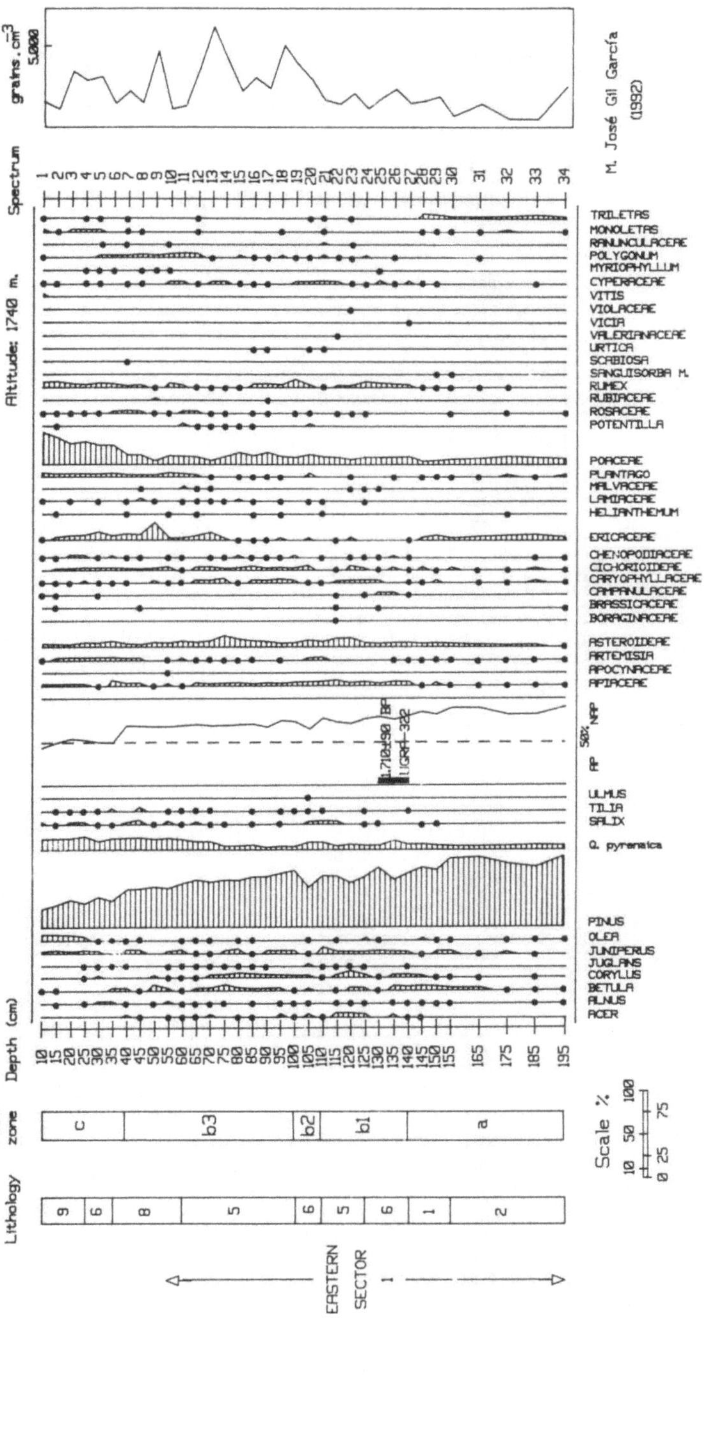

Figure 2-6. Pollinic diagram of Puerto de Morcuera (PM VIII)

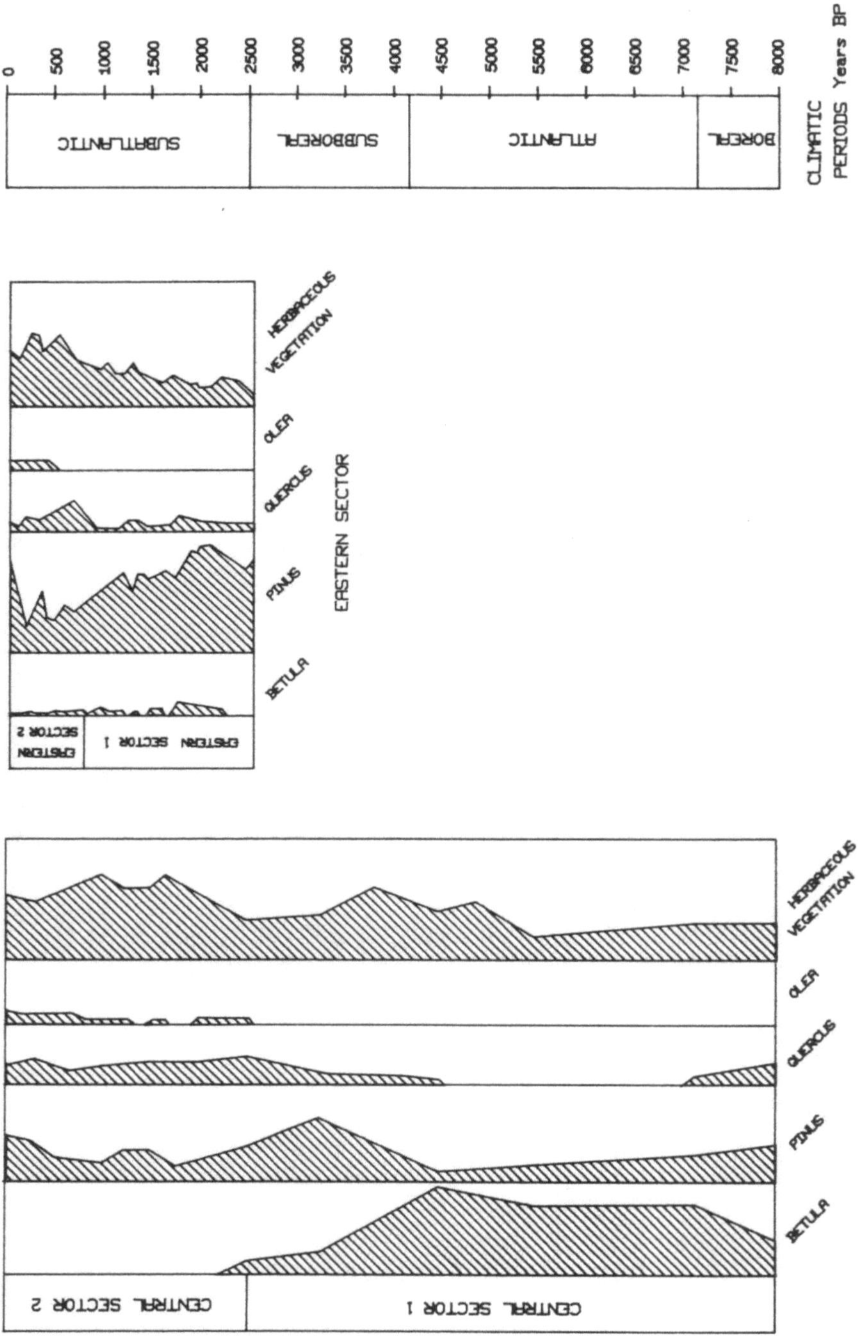

Figure 2-7. Vegetation evolution in the Holocene period

al., 1986) a typical mediterranean vegetation is developed. It is important to note the *Olea* presence in the lower topographic zones (Valle de Ambles- Dorado, 1993). The Subboreal period is not well detected in the study area, being its transitions with the other periods that determines it. The signs of certain cooling and aridity are detected in mountain areas (Gredos), where the *Betula* as well as the remaining components connected to wet environments decrease and *Pinus* dominate (Ruiz et al., 1992).

The Subatlantic period, warmer and drier, is determined, in the whole area, by the generalized A.P. descent and by the substitution of *Quercus*-p by *Quercus*-c, as well as by the presence of continuous curves of *Olea, Cerealia, Plantago, Rumex* and other elements, which accompagny cattle and agricultural activities. During this period, the climatic changes are less severe, and are detected at historical scale, but do not mark vegetation due to the alteration brought about by human intervention on the environment.

Nevertheless, by using these data carefully, the most significant climatic events during the Subatlantic climatic period which are detected in the Pollinic Diagrams are: The generalized A.P. descent and the increased development of *Quercus*, mainly the perennial ones, instead of other types of woody plants, are a response to warmer and drier conditions and, only when there is an intensification of the anthropic activity on the environment, are *Cerealia* and *Olea* developments that mark, together with small climatic changes, the vegetation evolution during this period. So, from the last 2000 B.P., "Iberia" is one of the most important provinces of the "Roman Empire", providing wood, cereals, wine, horses, metals to it. This leads to a big deforestation in mountain areas (altitude pastures - with the *Pinus* descent - as well as in the valley zones - cultivations instead of *Quercus*).

During the Middle Age (Bauer, 1980), the deforestation is total as a consequence of historical events: The Reconquest, the Invencible Army or the cattle hegemony over the agriculture (Mesta). From the middle of XVI to XVII (the last Austrias kingdom) there were *Pinus* reaforestation periods, alternated with cereal cultivations; the relict *Fagus* woods in the eastern sector (Montejo) are regenerated because of the better climatic conditions (Little Climatic Optimum) and there also was the Mesta development which favours the altitude shrub expansion. Little Ice Age is slightly detected in the Pollinic Diagrams of mountain areas, by the *Betula* percentage increase, because more cooler weather and snowing would result in an increase in raining. In the valley deposit (Valle de Ambles) the slight presence of *Betula* and *Quercus*-p and the absence of *Olea* indicate a greater development in favorable mountain areas during this period.

During the XVIII-XIX centuries, one of the most severe deforestations occurred, caused by the construction of the navy fleet and the desamortization even though the laws and the protection of some woods (Segovia), *Olea* and cereals cultivation in the valleys were more intense.

Finally, in the XX century, the *Pinus* reaforestation and the *Quercus*-p and *Quercus*-c improvement occurred. Given all these, we may conclude:

a) A vegetation evolution for the whole study area was found expressed by the substitution: Bt-p-Q-p+*Olea*, which was mainly controlled by the climatic changes.
b) Anthropic action was noticed relatively early in the lower zones, but its reflection in the altitude Pollinic Diagrams was not detected until 2000 B.P., as determined by the A.P. curve descent and by the anthropic taxa increase and development. Thus, during the Subatlantic climatic period the central region was characterized by a typical mediterranean vegetation, already established in lower zones since the Atlantic period with the *Pinus* dominance in central and eastern regions. In the western region *Pinus* was less represented and it even disappeared (Potugal, Estrela, Peneda) (Ruiz *et al.*, 1993); *Quercus* was dominant only in the most western region about at 1050 B.P., while in the rest of the central region it played a secondary role; *Quercus*-c was the dominant one in the eastern zone, and *Quercus*-p was in the central zone and in the low altitude deposits.
c) *Betula* was an accompanying element in altitude zones, mainly in the most western sector; in the central and eastern zones it was less developed but in the Little Ice Age changes it showed an increase.

The mediterranean wood indicates more oceanic conditions in the western sector, against the greater continentality reflected in the Pollinic Diagrams of the central and eastern sectors; only in the most eastern extreme, in small places, more oceanic vegetation can be found with *Betula* and *Fagus* (Gil *et al.*, 1993) because it is a wet winds collision zone. Although, the human activities (cutting, burning) and the soil use (pastures, cultivations) have resulted in environmental changes, climatic changes are also responsible to some extend for these environmental alteration (Little Climatic Optimum, Little Ice Age).

REFERENCES

Acaso E. (1983). Estudio del Cuaternario en el Macizo Central de Gredos. Universidad de Alcala de Henares (Tesis Doctoral).

Alia Medina M., Manendez Amor J. and Vidal Box C. (1957). Livret-guide de l'excursion C_3 and C_4 Guadarrama-Massif de Peñalara et variation El Escorial-Manzanares el Real. V Congrés Internacional I.N.Q.U.A., pp. 5-41. Madrid-Barcelona.

Atienza Ballano M., Gómez-Lobo Rodriguez A., and Ruiz Zapata B. (1990). Estudio polunico de un deposito localizado en la Garganta del Trampal (Sierra de Bejar. Avila). Actas de Gredos. Boletín Universitario 10, pp. 19-24.

Bauer E. (1980). Los montes de España en la Historia. Servicio de Publicaciones Agrarias. Madrid.

Dorado Valiño M. (1993). Evolucion de la vegetacion durante el Holoceno en el Valle de Amblés (Avila). Estudio Palinologico. Tesis Doctoral. Universidad de Alcala de Henares, pp. 212.

Garcia Anton M., Juaristi M., Ruiz Zapata B., and Sainz Ollero H. (1986). Contribución al conocimiento del paisaje vegetal en la Submeseta Sur Iberica. Analisis polínico de sedimentos hidroturbosos en el Campo de Calatrava. Quaternary Climate Western Mediterranea, Vol. 1, pp. 198-204.

Gil Garcia M.J. (1992). Dinamica de la Paleovegetación en el sector oriental del Sistema Central Español durante el Holoceno, en base al analisis polínico. Implicaciones climaticas. Tesis Doctoral.

Universidad de Alcala de Henares, pp. 221.

Gil Garcia M.J., Martin Arroyo T., Tomas Las Heras R., and Ruiz Zapata M.B. (1993). Reconstrucción del paleoambiente en el macizo de Ayllon (Sistema Central Español) durante el Holoceno a traves del analisis polinico. III Reuniao do Quaternario Iberico. Coimbra. Portugal (in press).

Goeury C. (1987). Acquisition, gestion et representation des donnes de l' analyse pollinique sur micro-ordinateur. Inst. Fr. Pondichery. Trav. Sec. Sci. Tech., 16: 405-416.

Grimm E.C. (1992). Graphic Software Systems (*Tilia*), Inc., P.O. Box 4900 Beaverton. Oregon 97005. USA.

Jimenez Ballesta A., Lõpez Martinez J., Lõpez Garcìa P., and Ibañez J.J. (1985). Contribución al conocimiento de las formaciones superficiales turbosas en las Sierrras de Guadarrama y Ayllon. Analisis polínicos. Actas I Reunión del Cuaternario Iberico, 1: 213-224. Lisboa.

Lopez Garcia P. (1982). Estudio palinológico del yacimiento de Lavapes. Museo de Pontevedra, pp. 3-7.

Pedraza J. (1978). Estudio Geomorfológico de la Zona de Enlace entre las Sierras de Gredos y Guadarrama (Sistema Central Español). Tesis Doctoral. Universidad Complutense de Madrid, pp. 540.

Ruiz Zapata M.B., Gil M.J., Dorado M., Andrade A., Atienza M., Gomez L. A., and Martin T. (1992). Evolución durante el Cuaternario Reciente de los Taxa arboreos mas representativos en el Sistema Central y Sistema Iberico. Actas III Cong. de Geol. de España y VIII Latinoamer.de Geol. Tomo, I: 554-559.

Ruiz Zapata M.B. and Acaso Deltell E. (1984).Clima y vegetación durante el Cuaternario en el Macizo Central de Gredos (Sistema Central Español). Geologia, T.,1: 723-740.

Ruiz Zapata M.B. and Acaso Deltell E. (1981). Contribución al estudio del cuadro Vegetal y climatico durante el Cuaternario reciente en el Macizo Central de Gredos (Avila). Bol.R. Soc. Esp. Hist. Nat., (G), 79:299-307.

Ruiz Zapata M.B., Garcia Anton M., and Acaso Deltell E. (1986). Datos polínicos para el conocimiento de la Vegetación en el Macizo de Peñalara (Sierra de Guadarrama). Actas de Palinologia. VI Simposio de Palinologia, pp. 351-354. Salamanca.

Ruiz Zapata M.B., Garcia Anton M., Vazquez G.R., Gil Carcia M. J., and Andrade Olalla A. (1988). Analisis polinico de dos turberas, localizadas en el Macizo de Peñalara (Sierra de Guadarrama. Madrid). II Congreso Geologico de España. Comunicaciones, Vol.1: 329-332. Granada.

Ruiz Zapata M.B., Correira A.I., Daveau S., and Lecompte M. (1993). Datos preliminares sobre la evolución de la vegetación en las Sierras del no de Portugal durante el Holoceno. Actas III Reuniao do Quaternario Ibérico, pp. 60

Van Den Brink L.M. and Janssen C.R. (1985). The effect of human activities during phases on the development of montane vegetation in the Serra da Estrela, Portugal. Rev. Palaeobot. Palynol., 44: 193-215.

Vázquez Gómez R. (1992). Evolución del Paisaje Vegetal durante el Cuaternario Reciente en la Zona Central y Oriental de la Sierra de Guadarrama, a partir del Análisis Polinico Tesis Doctoral. Universidad de Alcala de Henares, pp. 401.

CHAPTER 3

GLOBAL WARMING AND GAS TRANSFER AT THE AIR-WATER INTERFACE

IOANNIS K. TSANIS[1] AND FRANCOIS P. BRISSETTE[2]

[1] *Department of Civil Enginering, McMaster University Hamilton, Ontario, and*
[2] *École de Technologie Supérieure, Université du Quebec, Montreal, Quebec, CANADA*

3-1. INTRODUCTION

Importance of the Work

On the eve of the year 2000, humans are now facing a major environmental crisis. The atmosphere and oceans are used as dumps, forests are being logged at rates of hectares per second, natural resources are being depleted, the best agricultural top-soils are impoverished or eroded away, fish stocks are decimated and animal species are disappearing at an alarming rate. We are now witnessing the most dramatic changes to affect the earth since the dawn of civilization and it is becoming obvious that the magnitude of the stresses imposed on the earth far exceed its capacity to self-regulate, except perhaps in the context of Lovelock's (1979) "Gaia" hypothesis. Since the industrial revolution, humans have been blindly conducting a planetary-scale natural science experiment, and only recently have they been able to foresee some of the end results of this uncontrolled experiment.

Life as it is now known on earth is the result of a complex and fragile interaction between the earth's surface and its atmosphere. The earth's breathing and living life is concentrated in a remarkably thin layer of water and air between the earth's crust and the seemingly infinite outerspace. This thin layer, with all its diversity and complexity is the result of hundreds of millions of years of a not-fully understood interaction between the earth and the sun's energy, driving the interaction and equilibrium between the earth's water and atmosphere. Dramatically increasing environmental stresses and pollution are now putting serious strains on the earth's climate. The magnitude of the stresses and the results on the earth's atmosphere and climate are slowly starting to emerge. Of all the possible effects, the best-known is probably the global warming of the earth's climate, the so-called "greenhouse effect", which is the direct consequence of, until now, unrestrained

NATO ASI Series, Vol. I 36
Diachronic Climatic Impacts on Water Resources
Edited by Andreas N. Angelakis and Arie S. Issar

emission of gases in the atmosphere. The greenhouse effect is the result of enhanced conversion of solar radiation into heat in the lower levels of the atmosphere, due to the effect of greenhouse gases, mainly carbon dioxide (CO_2). The process involves selective transmission of shortwave radiations by the atmosphere which is absorbed by the earth's surface followed by infrared reradiation. An important part of the earth infrared reradiation is then absorbed and reradiated back to the surface by the greenhouse gases. The process has always occurred naturally due to the effect of water vapour but the rapid increase of carbon dioxide in the atmosphere is disrupting the equilibrium with the result that heat is presently accumulating in the lower levels of the atmosphere. All other factors being constant, if more heat is reradiated back toward the surface by increased greenhouse gases, the heat budget equilibrium will only be reestablished if the temperature gradient in the lower levels of the atmosphere increases. This can only be achieved through a temperature rise. Experts argue that this should cause an increase of the average earth's surface temperature between 1.5 and 4.5°C over the next 50 years, a very large and instantaneous change compared to the geological time scale (Regional Seas, 1990). This warming is predicted to be larger in middle and high than in low latitudes.

Direct effects of this global warming are very serious and potentially devastating. The icepack volume reduction could produce a sea level rise of 1.5 m, sufficient to flood 10- 15% of coastal areas, where a disproportionally large percentage of populations are concentrated. The sudden temperature change could be fatal to many animal species which are temperature sensitive, it would in turn affect animal species by disrupting living habitats and eventually inducing changes in the food chain. The extent of these changes is still unknown, as scientists are trying to better understand the physical processes linking the earth's energy budget to anthropogenically induced changes on the earth's surface and in the atmosphere. It is probably safe to say that science's great challenge of the 21st century will be to deal with environmental issues as humans are now only starting to assess long-term effects of sustained development growth.

Examination of the principal carbon reservoir sizes on the earth (Liss and Crane, 1983) indicate the smallness of the atmosphere as a reservoir for carbon relative to the land biosphere, sedimentary deposits, the oceans and fossil fuel reserves. Small transfer from the above reservoirs to atmosphere can have a large effect on the latter. The yearly exchange of the carbon between the land biosphere and the oceans and the atmosphere is estimated as 102 Gt Cyr^{-1}. Man-made CO_2 emissions are estimated to 5 Gt C yr^{-1} from fossil fuel burning and up to 2 Gt C yr^{-1} from deforestation and changes in land use. It is estimated that 2 to 3 Gt C yr^{-1} are taken by the ocean and 2.5 Gt C yr^{-1} remain in the atmosphere, leading to an observed increase in atmospheric CO_2 partial pressure.

The amount of CO_2 in the air has increased by about 30% over the last 150 years but the overall quantity is nevertheless small compared with the amount of CO_2 contained in the oceans (150 times more) and it is not well understood why the ocean has not taken up more of the anthropogenic CO_2, since presumably it was in an equilibrium state 150 years ago. The ocean is a natural desorber of CO_2 and clearly, important questions need to be resolved. Just how much CO_2 is or should be absorbed by the ocean, is this rate of absorption going to be proportional to the accelerated input? Which factors control the absorption of CO_2 by the ocean? The accommodating capacity of the ocean appears to be smaller compared to accelerated input and it seems that the oceans take up about half of CO_2 released by fossil fuel combustion (Regional Seas, 1990).

The overall problem of CO_2 cycling is complex and global in perspective as it is also linked to deforestation and increased development in third world countries. Still the ocean is the key link in the problem of CO_2 equilibrium, and knowledge of the physical parameters that control the air-water exchange of CO_2 and other gases is fundamental to an increased understanding that will give us the knowledge needed for the application to advanced coupled ocean-atmosphere climatic model which will be a fundamental tool in assessing the effect of anthropogenic changes on the earth's climate, and pave the way to better insights on the concept of sustainable growth, if such a thing is possible.

The gas exchange that occurs at the air-water interface is the result of complex physico-chemical processes. Beyond diffusion, Jahne *et al.* (1987) and others have shown that mass transfer is dramatically affected by the presence of waves. The transfer can be linked to the turbulent intensity in both the air and water phases, and processes such as wave breaking and the injection of air bubbles in the liquid phase, and conversely, spray of liquid droplets in the atmosphere also has an effect. This paper is an attempt at reviewing the literature on various topics linked to the problem of gas exchange at the air-water interface, so as to provide a general overview of the physical processes controlling the exchange and models presently used in calculating the mass transfer of CO_2 across the interface.

Methods for Evaluating Global CO_2 Exchange

Evaluation of the exchange coefficient for CO_2 between air and sea on a global scale requires knowledge of the transfer velocity, which is a function of wind speed and wave conditions, and CO_2 partial pressure difference between the atmosphere and the ocean. The CO_2 flux is given by the product of CO_2 partial pressure gradient, transfer velocity and solubility of CO_2 in seawater.

In the field there are a number of methods used to estimate the transfer velocity. These methods are (a) box method, (b) dissolved gas balance method, (c) micrometeorological techniques, (d) natural and bomb-produced ^{14}C, (e) the radon deficiency method and (f) sulphur hexafluoride. All these technologies are briefly reviewed by Liss and Merlivat (1986). Based on some field and laboratory measurements, they proposed an empirical relation between transfer velocity at 20°C and wind speed which has been used in many CO_2 flux estimation studies.

The other quantity which determines the CO_2 exchange flux is the partial pressure difference ΔpCO_2 between air and sea. The partial pressure of CO_2 at a given location in the surface ocean is determined by the complex interaction of several processes such as (a) upwelling, depth of mixed surface ocean layer, horizontal advection of waters with different history, (b) gas exchange with the atmosphere, and (c) chemical equilibrium with the different components of dissolved inorganic carbon and activity of the marine biosphere. The difference ΔpCO_2 on a global scale was measured amongst others by Keeling, (1968) and Takahashi *et al.*, (1983).

Etcheto *et al.*, (1991) measured the wind speed by a microwave radiometer SSM/I (Special Sensor Microwave Imager) which retrieves the wind speed at 19.5 m effective height. Then they converted these measurements to 10 m effective height assuming the atmosphere to be neutral and using 1.5×10^{-3} for the drag coefficient. The Liss and Merlivat relationship between the transfer velocity at 20°C and the wind speed at 10 m height above the ocean surface was then used to evaluate the exchange coefficient K (mol m^{-2} yr^{-1} μatm^{-1}) which is proportional to transfer velocity k (cm h^{-1}) (Etcheto and Merlivat, 1988). The measurements were binned in boxes 2.5° x 2.5° in latitude and longitude and the average transfer velocity over a given region was computed by interpolation and integration over the region. Knowledge of the air-sea difference of the CO_2 partial pressure permits the computation of the net flux exchange between air and ocean by interpolating and integrating over the globe the product of transfer velocity and interfacial CO_2 partial pressure difference.

Tans *et al.* (1990) showed that the available experimental data at the ocean global scale are still sparse and allows an estimate of the mean value and of the seasonal variation of CO_2 partial pressure flux only for two periods of the year in the northern and equatorial areas (there is no such information for the southern ocean). Etcheto *et al.* (1991) by using satellite wind speed measurements, found that the seasonal variation of the CO_2 exchange coefficient is large and should be taken into account in net flux computations between the ocean and atmosphere.

They found the main variation is north of 40° N, by a factor of 4 between July and January. In the zone 10° N-40° N, the variation is 2 while it is 1.5 at the corresponding south latitudes. Strong winds in autumn and winter cause higher values of exchange coefficients during these periods (the seasonal variation is a factor of 1.6, maximum in the winter).

The partial pressure of CO_2 in the ocean surface varies spatially and temporally by up to ±100 ppm (Regional Seas, 1990). Supersaturated waters with ΔpCO_2 of up to 100 ppm can be found close to equator. Unsaturated waters with less well known distribution of ΔpCO_2 (due to seasonal bias in the available measurements) can be found to the north and south regions.

Climate Change in the Mediterranean Sea

The climate of Mediterranean region is characterized by its winter-dominated rainfall, dry summers and local microclimates due to terrain changes, i.e., mountains and inland seas. If the temperature increased 1.5°C by the year 2025 (greenhouse effect), it is expected to lead to substantial changes in climate at the regional level over the next century. It will cause an increase in the evapotranspiration throughout the Mediterranean and decrease/increase in precipitation in the South/North, respectively. The frequency of floods, storms, and hot dry summers could increase. The Mediterranean region will be vulnerable, particularly through changes in rainfall and water supply that will have implications on agriculture and in domestic and industrial supply.

A fully detailed scientific study discussing the "Greenhouse Effect" impact on the Mediterranean region is given by Jeftic *et al.* (1993). The above study combines scientific/environmental studies and economic/demographic/social studies for the Mediterranean region.

Simulations of climate change. Wilson and Mitchell (1985) used the UK Meteorological Office 5-layer model to determine climate changes (mean annual cycle and the frequency of extreme events) over Western Europe. Although the broad features of the simulation were found satisfactory, the model produced too many cold episodes in spring, and an excessive number of wet days over northern Europe. Under the scenario of quadrupling CO_2 concentrations, the number of cold episodes was reduced and precipitation was less frequent in summer and autumn over much of Europe, and throughout the year in the south.

The above results should be regarded with caution in view of the model's deficiencies in its simulation of the present climate, the approximate nature of the specified changes in sea surface temperature and the short sample period used given the size of the area considered. However, the results provide a useful illustration of the general nature of changes that might be expected with increases in CO_2, even though the changes at a given location may not be correct. If regional data is important, then a finer resolution than the 500 km or more which is typical of many current climate models is required, and a period of integration longer than the three years used in this study is needed.

Palutikof *et al.* (1992) used the grid-point-output from four General Circulation Models (GCMs) to study the climate variables, mean sea level pressure (MSLP) and precipitation (based on the doubled CO_2 scenario). The simulated results were compared with observations for grid-points in the study region. Their direct scenarios indicate that the temperature change due to the greenhouse effect for the Mediterranean Basin should be similar to the global response. Precipitation is shown to increase in autumn and winter but decrease in summer and, particularly in the eastern Mediterranean, in spring also. The mean change is, in winter, around +3%/°C and in spring also. However, the confidence in the model scenarios of precipitation is low because of the uncertainty associated with the GCM results.

Wigley *et al.* approach (1994) was used to study seasonal and annual scenarios of the regional changes in temperature, precipitation and MSLP where the information from the four models is combined into a single scenario for each variable. The methods developed by Kim *et al.* (1984) and Wigley *et al.* (1990) were used in order to produce a set of high resolution scenarios based on the statistical relationship between grid-point GCM data and observations from surface meteorological stations. The temperature scenarios constructed were an improvement on the previous direct scenarios. The scenarios of precipitation due to added uncertainty introduced between the point precipitation anomalies and the regionally- averaged predictor variables, is not considered an improvement to the direct scenarios.

The occurrence of extremes (frequency and severity of droughts and floods) was studied by a two-part statistical model of precipitation which combines a first-order Markov chain (precipitation occurrence process) and a gamma function (distribution of precipitation amounts on wet days). The model was applied to daily precipitation records and the simulated results were compared with the observed ones. The implications for extreme events were examined by studying the change in the maximum and minimum precipitation amounts per wet day, and in the length of wet and dry spells after a change in the mean precipitation of 10% was applied. Changes in the order of 20% were not unusual and have particular implications in areas where water supplies are marginal.

Very recently (UNEP, 1993) a detailed scenario was constructed by the Climate Research Unit of the University of East Anglia, for the eastern Mediterranean and for the island of Rhodes, on the basis of GCM scenarios of greenhouse effects. These scenarios include an increase in mean annual temperature of 0.9°-1.4°C and 1.5°-3.3°C by the years 2050 and 2100 respectively; and a sea level rise of 20-30 cm and 60-100 cm by the years 2050 and 2100 respectively.

Annual and seasonal scenarios for both temperature and precipitation change were produced. For temperature, the greatest sensitivity to the greenhouse effect was found in the mainland areas to the north-east of the study region. Temperature increases less than the global mean temperature change were indicated for the south-west of the region. The scenarios of precipitation were difficult to evaluate and of low confidence.

Impact of climate change. The impact of climate change will be felt in society. The increase in temperature will increase evapotranspiration and decrease total runoff and infiltration. The ground water reserves will diminish due to lower infiltration resulting in lower levels of the aquifers and in drying up of springs. These factors will increase the vulnerability of the society to climatic stress, particularly in the South. These demographic and climate changes will increase the frequency of catastrophic events (Regional Seas, 1992).

Other effects of climate change in addition to deterioration of water resources will be on agriculture and aquatic ecosystems. Inadequate irrigation systems, soil degradation and salinization of irrigation water will have negative consequences on sensitive yield. A possible change of marine circulation will alter pollutant dispersal patterns and marine productivity. A sea level rise of 20 cm by 2025 will affect low-lying seas, deltas and coastal cities. These areas are already experiencing environmental problems due to development such as water pollution including salinization, deterioration of wildlife habitats and shoreline erosion. These problems will be increased by adverse socio-economic conditions, the effects of which will be superimposed upon those of climate change (Regional Seas, 1992).

Flora and fauna of the wetlands will adapt to the new environmental conditions while marine land weeds will benefit from the excess CO_2. Other species that don't have tolerance to high salinities will disappear. The shift to the north of the bioclimatic zonation will force migration of several species, and probably increase insect populations. The latter will set favourable conditions for an increasing health risk from agricultural pests, bacteria and diseases, especially in the swamps.

The major impacts in the study of the expected climatic changes on the Island of Rhodes are similar to those previously mentioned, and in short will occur due to: (a) temperature rise (rapid filling of the surface reservoirs, dry up of the springs, diminish the deterioration of ground water reserves due to salinization) and (b) sea level rise (soil erosion, coastal zone). In addition, the tourist season will be lengthened, the maquis ecosystem will be changed and incidence of fire will be increased.

A number of strategies should be formulated and preventive or reactive measures should be implemented for responding to the impacts of climate change. Some of these strategies are (Regional Seas, 1992):

a) Less urbanized beaches.
b) Replacement of extensive, uneconomical crops in low-lying lands with lagoons.
c) Storm impact maps for the "high risk" areas (coastal zone protection) - study coastal dynamics of erosion and deposition and the relative span of human installations.
d) Engineering solutions (building dikes and walls).
e) Conservation of soil, groundwater and wetland resources (environmental stability).
f) Control coastal development, land reclamation and groundwater exploitation.

Specific strategies for handling the impact of climate change on the Island of Rhodes are (UNEP, 1993):

a) A coastal zone management plan (land use zones, building standards etc.).
b) An integrated water management plan (water supplies, dams, boreholes in aquifers, avoidance of infiltration of saline waters into the aquifers, etc.).
c) Reforestation.
d) Evaluation of the consequences of lengthening the tourist season.

3-2. GAS TRANSFER MODELS

Fundamentals

Henry's law, expressed as:

$$P_g = HC_1 \tag{3-1}$$

where P_g is the gas partial pressure (Pa), and C_l is the gas concentration in the liquid expressed (mol/m^3) indicates that the quantity of gas that will dissolve in a liquid is proportional to its pressure in the gas phase. Henry's constant (Pa . m^3/mol) is often expressed in dimensionless form as:

$$C_a/C_i = H' = H/RT \tag{3-2}$$

using the gas law with T, the absolute temperature and R the gas constant (8.314 Pa m^3/mol/°K).

Henry's law provides a relationship to determine the equilibrium concentrations of a gas on each side of the gas-liquid interface. What it does not provide is information on how easily the equilibrium will be reached, or on how rapidly a gas will cross the interface. Depending on the gas solubility or reactivity within the liquid phase, gases are said to be "liquid phase controlled" or "gas phase controlled". For gases that have a low solubility in a liquid, the main resistance at the gas-liquid interface is in the liquid phase and the gas is under "liquid phase control" or "water phase control", since the liquid this review is concerned with is water (gases such as O_2, N_2, CO_2, CO). For gases that have a high solubility or high reactivity in a liquid, the main resistance to the transfer is in the gas-phase and the gas is under "gas phase control", or "air-phase control" (gases such as H_2O, SO_2, SO_3).

Central to all gas-transfer problems is the gas flux equation, (Fick's law) given for a one dimensional problem as:

$$F = -D\, \partial C/\partial z \tag{3-3}$$

where F is the gas flux or net rate of transport across any plane perpendicular to the z axis, $\partial C/\partial z$ is the concentration gradient at z, and D is a diffusivity coefficient.

The rate of mass accumulation is described by the diffusion equation

$$\frac{\partial C}{\partial t} = D \frac{\partial^2 C}{\partial z^2} \tag{3-4}$$

Fick's law can be rewritten in a discrete form as

$$F = k\Delta C \tag{3-5}$$

with

$$k = D/\Delta z \tag{3-6}$$

k is a mass transfer velocity across the layer of thickness Δ_Z (units of m/sec), or more precisely a measure of the gas flux per unit concentration difference, also called the exchange constant by Liss and Slater (1974), a term slightly misleading since k is not a constant and depends on many factors.

Basic Models

Gas transfer at the air-water interface is a complex problem involving several physical processes. Insights to the real phenomenon can be obtained by looking at simplified physical models. The aim of basic models (as well as any models) is to find a representation of the mass transfer velocity.

Film model. The film model, proposed by Whitman (1923), idealizes the air-water interface as a two-layer film system. The gas exchange is realized by molecular diffusion within the thin layers on each side of the interface, whereas the bulk of the water phase (and air phase) is considered to be well mixed. Since the transport through the thin films is the result of molecular diffusion, Fick's Law (Eq. 3-3) is applicable. If the process has reached a steady state, the flux will be constant across the interface and using the discrete form of Fick's Law (Eq. 3-5), the mass flux can be expressed as

$$F = k_a (C_a - C_{sa}) = k_w (C_{sw} - C_w) \tag{3-7}$$

where the subscripts "a" and "w" respectively refer to the bulk of the air and water phases, and "sa" and "sw" respectively refer to the air and water phases in the sub-layers on each side of the interface.

The film model clearly idealizes the air-water interface and as such, its applicability to real problems may be called into question but, nevertheless, as mentioned by Danckwerts (1970), the film model recognizes an essential feature of gas transfer, in that before the gas can be advected or convected in the liquid, it must first get into it by molecular diffusion. The model is also useful in helping visualize processes at the interface (Liss and Slater, 1974), and furthermore (at least up to the mid 70's), predictions based on the film model showed little or no difference from those derived using more sophisticated models (Liss and Slater, 1974 and Danckwerts, 1970).

As pointed out by Liss and Slater (1974) the film model can be especially useful in expressing the resistance to the gas transfer process, which is due to both phases across the interface. At the interface, the gas obey's Henry's Law (3-1) such that (3-7) can be rewritten as:

$$F = \frac{1}{R_w}\left[\frac{C_a}{H} - C_w\right] \tag{3-8}$$

where R_w is a measure of the ideal resistance to the gas transfer on each side of the interface such that:

$$R_W = \frac{1}{k_W} + \frac{1}{H\,k_a} = (r_W + r_a) \tag{3-9}$$

and r_a and r_W represent the contribution of the air and water phases to the total resistance to gas transfer in the system.

Surface renewal models. In the first of such models, Higbie (1935) assumed that the top layer was periodically replaced by fluid with the bulk concentration C_{w0}. The layer is replaced at a time interval T, after having absorbed a quantity of gas Q. The average gas flux $\overline{F}$ is then defined as:

$$\overline{F} = \frac{Q}{T} \tag{3-10}$$

The gas transfer is achieved through physical diffusion (no reactivity) and in such a case, the diffusion equation is solved under the boundary conditions ($z = z_0 \rightarrow C_w = C_a$) which leads to the solution

$$C_w - C_{w0} = (C_a - C_{w0})\left[1 - \operatorname{erf}\left(\frac{z}{2\sqrt{D\ t}}\right)\right] \tag{3-11}$$

The flux of gas at $z = z_0$ is obtained from (3-11) and (3-3) and can be shown to be equal to:

$$F = (C_a - C_{w0})\ \sqrt{D/(\pi\ t)} \tag{3-12}$$

The mass transfer velocity k is directly obtained from the average flux $\overline{F}$ during a time interval T and is equal to:

$$k = \frac{\overline{F}}{C_a - C_{w0}} = 2\ \sqrt{D/(\pi\ T)} \tag{3-13}$$

The Higbie model specifies the same time of exposure T for all elements of the surface, an unrealistic assumption (Danckwerts, 1970). To overcome this problem, Danckwerts (1951) proposed a distribution in which the fraction of the surface renewed as a function of time Fr(t) is given by the exponential function:

$$F_r(t) = 1 - e^{-st} \tag{3-14}$$

where S is the surface renewal rate (for values of S<<1, S is equal to the fraction of the area of surface which is replaced in unit time). This means that on the surface, the distribution of "ages" D(t) (the fraction of the surface which has been exposed for a time t) is given by:

$$D(t) = S\,e^{-St} \quad \text{with} \int_0^\infty S\,e^{-St}\,dt = 1 \tag{3-15}$$

The average flux $\overline{F}$ is then simply given through (3-12) averaged over all elements of the surface (of all ages). The mass transfer velocity k is then given by:

$$k = \sqrt{DS} \tag{3-16}$$

In both of these renewal models, all the hydrodynamic properties of the system are accounted for by one single parameter: the time of exposure T (units of time) in the Higbie model and the surface renewal rate S (units of reciprocal time) in the Danckwerts model. The Danckwerts (1951) model implies that the surface renewal rate S is equal to $4/\pi$ times the inverse of the exposure time T of the Higbie (1935) model ($S = 4/\pi T$).

Rigid wall models. These models (also called boundary layer models (Liss and Merlivat, 1986)) link gas transfer to the exchange of momentum over a hydrodynamically smooth surface, allowing use of existing knowledge on transfer over rigid, smooth surfaces (Deacon, 1977). Both Münnich and Flothmann (1975) and Deacon (1977), on the basis of the universal velocity profile over a smooth surface, derived models predicting the mass transfer velocity k with a relationship of the form:

$$k = \frac{1}{\beta}\left[\frac{D}{v}\right]^n u_{*w} \tag{3-17}$$

where v is the water kinematic viscosity and u_{*w} is the friction velocity in the water phase. Deacon's model is expressed as:

$$k = 0.082 \sqrt{\rho_a / \rho_w} \left(\frac{v}{D}\right)^{-\frac{2}{3}} u_{*a} \tag{3-18}$$

where ϱ is the density and the subscripts have their usual meanings.

General expression of basic models. The film, renewal and rigid wall models represent the three classes of models outlined by Liss and Merlival (1986). All of these models can be represented in a simple form by (Jähne *et al.*, 1984):

$$k = \beta^{-1} Sc^{-n} u_{*w} \tag{3-19}$$

where the Schmidt number $(Sc = v/D)$ is the ratio of the molecular transport properties (Jähne *et al.*, 1987) with v representing momentum and D the diffusion. For Deacon's rigid wall model the values of β and n are found to be 12 and 2/3 respectively. Münnich and Flothmann (1975) fitted the sub-logarithmic part of the universal velocity profile with a surface-renewal model (Jähne *et al.*, 1979) and found respective values of 16 and 1/2. Finally, a similar treatment with the film model yielded values of (Jähne *et al.*, 1984) 11 and 1. A comparison of the models is shown in Jähne *et al.* (1984) where it was found that data for rough surfaces were better represented by surface renewal models.

Combinations of the film and surface renewal models have also been presented by Dobbins (1956); Toor and Marchello (1958); Harriot (1962) and Marchello and Toor (1963). Typically, these incorporate both models as end members of a continuous spectrum of the combination of both models.

Other Models

The surface renewal theory of Danckwerts (1951) has probably received the most attention and considerable work has been put into conceptual theories that identify as the source of renewal either small eddies and large eddies (the so-called eddy- structure models) or eddy diffusivity (Gulliver, 1991). Both models can be looked at by considering the general mass transport equation. For a mean flow velocity U in the x-direction, the mass transport equation can be written as

$$\frac{\partial C}{\partial t} + (U+u)\frac{\partial C}{\partial x} + v\frac{\partial C}{\partial z} + w\frac{\partial C}{\partial z} = D\left\{\frac{\partial^2 C}{\partial x^2} + \frac{\partial^2 C}{\partial y^2} + \frac{\partial^2 C}{\partial z^2}\right\} \tag{3-20}$$

where u, v and w are the fluctuating velocity component and D is the molecular diffusion coefficient. A characteristic thickness of the concentration boundary layer at the interface can be defined as (Hanratty, 1991);

$$\delta = D/k \tag{3-21}$$

For typical values of the diffusivity D the thickness δ is quite small and the derivatives in the x and y directions are small. Then:

$$\frac{\partial C}{\partial t} + w\frac{\partial C}{\partial z} = D\frac{\partial^2 C}{\partial z^2} \tag{3-22}$$

From (3-22) it can be seen that mass transfer across the air-water interface is controlled by the diffusivity of a given gas and by velocity fluctuations normal to the interface. For a mobile surface, the velocity fluctuation can be represented as $w = \beta(x,y,t)z$ (McCready *et al.*, 1986) and as such, β becomes the variable of interest, to be linked with properties of the turbulent system. Recently, new approaches have looked at the problem of mass transport by trying to solve numerically the time dependent mass transport equation (Eqs. (3-20) or (3-22)) (McCready *et al.*, 1986; Campbell and Hanratty, 1983 and Hanratty, 1991). Such approaches have a definite potential to applications of mass transfer problems in turbulent flows, if the near-surface turbulence can be measured accurately (Gulliver, 1991). Earlier approaches have focused on steady state solutions of the mass transport equations using idealized turbulence structure models (eddy structure models) or eddy diffusivity concepts.

Eddy structure models. Eddy structure models are directly related to the surface renewal model of Danckwerts (1951). They try to express the renewal time as a ratio of a length and a velocity scales, characteristics of the flow turbulence structure. These models idealize the turbulence and ignore damping effects close to the interface in order to select appropriate scales. The surface renewal time S can be expressed as the ratio of a characteristic velocity scale V_c and a characteristic length scale L_c ($S = V_c/L_c$) such that:

$$k = \sqrt{D\ V_c / L_c} \tag{3-23}$$

Two main models have been put forward to identify the velocity and length scales: the large eddy model (Fortescue and Pearson, 1967) and the small eddy model (Lamont and Scott, 1970).

Large eddy model. Fortescue and Pearson (1967) assumed that the mass transfer toward the turbulent liquid was controlled by large scale eddies in the turbulent flow. Associated with the large scale flow, the velocity scale was taken to be that of the bulk of the turbulent flow field. Assuming that the large eddies were rounded cells of dimension Λ they solved numerically the steady state mass transport equation and found:

$$k = 1.46 \sqrt{D\ U_{rms} / \Lambda} \tag{3-24}$$

where U_{rms} is the root-mean-square velocity of the turbulence fluctuations and Λ is a characteristic lengthscale of the eddies. This model identifies the velocity scale as Urms and the characteristic length scale as being very close to $\Lambda/2$.

Small eddy model. On the other hand, Lamont and Scott (1970) assumed that mass transfer by molecular diffusion was controlled by smaller eddies. Idealizing the motion of the small eddies, the mass transfer velocity was solved analytically as:

$$k = 0.4 \sqrt{D \varepsilon^{1/2} / \nu^{1/2}} \tag{3-25}$$

where ε is the rate of dissipation of energy. For the small eddy model, the characteristic velocity and length scales can be identified as the Kolmogorov characteristic micro-scales in the dissipation range defined as: $L_c = (\nu^3/\varepsilon)^{1/4}$ and $V_c = (\nu\varepsilon)^{1/4}$.

The rigorous application of either the small or large eddy models would require precise measurements of turbulent flow parameters, such as dissipation rate, which are rather difficult to obtain. In order to ease the application of eddy structure models, Brtko and Kabel (1978) developed methods to approximate the necessary input parameters. They also derived guidelines for the applicability of both models as a function of a turbulence based Reynolds number.

Eddy diffusivity models. In these models, the turbulent diffusivity is taken into account in the form of a turbulent eddy diffusivity term D_t likened to the molecular diffusivity D. In such cases, the fluctuating velocity components are not taken into account and (3-20) can be rewritten

$$\frac{\partial C}{\partial t} + U \frac{\partial C}{\partial x} = \frac{\partial}{\partial z} \left((D + D_t) \frac{\partial C}{\partial z} \right) \tag{3-26}$$

For mobile interfaces, the turbulent eddy diffusivity D_t is usually represented as:

$$D_t \propto Sc\, (y u_* / \nu)^2 \tag{3-27}$$

which yields

$$k \propto u_*\ Sc^{-1/2} \tag{3-28}$$

As pointed out by Hanratty (1991) eddy diffusivity models can be related to eddy structure models by expressing the eddy diffusivity as the product of a characteristic velocity scale and a characteristic length scale. Eddy diffusivity models have been proposed by King (1966),

Lamurelle and Sandall (1972), Prasher and Frickie (1974) and Henstock and Hanratty (1979) amongst others. A review of such models is presented in Theofanous (1984).

Gas-transfer Models: Conclusion

The described models (and others) all try to express gas transfer rates in terms of known physical parameters. All models, in order to calculate mass transfer, require the determination of one or many of the parameters which control the turbulent system. From the basic models which require the thickness of the film layer or a renewal rate, to the solution of the three-dimensional time-dependent transport equation which requires a representation of the turbulent velocity field, an adequate application and/or verification and comparison of any model demands knowledge about the physical process operating at the interface.

3-3. EXPERIMENTAL DATA

The Effect of Waves

It has been known for quite some time that waves have a significant enhancement effect on gas transfer. MacIntyre (1971) demonstrated a 3.5 times increase in the gas transfer velocities simply due to the surface dilation associated with fully developed capillary ripples. Following various flume studies, the influence of waves on mass transfer rates has been noted by many authors (Kanwisher, 1963; Broecker *et al.*, 1978; Flothman *et al.*, 1979; and Jähne *et al.*, 1979). In this last study, the authors, in trying to explain discrepancies between the different experiments, clearly stated that in addition to wind speed or shear velocity, wave spectral characteristics would have to be taken into account. Jähne *et al.* (1987) compared experimental data from various authors (Liss *et al.*, 1981 and Merlivat and Memery, 1983; McCready and Hanratty, 1984) and noticed variations up to a factor of four in gas transfer for similar shear velocity and independently of the scale of the wave field. This fact, coupled with new experimental data (including wave slope data) prompted them to propose the mean square slope of the waves as the best parameter to characterize the water surface for gas transfer purposes.

Liss and Merlivat (1986), produced an idealized plot of the mass transfer velocity for water phase controlled gases, based on measurements made in Rockland Lake by Wanninkhof *et al.* (1985) up to 8 m/s and on wind tunnel observations made by Broecker and Siems (1984) and Merlivat and Memery (1983) (see Fig. 3-1). At low wind speed (<3.6 m/sec) which is called the smooth surface regime by Liss and Merlivat (1986) the gas transfer velocity is proportional to $Sc^{-2/3}$ and directly proportional to the water phase shear velocity as predicted by the rigid-wall model of Deacon (1977). This regime is characterized by the absence of a well developed wavefield, which emphasizes the appropriateness of Deacon's model. Gas transfer is then suddenly enhanced at a speed of roughly 3.6 m/sec (Liss and Merlivat, 1986) which corresponds to a rough sea, with little breaking. In that regime (rough surface regime) the gas transfer velocity is proportional to $Sc^{-1/2}$. Furthermore, in that regime, the proportionality constant ß (Equ. 3-33) is not constant but depends on the wavefield (Jähne, 1991b; Broecker *et al.*, 1978 and Broecker *et al.*, 1986). Finally, at wind velocities greater than about 13 m/sec, a second enhancement is observed (breaking wave regime) which is linked to the widespread appearance of wave breaking. This enhancement is thought to be caused by bubble injection and intense mixing caused by wave breaking (Kanwisher, 1963 and Woolf and Thorpe, 1991). Based on this idealized figure, Liss and Merlivat (1986) proposed the following relationship for the variation of gas transfer velocity k in cm/hours (for S≈600) as a function of the wind velocity U (m/sec) at a 10 m height:

$$k = 0.17\,U \qquad U \leq 3.6 \qquad (3\text{-}29a)$$

$$k = 2.85\,U - 9.65 \qquad 3.6 < U - 13 \qquad (3\text{-}29b)$$

$$k = 5.9\,U - 49.3 \qquad U > 13 \qquad (3\text{-}29c)$$

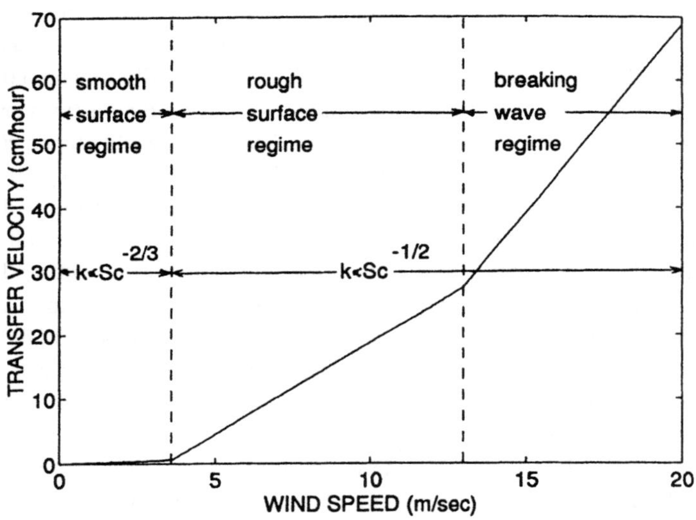

Figure 3-1. Idealized plot of mass transfer velocity for a Schmidt number of 600. The plot is based on the parameterization of Liss and Merlivat (1986)

Other observations of wind speeds up to 22 ms^{-1} are in good agreement with this relation: measurements in Cornish lakes by Upstill-Goddard *et al.* (1990), measurements in the North Sea by Watson *et al.* (1991) and measurements in the very large Delft wind tunnel by Wanninkhof and Bliven (1991). Such relationships are what could be considered state of the art in modelling air-water exchange (Jähne, 1991b) and have been used for obtaining global exchange values for carbon dioxide (Erickson, 1989 and Etcheto *et al.*, 1991) and dimethyl sulfide (Erickson, 1990) (also see Gosink, 1992). However, such results should be treated with caution since the wind speed dependence is clearly not the sole factor that should be taken into account (Jähne, 1991b). Wave roughness, which is not only a function of wind speed but also of fetch and time, has to be incorporated in the gas transfer representation as the apparent representation of the wind disturbed water surface (Kitaigorodskii and Donelan, 1984). Following experimental results that indicated similar gas transfer velocities for similar shear velocities and different wave structures (compared from wave height spectra), McCready and Hanratty (1984) argued that only high-frequency waves are important in the gas-transfer process. They also argued that since the boundary layer is so thin, the smallest waves should be most effective in transferring mass. Breaking of even the smallest waves should be sufficient to disturb the boundary layer. This is in accordance with the theoretical work of Coantic (1986) which proposed a model of gas transfer enhancement due to the interaction between capillary waves and the wind-induced shear current.

On the other hand, Kitaigorodskii (1984) related gas transfer to the dynamics of large waves with emphasis on wave breaking. More recent work by Jähne (1991b) provided significant insights on the phenomenon. Following experimental work in 6 different flumes with fetches varying from 1.5 to 250 meters, he observed that the air-sea exchange enhancement was not closely linked with the spectra densities of small-scale water surface waves and also that the modification of wind-waves by mechanically produced waves did not influence the transfer rates significantly. Significant enhancement was observed only if near unstable waves were generated, producing wave breaking. These results strongly suggest that the stability of the wave field might be the key point in gas-transfer enhancement. In such a case, the slope spectrum (or the mean square slope) might be the most important parameter in describing the wave field, and not the wave height spectrum. The importance of the mean square slope has also been outlined by Huang *et al.* (1981) for empirical wind wave studies. However, Jähne noted that the randomness of the wave field also has a definite effect since mechanically adding steep regular waves did not produce any significant enhancement of the mass transfer velocity. These last results were contradicted by data obtained by Daniil and Gulliver (1991). Nevertheless, clearly, the mechanisms by which turbulence is induced by wave action in the surface boundary layer are not well known. Wave breaking, direct dissipation in the boundary layer and interactions with the wind-induced shear-current all have potential effects. Ocampo Torres *et al.* (1994) in laboratory measurements of mass transfer of CO_2 and water vapour for smooth and rough flow conditions found at wind speeds of 2 to 3 m/s a characteristic common point in the transfer mechanism for both gas and aqueous phase controlled constituents that corresponds to the occurrence of the first wind generated wavelets. At higher winds the transfer rate of CO_2 increases more rapidly than that of H_2O, reflecting the increased importance of wave-generated turbulence for the ventilation of water phase limited compounds at the air-water interface. The principal resistance to transfer of CO_2 across the air-water interface occurs in the thin (of the order of 100 μm) diffusive boundary layer beneath the surface. Therefore, the occurance of the first wind generated wavelets can cause an enhancement of the mass transfer rate (due to breaking of capillary waves that have wavelengths 10-100 μm).

Other Sources of Near-Surface Turbulence

Ultimately, gas transfer at the air-water interface must be achieved through molecular diffusion. But near-surface turbulence plays a fundamental role in diffusing the gas into the bulk of the water and as such, turbulent processes dominate the global exchange rates. As outlined by Hanratty (1991) four sources of near-surface turbulence can be identified: shear layer turbulence, wave-induced turbulence, wall turbulence and gas-phase turbulence, the first two being the more important. Wall turbulence can be potentially important in open-channel flows or in flumes as turbulent bursts originating from boundaries can impact on the free-surface (Rashidi and Banerjee, 1988 and Rashidi and Banerjee, 1990). However, the effect should be negligible when wind induced shear stress is large compared to stresses at the boundary. Wall turbulence is absent in cases of deep bodies of water (such as the ocean) but should be considered in shallow (or narrow) open-water channel flows and some experimental setups. Gas phase turbulence can induce shear stress fluctuations at the interface which can in turn cause turbulent fluctuations in the water phase. Nevertheless, this effect should not be significant (Hanratty, 1991). This leaves two main sources of near-surface turbulence: wave induced turbulence and shear-layer turbulence. When the wind blows on top of a still surface of water, it creates a shear layer with a typical surface velocity of about 3% the wind velocity. For sufficiently high enough shear stresses on the surface, the shear layer is turbulent (Rashidi and Banerjee, 1990). Finally, as discussed in the preceding section, waves, through physical processes not well known, generate near-surface turbulence. The enhancement of mass transfer observed at the onset of wave generation makes the study of wave induced turbulence the most promising approach in understanding the effect of air velocity on gas transfer (Hanratty, 1991).

The effect of turbulence on gas transfer in conditions of no shear stress at the interface has been studied by a few authors with the use of grid-stirred tanks. The advantage of using such tanks is that they produce horizontally near-homogeneous turbulence (Hopfinger and Toly, 1976) and a well-controlled turbulent environment. Dickey *et al.* (1984) and Pankow *et al.* (1984) performed such experiments and used visualization techniques in order to evaluate turbulent parameters. They found reasonable agreement of their mass transfer measurements with eddy-structure models. In a more recent investigation, Chu and Jirka (1992) obtained simultaneous measurements of turbulent parameters and concentration very close to the surface and identified large eddies are being responsible for gas transfer. The extent to which grid-stirred turbulence approximates wave induced turbulence in a strong shear layer is unknown.

3-4. DISCUSSION AND NEEDS FOR FUTURE RESEARCH

Present Knowledge

Results from field measurements based on field techniques are not consistent. Broecker and Peng (1974) used the ^{14}C method and found that transfer velocity had a value of 20 cm h^{-1}. Estimates by Erickson (1989) by using the Liss-Merlivat relation (1986) and global winds were 1.8 times lower than the latter. Peng *et al.* (1979) and Smethie *et al.* (1985) by using the radon deficiency method failed to show a close dependence between the transfer velocity and wind speed. During the FGGE experiments with constantly blowing winds, Kromer and Roether (1983) show that the gas exchange rates are considerably lower than the rates measured during the JASIN experiment under highly variable winds.

Recent theoretical and experimental work has helped to identify several important aspects of gas-transfer at the air-water interface. The change of behaviour in mass transfer at the onset of waves has been clearly identified. The $Sc^{-2/3}$ dependency of gas transfer for smooth surfaces (described by rigid-wall models) as well as the $Sc^{-1/2}$ dependency (surface renewal models) for rough surfaces is now well established. The importance of waves has been outlined in various experiments and is now clearly recognized as a factor of prime importance in the control of gas transfer. Enhancement of gas transfer does not seem to be simply directly related to the spectral density of either capillary waves or gravity waves but rather to the overall stability of the wavefield. Despite those significant advances, much is still left unknown. The details of the physics at the interface are not well understood, and despite the fact that the sources of near-surface turbulence have been identified, little is known about the mechanics of the various phenomena. For example, it is not clear if waves directly dissipate to near-surface turbulence or if they interact with the turbulent shear current. The relative contributions of large and small-scale turbulent motions to gas transfer is also largely unknown and the importance of bubble production and injection in the water phase through wave breaking remains to be assessed.

So far experimental work has focused on measuring bulk liquid phase concentrations of various gases, and although such work has succeeded in identifying the various physical parameters controlling the gas exchange at the interface, comparison between experiments is difficult, important discrepancies are observed and extension of the results to field applications is at best questionable. At present, it is reasonable to say that an emphasis should be put on the physics of the interface in order to improve our basic knowledge of what exactly happens in or close to the thin surface boundary layer, and of the turbulence mechanisms which operate and are responsible for the turbulent diffusion of gases after they have been molecularly diffused through the interface. This information is critical for any application of mass transfer models to real-life problems. Presently, despite all the advances in the air-water interaction field and a plethora of proposed models, it is still not clear if large scale turbulent motions (large eddies) or eddies in the dissipation range control mass- transfer of gases at the air-water interface.

As mentioned by Wolff *et al.* (1991), and in the preceding sections, gas transfer is controlled to a large extent by the presence of surface waves. Surface waves contribute to the turbulence regime close to the interface through physical processes which are not well understood. One of the main problems is the scarcity of information on the velocity field and gas concentration close to the interface. Since the boundary layer is very thin (10- 300 μm), it is extremely difficult to obtain concentration measurements near the surface using conventional techniques and as such, most previous investigations have relied on bulk concentration measurements, giving no detailed information close to the interface. This is where visualization and image analysis techniques can potentially play an important role. Considerable work has been done on the chemical aspects of gas transfer processes, measuring gas concentrations in various experimental conditions. Such studies have been fundamental in identifying the parameters responsible for gas transfer enhancement. On the other hand, comparisons between data obtained from various experimental setups (and different physical conditions) are hard to make and differences are often significant (see Jähne *et al.*, 1987 for example) and so, extension of laboratory data to field applications and advanced coupled ocean-atmosphere models will at best give an order of magnitude estimation. Near-surface turbulence which is ultimately responsible for the gas transfer enhancement must be investigated. The main problem in looking at near-surface turbulence is the thickness of the boundary layer which precludes the use of intrusive methods of measurement. In such a case, visualization techniques, coupled with digital image processing have the potential to yield significant new insights to the physics operating at the air-water interface.

Experiments on Physical Processes at the Interface

So far, little visualization work has been made which had mass transfer applications as its main scope. Two main types of visualization experiments have been performed so far with application to gas-transfer at the air-water interface. The first type involves experiments aiming principally at obtaining quantitative information on near-surface gas concentrations, and the second type involves experiments designed to obtain measurements of the near-surface turbulence structure and flow field.

In the first type of visualization experiments, Jähne (1991a) discussed two approaches that would allow for a detailed experimental investigation of the gas transfer process. The first approach named the "controlled flux technique" is not strictly a visualization technique but its scope is the same. The principle of the technique consists in controlling the tracer flux (heat in this experiment) across the interface. By measuring the concentration right at the surface in the upper portion of the boundary layer and by turning the tracer flux on and off, Jähne was able to measure the concentration difference between the surface and bulk concentration from which he was able to determine the characteristic time scale needed for a tracer to cross the boundary layer.

The second approach uses fluorescence and digital image processing as a mean of obtaining concentration profiles across the boundary layer. The idea of using fluorescence to evaluate mass transfer rates has been originally used by Pankow *et al.* (1984) when they used a dye whose fluorescent characteristics were Ph dependent. The setup allowed them to measure CO_2 mass transfer rates in a small cell apparatus with grid-generated turbulence. The small size of the cell (10 cm) did not allow them to induce large levels of turbulence but they nevertheless outlined the potential of the fluorescence technique in making non-intrusive measurements of concentration close to the surface. The fluorescence method presented by Jähne bypasses the drawbacks of the controlled flux technique which provides only point measurements and uses heat as a tracer which has a molecular diffusivity much higher than typical gases. Two different techniques were used. In the first one, an acid or alkaline layer was generated within the boundary layer using a reactive gas as a tracer. The thickness of the boundary layer is then related to the intensity of a fluorescent pH indicator. Photographs of the water surface can then be taken which outline the spatial distribution of the thickness of the boundary layer in various wind conditions. In the second technique, concentration profiles are directly obtained using a tracer whose concentration directly affects the fluorescence intensity. This technique allowed for the first time concentration profiles to be measured within the boundary layer. A fluorescence technique was also used by Wolff *et al.* (1991) using pyrenebutyric acid, a fluorescent dye whose fluorescent intensity is inversely proportional to oxygen concentration. Comparison of the data with oxygen concentration measured directly with an oxygen sensor outlined the validity of the method in predicting near-surface gas concentrations. Similar conclusions were also obtained by Gardin *et al.* (1989) in a comparable experiment.

In flow structure and turbulence visualization work related to mass transfer, Dickey *et al.* (1984) used a streak photography technique to obtain instantaneous fluid velocity vectors. Using these velocity vectors, they evaluated turbulent velocities, length scales and dissipation rates. Gas transfer rates were measured concurrently. Their data was found to be consistent with surface renewal methods, but the use of grid-generated turbulence introduces a simplification of the problem as the effect of waves is neglected. Su *et al.* (1984) looked at the behaviour of wave breaking and air entrained using 35mm photography with an underwater stroboscopic light source. The images clearly demonstrate that air bubbles are carried deep underwater by breaking waves but only a qualitative discussion of the phenomenon is presented.

Tsanis and Donelan (1989) presented what could possibly be the only visualization work so far on the structure of surface drift and wave induced turbulence (see Figure 3-2). Using a shadowgraph setting, they looked at the mechanisms of near-surface turbulence. The thickness of the mixing layer was measured at various fetches for different wind velocities, but no attempt was made to look more closely at the finer structure of the turbulent field. Wierzimok and Jähne (1989 and 1991) used small polystyrol particles to visualize the flow field below a wind-disturbed water surface. Using a particle tracking algorithm, they were able to obtain both the Eulerian and Lagrangian velocity fields. Finally, present work by the authors, using a laser-induced fluorescence technique is showing great potential in resolving the finer details of near surface turbulence (see Figure 3-3).

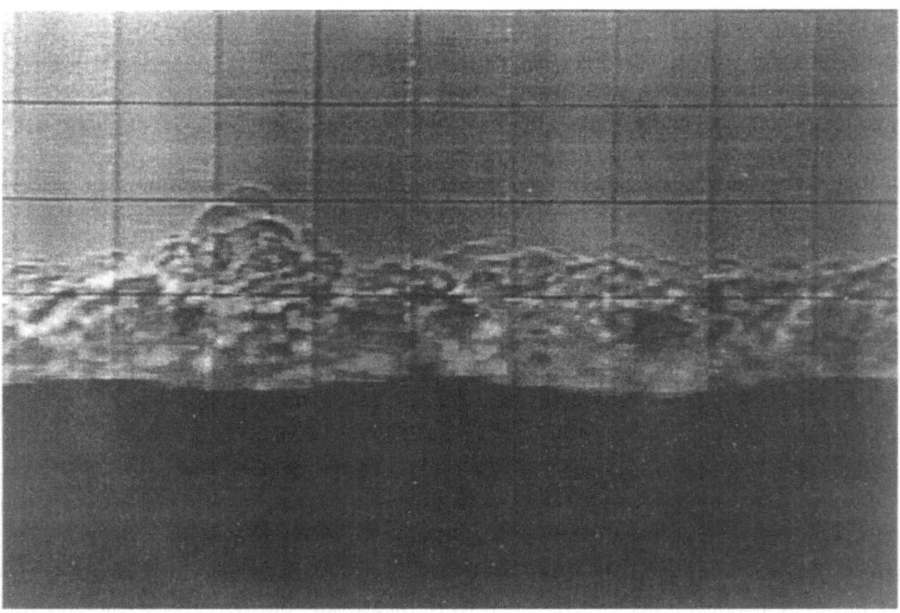

Figure 3-2. Shadowgraph image of the mixing layer in the presence of waves. The grid squares are 2.5 x 2.5 centimeters. Wind velocity = 20 m/sec, fetch length = 11.75 m From Tsanis *et al.* (1989)

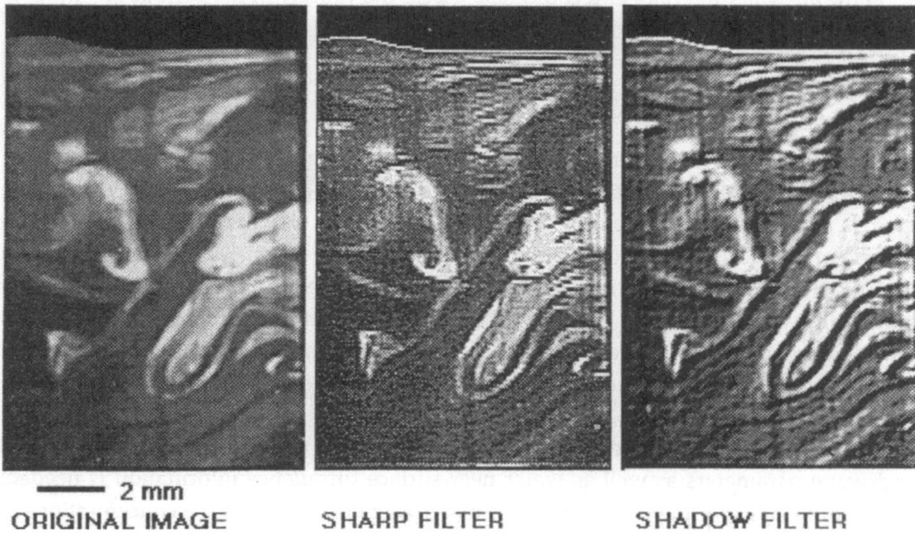

Figure 3-3. Turbulence structure close to the interface obtained using a laser-induced fluorescence technique. The image represents a surface of approximately 1 x 2 centimeters (current experiments by Brissette and Tsanis)

The methods discussed above clearly outline the potential of visualization tools in yielding new information on the transfer process at the interface and perhaps more importantly, on the physical processes that control the mass transfer. This area of research is relatively new and so far, few quantitative results have been obtained. As mentioned by Jähne (1992) presently the lack of detailed measurements about the mass transfer is a major problem and as such all conceptual models or theories remain speculative as they cannot be compared against experimental or field data. As such it is believed that the main need of future gas-transfer work should be aimed at gaining a better understanding of the physical processes at the air-water interface.

The main focus should be directed toward the characterization of near-surface turbulence in various conditions, for wind waves as well as for mechanically generated waves. Steep and less steep waves as well as breaking conditions should be looked at. Large scale breaking waves can be generated by combining mechanically generated steep waves and wind action. Wave information should be obtained including slope spectra. Because of the importance of obtaining information in the ultra-gravity and capillary range, non intrusive methods should be used (such as used by Kahma and Donelan, 1988). Available information on the variations in the thickness of the boundary layer such as obtained by Jähne (1992 and 1991a) could then possibly be linked to known turbulent processes.

The above observations would provide significant information on the processes that act at the air-water interface and could help assess the validity of the various gas-transfer models.

3-5. CONCLUSIONS

Adequate knowledge of mass transfer of greenhouse gases at the air-water interface is of prime importance for the application of advanced coupled atmospheric-oceanic models and for the assessment of climatic impacts in the Mediterranean region and in the rest of the world. Presently, state-of-the-art modelling relies on the parameterization of mass transfer velocity as a sole function of wind speed, a parameterization clearly inadequate considering the considerable scatter of existing experimental data and a known dependency on wave conditions. Results from laboratory experiments cannot directly be transferred to the oceanic environment without knowing more about the mechanisms of gas transfer.

An appropriate parameterization of the mass transfer velocity as a function of wind and wave parameters as well as water near-surface turbulence information is needed. The traditional experimental approach of measuring bulk liquid phase concentrations of various gases has succeeded in identifying the various physical parameters controlling the gas exchange at the interface but has failed in giving any quantitative information about the details of the physical processes acting at the interface. More details about the mechanisms can only be revealed by studies which need elaborate visualization tools and image processing facilities. These studies should include the energy balance in wind waves, especially the dissipation term, the turbulent flow in the water phase close to the interface and the distribution of the gas tracer concentration in the boundary layer. Only then will it be possible to fully assess current gas transfer models and better evaluate the role of the oceans in regulating climatic changes.

In field measurements the spatial and temporal observational records of the partial pressure gradient of CO_2 between ocean and atmosphere, should be improved. The recent estimates are accurate on a global scale within an order of magnitude due to scarcity or no existence of the data. Estimates of the CO_2 flux based on the available data set can provide an estimated mean value and the seasonal variation of ΔP_{CO_2} for two periods of year in the northern and equatorial oceans. No such information exists for the southern ocean.

The large scale atmosphere General Circulation Models (GCMs) can be accepted with some confidence as predictors of future climates on a global scale. Due to their coarse resolution and highly smoothed orography the GCMs are not applicable for predicting any future changes in climatic variables at a smaller regional scale. The accuracy of any regional scenarios so constructed is constrained by the reliability of GCM results. The GCMs are unable to reproduce accurately the characteristics of present-day regional climates, and this must cast doubt on the predictions of the future. The scenarios examined until recently can only be taken as an indication of the range of possible changes that might occur as a result of greenhouse warming. The quality of these scenarios will improve as more accurate GCM predictions become available in the future.

The relation between the frequency of extreme events (high temperature, high and low precipitation events) with the mean climatic conditions should be studied to help in predicting probabilities of occurrence. Finally, due to complexity of the processes involved, a systems analysis seems to be the best approach to study the implications of climate impacts in most regions.

REFERENCES

Broecker H-C., Petermann J., and Siems W. (1978). The influence of wind on CO_2 exchange in a wind-wave tunnel, including the effects of monolayers. J. Mar. Res., 36: 595-610.

Broecker H-C. and Siems W. (1984). The role of bubbles for gas transfer from water to air at higher windspeeds. Experiments in the Wind-Wave facilities in Hamburg. *In*: Gas Transfer at Water Surface. Brutsaert, W. and Jirka, G.H. (Eds.), Don Reidel, pp. 229-236.

Broecker W.S., Ledwell J.R., Takahashi T., Weiss R., Merlivat L., Memery L., Peng T.H., Jähne B., and Münnich K.O. (1986). Isotropic versus micrometeorologic Ocean CO_2 fluxes: a serious conflict. J. Geophys. Res., 91: 10517-10527.

Broecker W.S., and Peng T.-H. (1974). "Gas exchange rates between air and sea", Tellus, 26: 21-35.

Brtko W.J., and Kabel R.L. (1978). Transfer of gases at natural air-water interfaces. J. Phys. Oceanogr., 8: 543-556.

Campbell J.A., and Hanratty T.J. (1983). Mechanism of turbulent mass transfer at a solid boundary. AIChE Jour., 29: 221.

Chu C.R. and Jirka G.H. (1992). Turbulent gas flux measurements below the air-water interface of a grid-stirred tank. Int. J. Heat Mass Transfer, 35(8): 1957-1968.

Coantic M. (1986). A model of gas transfer across air-water interfaces with capillary waves. J. Geophys. Res., 91(C3): 3925-3943.

Danckwerts P.V. 1970. Gas-liquid reactions. McGraw-Hill, London, pp. 400.

Danckwerts P.V. (1951). Significance of liquid-film coefficient in gas absorption. Indust. and Eng. Chem., 43(6): 1460-1467.

Daniil E.I. and Gulliver J.S. (1991). Influence of waves on air-water gas transfer. ASCE J. of Env. Eng., 117(5): 522-540.

Deacon E.L. (1977). Gas Transfer to and across an air-water interface. Tellus, 29: 363- 374.

Dickey T.D., Hartman B., Hammond D., and Hurst E. (1984). A laboratory technique for investigating the relationship between gas transfer and fluid turbulence. *In*: Gas Transfer at Water Surfaces, Brutsaert, W., and Jirka, G.H. (Eds.), Don Reidel, pp. 93- 100.

Dobbins W.E. (1956). The nature of the oxygen transfer coefficient in oxygen aeration systems. *In*: Biological Treatment of Sewage and Industrial Wastes. McCabe J. and Eckenfelder, W.W. (Eds.), pp. 393.

Erickson D.J. II. (1989). Variations in the global air-sea transfer velocity field of CO_2. Global Biogeochem. Cycles, 3: 27-41.

Erickson D.J. III. (1990). Global ocean-to-atmosphere dimethyl sulphide flux. J. Geophys. Res., 95: 7543-7552.

Etcheto J., Boutin J., and Merlivat L. (1991). Seasonal variation of the CO_2 exchange coefficient over the global ocean using satellite wind speed measurements. Tellus, 43(b): 247-255.

Etcheto J. and Merlivat. L. (1988). Satellite determination of the carbon dioxide exchange coefficient at the ocean-atmosphere interface: a first step. J. Geophys. Res., 93: 15669-15678.

Flothman D., Lohse E., and Münnich K.O. (1979). Gas exchange in a circular wind/water tunnel. Naturwissenschaften, 66: 49-50.

Fortescue G.E., and Pearson J.R.S. (1967). On gas absorption into a turbulent liquid. Chem. Eng. Science, 22: 1163-1176.

Gardin P., Serieys J.C., Cunin J.C., and Cognet G. (1989). Concentration measurements by means of fluorescent particles. Proc. 5th Int. Symp. Flow Visualization, Prag., pp. 23-28.

Gosink T.A. (1992). Air-water gas exchange. An empirical approach. The Science of the Total Environ., 112: 221-231.

Gulliver J.S. (1991). Introduction to air-water mass transfer. *In*: Air-Water Mass Transfer, 2nd Int. Sympos. on Gas Transfer at Water Surf., Wilhelms, S.C., and Gulliver, J.S. (Eds.), pp. 1-7.

Hanratty T.J. (1991). Effect of gas flow on physical absorption. *In*: Air-Water Mass Transfer, 2nd Int. Sympos. on Gas Transfer at Water Surf., Wilhelms, S.C., and Gulliver, J.S. (Eds.), pp. 1-7.

Harriot P. (1962). A random eddy modification of the penetration theory. Chem. Eng. Sc. 17: 149-154.

Henstock W.H. and Hanratty T.J. (1979). Gas absorption by a liquid layer flowing on the wall of a pipe. AIChE Jour., 25: 122.

Higbie R. (1935). The rate of absorption of a pure gas into a still liquid during short periods of exposure. Am. Inst. Chem. Eng. Trans., 31: 365-390.

Hopfinger E.J. and Toly J.A. (1976). Spatially decaying turbulence and its relation to mixing across density interface. J. Fluid Mech., 78: 155-175.

Huang N.E., Long S.R., and Bliven L.F. (1981). On the importance of the significant slope in empirical wind-wave studies. J. Phys. Oceanogr., 11: 569-573.

Jähne B., Münnich K.O. and Sigenthaler U. (1979). Measurements of gas-exchange and momentum transfer in a circular wind-water tunnel. Tellus, 231: 321-329.

Jähne B., Huber W., Dutzi A., Wais T., and Ilmberger J. (1984). Wind/Wave tunnel experiment on the Schmidt number and wave field dependence of air-water gas exchange. *In*: Gas Transfer at Water Surface, Brutsaert, W., and Jirka, G.H., (Eds0, Don Reidel, pp. 283-302.

Jähne B., Münnich K.O., Bosinger R., Dutzi A., Huber W., and Libner P. (1987). On the parameters influencing air-water gas exchange. J. Geophys. Res., 92(C2): 1937-1949.

Jähne B. (1992). Imaging of gas transfer across gas/liquid interfaces, Proc. of the Int. Seminar on Imaging in Transport Processes, May 25-29, 1992, Athens, Greece.

Jähne B. (1991a). From mean fluxes to a detailed experimental investigation of the gas transfer process. *In:* Air-Water Mass Transfer, 2nd Int. Sympos. on Gas Transfer at Water Surf., Wilhelms, S.C., and Gulliver, J.S. (Eds.), pp. 244-256.

Jähne B. (1991b). New experimental results on the parameters influencing air-sea gas exchange. *In:* Air-Water Mass Transfer, 2nd Ints. Sympos. on Gas Transfer at Water Surf., Wilhems, S.C., and Gulliver, J.S. Eds., pp. 582-592.

Jeftic L., Milliman J.D. and Sestini G. (1993). Climatic Change and the Mediterranean. Environmental and Societal Impacts of Climatic Change and Sea-Level Rise in the Mediterranean Region, United Nations Environment Programme (UNEP). Edward Arnold, Hodder & Stoughton Publishers.

Kanwisher J. (1963). On the exchange of gases between the atmosphere and the sea. Deep-Sea Res., 10: 195-207.

Kahma K.K. and Donelan M.A. (1988), A laboratory study on the minimum wind speed for wind wave generation. J. Fluid Mech., 192: 339-364.

Keeling C.D. (1968). Carbon dioxide in surface ocean waters: 4. Global distribution. J. Geophys. Res. 73: 4543-4553.

Kim J.W., Chang J.T., Baker N.L., and Gates W.L. (1984). The statistical problem of climate inversion: determination of the relationship between local and large-scale climate. Monthly Weather Review 112: 2069-2077.

King C.J. (1966). Turbulent liquid phase mass transfer at a free gas-liquid interface. Indust. Eng. Chem. Fund, 5(1): 1-8.

Kitaigorodskii S.A. (1984). On the dynamical theory of turbulent gas transfer across an air-sea interface in the presence of breaking wind-waves. J. Phys. Oceanogr., 14: 960-927.

Kitaigorodskii S.A. and Donelan M.A. (1984). Wind-wave effects on gas transfer. *In:* Gas Transfer at Water Surfaces, Brutsaert, W., and Jirka, G.H., (Eds), Don Reidel, pp. 147-170.

Kromer B. and Roether W. (1983). "Field determination of air-sea gas exchange by the radon deficit method during JASIN 1978 and FGGE 1979", Meteor Forsch. Eng., A/B, 24: 55-75.

Lamont J.C. and Scott D.S. (1970). An Eddy Cell Model of Mass Transfer into the Surface of a Turbulent Liquid. AIChE Jour., 16(4): 513-519.

Lamurelle A.P., and Sandall O.C. (1972). Chem. Eng. Sci., 27, 1035.

Liss, P.S. and Crane, A.J., (1983). Man-made Carbon Dioxyde and Climate Change.. A Review of Scientific Problems. Geo. Books, Norwich, England, pp. 127.

Liss P.S. and Merlivat L. (1986). Air-sea exchange rates: Introduction and synthesis. *In:* The Role of Air-Sea Exchange in Geochemical Cycling, Ed. P. Buat-Menard, Reidel, pp. 113-128.

Liss P.S., Balls P.W., Martinelli F.N., and Coantic M. (1981). The effect of evaporation and condensation on gas transfer across an air-water interface. Oceanog. Acta, 4: 128-138.

Liss P.S., and Slater P.G. (1974). Flux of gases Across the air-sea interface. Nature, 247 : 181-184.

Lovelock J.E. (1979) Gaia — A New Look at Life on Earth. Oxford University Press. MacIntyre, F., 1971: Enhancement of Gas Transfer by Interfacial Ripples. The Physics of Fluids, 14(8): 1596-1604.

McCready M.J. and Hanratty T.J. (1984). A comparison of turbulent mass transfer at gas- liquid and solid-liquid interfaces. *In:* Gas Transfer at Water Surface, Brutsaert, W., and Jirka, G.H., (Eds), Don Reidel, pp. 283-292.

McCready M.J. Vassiliadov E., and Hanratty T.J. (1986). Computer simulation of turbulent mass transfer at a mobile interface. AIChE Journal, 32:1108.

Marchello J.M. and Toor H.L. (1963). A mixing model for transfer near a boundary. Ind. Eng. Chem. Fundam., 2: 8-12.

Merlivat L. and Memery L. (1983). Gas exchange across an air-water interface: experimental results and modelling of bubble contribution to transfer. J. Geophys. Res., 88: 707-724.

Münnich K.O. and Flothmann D. (1975). Gas exchange in relation to other air/sea interaction phenomena. SCOR Workshop on air/sea interaction, Miami 8-12 Dec. (prepared for the Ocean Science Board, Nat. Res. Council, Wash., D.C., 1977).

Ocampo-Torres F.J., Donelan M.A., Merzi N., and Jia F. (1994). Laboratory measurements of mass transfer of carbon dioxide and water vapour for smooth and rough flow conditions. Tellus, 46B: 16-32.

Pankow J.F., Asher W.E., and List J.E. (1984). Carbon dioxide transfer at the gas/water interface as a function of system turbulence. *In:* Gas Transfer at Water Surfaces, Brutsaert, W., and Kirka, G.H., Eds., Don Reidel, 101-111.

Palutikof J.P., Guo X., Wigley T.M.L. and Gregory J.M. (1992). Regional changes in climate in the mediterranean basin due to global greenhouse gas warming. MAP Technical Reports Series No. 66, Mediterranean Action Plan, MED POL, (United Nations Environment Programme) UNEP, Athens.

Peng T.-H., Broecker W.S., Mathieu G.G., Li Y.-H., and Bainbridge A.E. (1979). Radon evasion rates in the Atlantic and Pacific Oceans as determined during the GEOSECS Program. J. Geophys. Res., 84:2471-2486.

Prasher B.D. and Frickie A.L. (1974). Ind. Eng. Chem., Process Des. Develop., 13: 336.

Rashidi M. and Banerjee S. (1988). Turbulence structure in free surface channel flows. Phys. Fluids, 31: 2491.

Rashidi M. and Banerjee S. (1990). Streaks characteristics and behaviour near wall and interface in open channel-flows. ASME Jour. Fluids Eng., 112: 164.

Regional Seas (1992). Potential impacts of expected climate change on coastal and near- shore Environment. UNEP Regional Seas Reports and Studies No. 140, UNEP, Athens.

Regional Seas (1990). GESAMP: Pollutant modification of atmospheric and oceanic processes and climate: Some aspects of the problem, UNEP Regional Seas Reports and Studies No. 117, UNEP, Athens.

Smethie W.M., Takahashi T., Chipman D.W., and Ledwell J.R. (1985). Gas exchange and CO_2 flux in the Tropical Atlantic Ocean Determined from ^{222}Rn and pCO_2 Measurements. J. Geophys. Res., 90: 7005-7022.

Su, M-Y. Green A.W., and Bergin M.T. (1984). Experimental studies of surface wave breaking and air entrainment. *In:* Gas Transfer at Water Surfaces, Brutsaert, W., and Jirka, G.H., (Eds.), Don Reidel, pp. 211-219.

Takahashi T., Chipman D., and Volk T. (1983). Geographical, seasonal and secular variations of the partial Pressure of CO_2 in surface waters of the north atlantic ocean: The Results of the North Atlantic TTO Program, pp. II.123-II.145. *In:* Proceedings of CO_2 Conference: Carbon Dioxide, Science and Consensus, Berkeley Springs, West Virginia, U.S.A., September 19-23, 1982, CONF-820970, U.S. Dept. of Energy, available from NTIS, Springfield, Va, U.S.A.

Tans P.P., Fung I.Y., and Takahashi T. (1990). Observational constraints on the global atmospheric CO_2 budget. Science 247: 1431-1438.

Theofanous T.G. (1984). Conceptual models of gas exchange. *In:* Gas Transfer at Water Surfaces, Brutsaert, W., and Jirka, G.H., Eds., Don Reidel, pp. 217-281.

Toor H.L., and Marchello J.M. (1958). Film-penetration model for mass and heat transfer. AIChE Jour., 4: 97-101.

Tsanis I.K., Merzi N., and Donelan M.A. (1989). Near surface wave turbulence via a shadowgraph system. Proc. 5th Int. Symp. Flow Visualization, Prag, pp. 915-920.

UNEP (1993). Implications of expected climatic changes on the island of Rhodes. Mediterranean Action Plan (MED POL), United Nations Environmental Programme.

Upstill-Goddard R.C., Watson A.J., Liss P.S., and Liddicoat M.I. (1990). Gas transfer velocities in lakes measured with SF_6. Tellus, 42B: 364-377.

Wanninkoff R., Ledwell J.R., and Broecker W.S. (1985). Gas Exchange Wind Speed Relation Measured with Sulfur Hexafluoride on a Lake. Science, 227: 1224-1226.

Wanninkhof R. and Bliven L.F. (1991). Relationship between gas exchange, wind speed and radar backscatter in a large wind-wave Tank. J. Geophys. Res. 96.

Watson A.J., Upstill-Goddard R.C., and Liss P.S. (1991). Air-sea gas exchange in rough and stormy seas, measured by a dual tracer technique. Nature, 349:145 - 147.

Whitman W.G. (1923). The two-film theory of gas absorption. Chem. Metallur. Eng., 29(4): 146-148.

Wierzimok D. and Jähne B. (1989). 3-dimensional particle tracking beneath a wind- stressed wavy water surface with image processing. Proc. 5th Int. Symp. Flow Visualization, Prag, pp. 943-956.

Wierzimok D. and Jähne B. (1991). Measurements of wave-induced turbulent flow structures using digital image sequence analysis. *In:* Air-Water Mass Transfer, 2nd Int. Symp. on gas transfer at water surf., Wilhelms, S.C. and Gulliver, J.S. (Eds.), pp. 200-209.

Wigley T.M.L., Jones P.D., Briffa K.R., and Smith G. (1990). Obtaining sub-grid-scale information from coarse-resolution general circulation model output. J. Geophys. Res., 95(D2): 1943-1953.

Wigley T.M.L., Santer B.D., Schlesinger M.E., and Mitchell J.F.B. (1994). Developing climate scenarios from equilibrium GCM results. Climatic Change (submitted).

Wilson C.A. and Mitchell J.F.B. (1985). Simulated Climate and CO_2-Induced Climate Change over Western Europe. Climate Change, 10: 11-42.

Woolf D.K. and Thorpe S.A. (1991). Bubbles and the air-sea exchange of gases in near- saturation conditions. J. Mar. Res., 49: 435-466.

Wolff L.M., Liu Z-C., and Hanratty T.J. (1991). A fluorescence technique to measure concentration gradients near an interface. *In:* Air-Water Mass Transfer, 2nd Int. Symp. on gas transfer at water surf., Wilhelms, S.C. and Gulliver, J.S., (Eds), pp. 210-218.

<u>*CHAPTER 4*</u>

CLIMATE CHANGES DURING THE HOLOCENE IN THE MEDITERRANEAN REGION

ARIE S. ISSAR AND DORIT MAKOVER-LEVIN

The Jacob Blaustein Institute for Desert Research,
The Ben Gurion University of the Negev,
Sede Boker Campus 84993, ISRAEL

4-1. INTRODUCTION

The information about the paleo-climate of the Mediterranean region is of special importance for investigating the possible impact of future climate change on the hydrological cycle. This is due to the location of this region inbetween two climatic belts: the northern zone dominated by the westerlies which bring humidty from the Atlantic ocean, and the southern area dominated by the Sahara desert belt. During winter the westerlies humid belt moves to the south, while during the summer the Sahara belt moves to the north.

As has already mentioned the contemporary climate of the Mediterranean region is influenced by the the seasonal north-south movement of the climate belts, thus it is characterized by cold wet winters and warm dry summers. The sea itself serves as a passage way of maritime polar air masses. It contributes moisture from its relatively warm surface to air masses brought in by the cold fronts and cyclonic disturbences, which travel southward and westward. These fronts usually result in rainstorms which are the main source of the humidity in this region. During the summer the region is dominated by the Sahara or the Azores high pressure belt causing drought.

The southern part of the region is influenced by the easterlies and monsoonal regimes and may be reflected in the deposits in this sea, especially off the delta of the Nile, which gets its water from the tropical and subtropical belt south to the Sahara. Another reason for the interest in the climate changes of this region is the fact that the countries around the Mediterranean have the longest recorded history, which enables to investigate the impact of climate changes on the human socio-economic systems.

4-2. CLIMATE CHANGES IN THE MEDITERRANEAN REGION DURING PLEISTOCENE-HOLOCENE TRANSITION PERIOD.

Leroi-Gourhan (1980) investigating pollen spectra, in the Middle East, found that

NATO ASI Series, Vol. I 36
Diachronic Climatic Impacts on Water Resources
Edited by Andreas N. Angelakis and Arie S. Issar

fluctuations of tempratures as well as wet and dry phases existed during the Lower and Middle Würm. The cold-wet maximum seems to be dated around 45,000 B.P. On the other hand droughts characterizes the coldest Würmian phase, which probably explains the scarcity of archaeological occupation between 23,000 to 19,500 B.P. The late glacial shows some improvements in climate dated to 17,000, 13,500 and 12,000 B.P. Then a richer and more diversified flora marks the beginning of the Holocene.

In the southern part of Israel, layers of loess were deposited during most of the Last Glacial period (Issar and Bruins, 1983). The uppermost stratigraphic paleosol with calcic horizon (Bruins, 1976 and Bruins and Yaalon, 1979) of the northern Negev, was formed around 13,000 B.P., according to radiocarbon dating by Goodfriend and Magaritz (1988). In northern Sinai and western Negev, the loess layers are overlain by sand dunes indicating most probably an increae in the transport of sands by the Nile due to an increase in its flow (Issar, 1992).

Vegetational and climatic interpretation of palynological data (Van Zeist, 1980) show the main tendencies in the Late Quaternary vegetational development in the Near East. Accordingly the period between 24,000 and 14,000 B.P. was characterized by colder temperatures, and was markedly drier than today, (even though pollen assemblage show also fluctuations and regional differences in humidity). From 14,000 to 10,000 B.P., one can observe an increase in temperatures. Many sites suggest a distinct rise in humidity around 14,000 B.P.

Continuous pollen diagrams for the last 3.5 M.Y. from boreholes penetrating the entire Quaternary sequences of the Hula and Dead Sea lakes, show that the Dead Sea served as a continental base level throughout this period, thus, ensuring a complete, continuous deposition. The glacial phases are manifested by periods of somewhat lower temperatures and higher rainfall, some in summer time. The interglacials are hot and dry, with Saharean conditions prevailing. The interstadials are similar to the present climate, characterized by a short, rainy winter and a dry, hot, long summer. It is possible that short dry phases would occur in Israel at peaks of the glacials, but in general the periods recorded by low sea levels had a wet climate (Horowitz, 1979).

The Dead Sea clearly shows a humid period in the terminal Pleistocene with lisan-type greenish-gray laminated clay sediments (Neev and Emery, 1967). The paleo Dead Sea (Lake Lisan) had a much greater areal extent and reached in fact its maximum level during this period from about 13,000-11,000 B.P. (Begin *et al.*, 1985). In northern Sinai Goldberg (1977) found evidence for a fresh water lake which occupied a large region at Gebel Maghara during that period. Abundant arboreal pollen from this period was found in the Central Negev. Issar and Eckstein (1969) describesd lacustrine layers in Wadi Feiran in Sinai. The upper layers were found to be *ca*. 20,000-24,000 years old.

A study of the lake levels and climate for the last 30,000 years in the Fucino lake, Central Italy, has been carried out by Giraudi (1989). Geological and geomorphological mapping identified delta and lacustrine sediments, as well as shoreline terraces, wave cut terraces and soft-rock pediments. Each provides evidence of ancient levels of Lake Fucino during the late Pleistocene and Holocene. The following changes can be recognized:

a) A slight rise in lake level dating back to an indefinite period before 33,000 years ago;
b) A strong rise in lake level during the period between about 30,000 and 20,000-18,000 years ago;

In conclusion it can be said that during Upper Pleistocene the glacial period was chracterized in this region by colder and more humid winters than at present, while during the postglacial period the region was warmer and droughts were more abundant. Various observations suggest, however, that during the peak of the glacial period, colder conditions spelled also dry conditions, apparently because the region came under the influence of the continental high pressure zone of Central Europe. During the peak of the interglacial period the southern part of the region may have enjoyed some summer rains due to the northward migration of the monsoonal belt.

4-3. CLIMATE CHANGES IN THE MEDITERRANEAN REGION DURING THE HOLOCENE

Climate Change and Settlements during the Neolithic Period (*ca.* 10,300 B.C. - 6,500 B.P.)

Thunell and Williams (1983 and 1989) who investigated the paleotemperature and paleosalinity history of the Eastern Mediterranean during the Late Quaternary, came to the conclusion that the Mediterranean isotopic signal is a complex record of regional temperatures and salinity changes superimposed on the change of isotopic composition due to the melting of the global ice. Hydrographic conditions in the Mediterranean at 8,000 BP must have been considerably different than at 18,000 B.P. and than today. The water balance became positive as precipitation and run-off exceeded evaporation. Salinities were considerably lower at 8,000 B.P. and the west-east (increasing) salinity gradient was reversed to an east-west gradient. This is supported by east African climate records which indicate the onset of very humid conditions occurred at *ca.* 12,500 B.P., with maximum wet conditions between 10,000 - 8,000 BP. This was also a time of intensified African monsoons and increased Nile discharge. According to Herman (1989) the record of past environments is more complete and the revolution of climatic events is higher in the Mediterranean than in the major ocean basins.

From the ratio of $^{18}O/^{16}O$ of the epipelagic foraminifera Globigerinoides rubes in the Mediterranean, Rossignol-Strick *et. al.* (1982) found that the oxygen isotopic composition decrease through several large shifts to minimal values between 8,000 and 6,000 B.P. Luz *et al.* (1980) finds $\delta^{18}O$ minimum at about 8,700-6,500 B.P. Rossignol-Strick *et al.* (1982) and Luz *et. al.* (1980) explained part of the decrease in ^{18}O by the fresh water influx which formed low salinity surface water (and also explained the sapropelic layer). The sharp depletion signaled lowering of salinity and represented the heavy Nile floods which correlate with the period from 8,000-7,000 B.P. This period was mainly a rainy period in Africa. It seems that the Nile River is the major source responsible for such a discharge of freshwater. Sapropels contain valuable records of changing oceanographic conditions in the eastern basin of the Mediterranean (Rossignol-Strick *et al.*, 1982, Williams *et al.*, 1980). Deposition of the most recent sapropel (9,000-7,000 B.P.) on the eastern Mediterranean Sea floor seems related to a circulation reversal of water (flowing out of the Mediterranean to the Atlantic during this period), which would have caused the eastern Mediterranean to become a nutrient trap, thus enhancing productivity.

In a recent work by Luz (1991) he pointed out an alternative explanation for the isotopic depletion. He claimed that the high influx of low salinity water entered the Mediterranean Sea from the Black Sea when the rising sea surface reached the level of the Bosporus. Issar (1992) suggested that the two isotopic signals described by Rossignol-Strick *et al.* (1982) and Luz *et al.* (1980) may be correlated with the two phases of global warming which occured during the Neolithic, (PPNB and PNB) as will be discussed later on.

Sea levels during the Neolithic period show a continuing rise from very low levels compared to the present sea level. The curve of ancient sea level of Galili *et al.* (1988) reveals that between 8,000-6,000 B.P. the sea level rose at a mean annual rate of 5.2 mm. Galili *et al.* (1985, 1988) who investigated the Holocene sea-level changes, based on submerged archaeological sites off the northern Carmel coast in Israel report that the northern section of the Carmel coastal plain between Atlit and Haifa has so far revealed the largest number of submerged prehistoric sites along the east Mediterranean. They indicate a continuous marine transgression between 8,000 and 1,500 years B.P. The sites are embedded in the upper part of a marshy clay that fills the trough between the coastal ridge and a ridge now submerged some 1,000-1,500 m. in the sea to the west. The submerged prehistoric sites belong to two main chronological units: Pre Pottery Neolithic B (*ca.* 8,000 B.P.) at depths of 12- 8 m, and Late Neolithic (*ca.* 6,500 B.P.) at depths of 5-0 m. A continuous Holocene rise of sea-level from -14 or -15 m at 8,000 B.P. to the present level was definitely reached by 1,500 B.P. and possibly by 2,000 B.P. According to their evidences no tectonic movements have occurred in the area during the last 8,000 years.

Crown (1972) and Issar (1992) suggested that during part of the Neolithic the increased humidity in southern Israel and Sinai was a function of a global warm climate and was thus connected with the migration of the monsoon belt northward. Issar (1991) also suggested that during the lowest and the middle part of the Neolithic (PPNA and PNA) humidity was associated with cold climate due to southern shift of the cyclonic belts. The present authors suggest to use of the following criteria in order to discern between the warm-wet periods and the cold-wet periods. The warm-wet periods were characterized by a the spread of settlements in the desert areas due to the abundance of monsoonal rain storms. This amelioration of the climate was restricted to the south and did not cause a rise in the level of the Dead sea, which is mainly fed by the Jordan. The ^{18}O composition of the lake sediments was rather haveay. Such a warm-humid period was that of the upper Pre-Pottery Neolithic (PPNB) from *ca.* 9,500 to 8,500 B.P. On the other hand a cold and humid period was manifested by a rise in the level of the Dead Sea as well as depletion in the ^{18}O in the lacustrine sediments in the region and the spread of settlements into the desert (Fig. 4-1). This occured during the lower part of the Pre-Pottery Neolithic (PPNA) *ca.* 10,000 to 9,500 B.P. as well as during the lower Pottery Neolithic (PNA) from ca. 8,500 to 7,500 B.P. A change towards a dry and warm climate occurred during the upper Pottery Neolithic (PNB) from *ca.* 7,500 to 6,500 B.P.

Goodfriend (1986) found evidence for humid conditions for the period 10,000-8,000 B.P. in the Negev in the very small size of fossil land snail shells (*Trochoidea seetzeni*), being about 20-25% smaller in diameter than modern ones. The analysis of ^{13}C of organic matter in Early Holocene land snails from the northern Negev gives paleo-geographic evidence of the former distribution of xeric C4 semi-shrubs (enriched in ^{13}C relative to C3 plants). This suggests an approximate doubling of rainfall in the northern Negev during the Early Holocene as compared to the present (Magaritz and Goodfriend, 1987). Faunal remains from the site of Gilgal (PPNA) in the lower Jordan valley (Tchernov, 1982) and from Wadi Tbeik (Pre-Pottery Neolithic B) in southern Sinai (Tchernov and Bar-Yosef, 1982) suggest the presence of fresh water resources at these places in these early Holocene periods.

From the Pre-Pottery Neolithic A, *ca.* 10,000-9,500 BP (PPNA), there are almost no remains in the Negev (Avner *et al.*, Manuscript). Bar Yosef (1986 a and b) claimed that even though the farming communities of the PPNA period expanded towards the northern part of Israel and to the Jordan Valley, there are evidences that communities of hunters and gathers continued to exist also in the marginal regions, such as the Negev mountains.

Bar Yosef (1986 a) suggested that during the Pre-Pottery Neolithich B (PPNB) (9,500-8,500 B.P.) the climate favored and attracted hunters and gathers to occupy more desert areas. The more well known sites from that period are Sde-Divshon, and some sites from Ramat Mitrad in the central Negev, and Nahal Lavan and Halutza from the western Negev (Gopher, 1981). There are PPNB sites in the eastern Uvda Valley, Ein Qtura (Avner *et al.*, Manuscript). Horowitz (1977) pointed out that according to the pollen spectrum found in the sediments of the PPNB (Sde Divshon), this period was more humid than today and enabled agriculture in the area. According to Bar Yosef (1986a) the economy of the PPNB sites in fertile areas was based on legumes and cereal cultivation together with hunting and herding.

Most of the Neolithic sites were covered by alluvial deposits, with large stones that indicate severe flooding. The Neolithic sites were located close to water resources and low areas which required the building of a protective system from flood and mudflow damage.

A moister climate has been suggested by Neev and Emery (1967) from the beginning of the Holocene till 7,000 B.P., while from about 7,000 B.P. the climate became drier and the driest period in the Holocene occurred between 6,500-5,500 B.P. This conclusion is based on their study of the sediments of the Dead Sea, where greenish-gray laminated marls were deposited, indicating a humid period, covered by a salt layer. Issar (1992) suggests that the lower and middle part of the Neolithic period (PPNA cool and humid, PPNB warm and humid and PNA cool and wet which was ended at 7,500 B.P.). It was followed by a warm dry period of PNB, which continued until *ca.* 6,500 B.P. This was follwed by the cool wet period of the Chalcolithic. Issar's suggestion is based on the interpretation of the archeological observations discussed above as well as on the report of high water level of the Dead Sea reported by Frumkin *et al.* (1991) at about 8,000 B.P., and a high stand of the Dead Sea at about 6,500 B.P. reported by Goodfriend *et al.* (1986). The analysis of ^{13}C of organic matter in Early Holocene land snails also support the above findings by showing approximate doubling of rainfall in the Northern Negev during the Early Holocene as compared to the present (Magaritz and Goodfriend, 1987).

Van Zeist and Bakker-Heeres (1979) investigated the economic and ecological aspects of the plant husbandry of Tell Aswad situated some 30 km E.S.E. of Damascus. Mean annual rainfall is below 200 mm at present. The subsoil in the Aswad area consists of lacustrine sediments deposited in a lake which up to Late Pleistocene times must has covered a largre part of the Damascus basin. The natural vegetation is a steppe. It is likely that in early Neolithic times the lake and the marshes extended up to the site of Aswad.

In northern Greece, a pollen diagram described by Wijmstra, (1969) shows at *ca.* 8,000 B.P. a marked decline in oak and an increase in pine and artemisia parallel to a general decrease in arboreal pollen values and an inecrease in non arboreal pollen values. This trend is reversed to the general trend in the pollen ratios, which one can see in the same diagram after *ca.* 14,600 B.P. and after 7,850. To the opinion of the present authors this shift represents the PNA cold period, which in northern Greece caused partial return to the steppe conditions of the Late Glacial (See also Bottema, 1987).

In Anatolia an anlysis of the ^{18}O ratios in the carbonate sediments of Lake Van (Schoell, 1978) show rather low values at the base of the Neolithic, a marked increase at *ca.* 9,000 B.P. followed by a depletion at *ca.* 8,000 B.P., followed by a period of heavier ^{18}O composition, which terminates at *ca.* 7,000 B.P. (Fig. 4-1). This correlates well with the climate changes during the Neolithic as already suggested in this chapter.

Degens *et al.* (1984) report on low levels of Lake Van and high salinities from the beginning of the Holocene to *ca.* 6,300 B.P. Lake Van's levels show a rapid rise and its

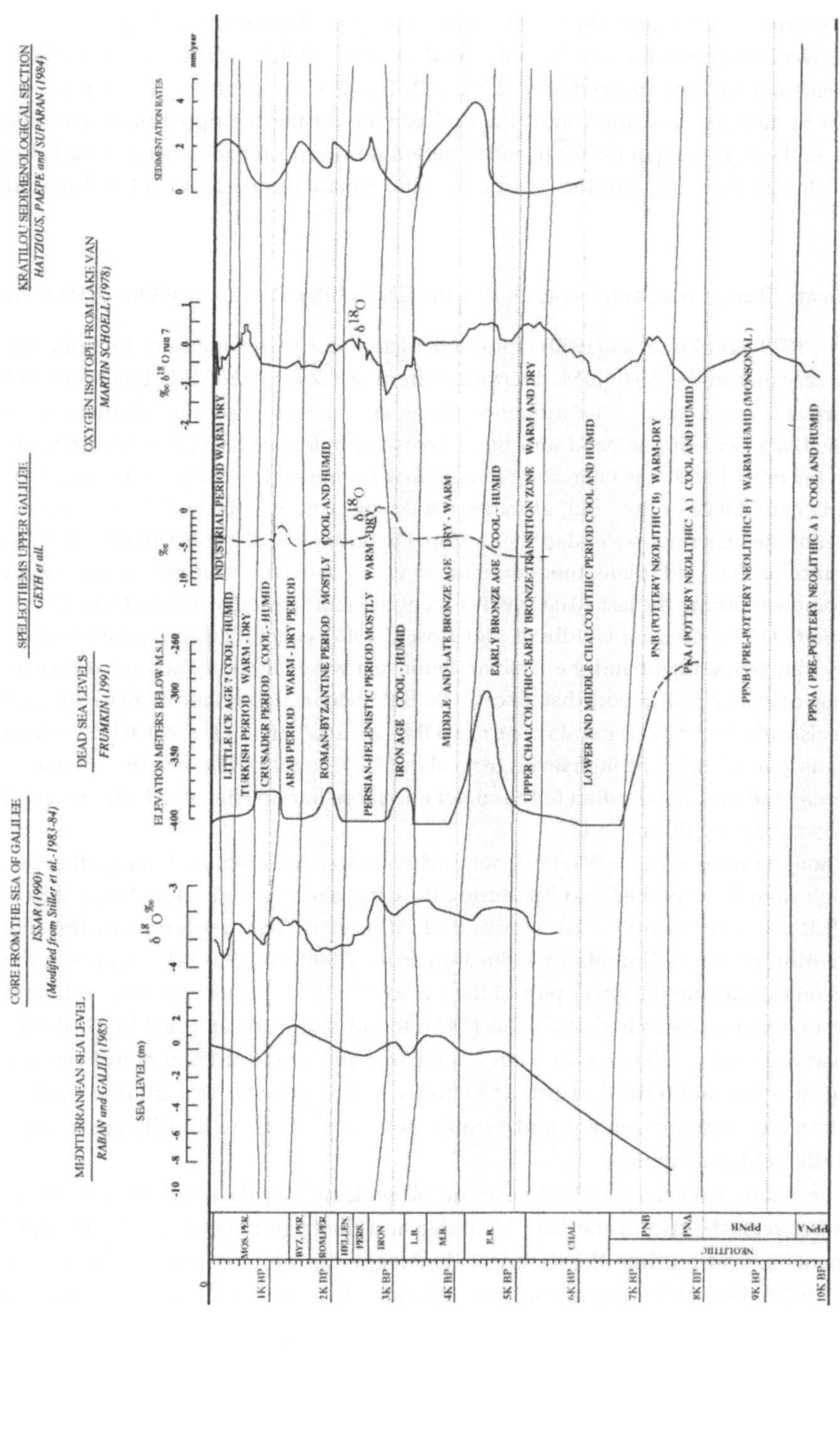

Figure 4-1. Climatic changes in the Mediterranean region during the Holocene

salinity decreased at the transition from the Neolithic period to the Chalcolithic period.

The Neolithic period in the Levant ended at *ca.* 6,500 B.P., while the Chalcolithic settlements in Israel commenced at *ca.* 6,200 B.P. This gap in settlement is explained by the present authors, as a function of the arid climate during the uppermost Neolithic period. At the beginning of the Chalcolithic period it is most probably that a remarkable climate change occurred, and the warm and dry climate turned to a cold and humid one.

Climate Change and Settlements During the Chalcolithic Period (*ca.* 6,500 - 5,200 B.P.)

Luz (1979) and Luz *et al.* (1980) report a depletion in ^{18}O in planktonic foraminifera in deep sea cores in the Eastern Mediterranean from *ca.* 6,200 to 5,300 B.P. It is suggested to regard this as a result of the influence of colder water entering the Mediterranean from the Black Sea, on one hand and the decrease of inflow of heavier water from the Nile on the other hand. It is interesting to note that Rasmussen (1991, p. 373) found, that a peak in abundance of the 'cold' affinities planktonic species appeared "when vertical turnover of the water masses ended the sapropel deposition at about 6,000 B.P.". But the appearance of the 'cold' planktonic foraminifer was a short one and the 'warm' fauna was prominent during the last 10,000 years except the short period around 6,000 B.P.

Roberts (1982, cited in: Goldberg and Rosen, 1987) concluded that lake levels in Turkey, Iran, Syria, Egypt and the Arabian Peninsula were high until about 5,000 years ago. Degens *et al.* (1984) report that since 7,000 B.P. there is a dramatic increase in Lake Van levels with downwards trends at *ca.* 6,000 B.P., *ca.* 5,000 BP (and *ca.* 4,000 B.P., which will be discussed later). The high levels from about 7,000 to 6,000 BP are to the opinion of the present authors an indication to the cooler and more humid climate which prevailed in the Levant during this period.

Although Frumkin *et al.* (1991) did not find evidence, in the Mount Sdom salt caves, for a high water level of the Dead Sea during the Chalcolithic, a high shore line, dated at 6,500 B.P., was reported by Goodfriend *et. al.* (1986). The depletion in the ^{18}O composition of the sediments of Lake Van from 7,000 to 6,000 B.P. supports the conclusion that during the lower part of the Chalcolithic the climate was more cold and humid. In another research, Goodfriend (1990) found that between 6,500 to 3,000 B.P., there was a shift of *ca.* 20 km to the south of the transition zone of 150 mm isohyet. The present authors maintain that this conclusion is too general and involves a few oscillations, its commencement and termination (as well as its middle part) were, apparently, cold and humid.

After a gap of several hundred years in the settlement of the southern part of the Levant between the end of the Late Neolithic and the beginning of the Chalcolithic, settlements spread out into the desert of the Negev during most of the Chalcolithic (Cohen, 1989). One of the prominent manifestation of this culture in the Levant is the

"Ghasulian Culture" on the name of site Tell Ghasul on the eastern side of the Jordan rift valley not far from Jericho. The "Ghasulian Culture" started during the third quarter of the 6th millenium B.P. (Hennessy, 1969 and Ussishkin, 1986). The center of the Chalcolithic settlement in the Negev was in the Beer-Sheva Valley. The archaeological remains from the Beer-Sheva area point to the importance of agriculture to the Chalcolithic society (Avner *et al.*, Manuscript). Cohen (1989) claimed that the settlements in the Negev mountains were temporary and semi-nomadic settlements, based on pastoral grazing and transportation of copper from the Feiran area and the Timna Valley.

Tsukada (cited in: van Zeist and Bottema, 1982) analyzed the Hula core. The oak pollen increases between *ca.* 6,400 and 4,300 B.P. is in correlation with Horowitz's (1980) interpretation of a more humid climate. The Tsukada pollen curve shows a slight decrease in oak pollen *ca.* 6,500 B.P. It might be correlated with the dry spell, which was mentioned earlier between the end of the Neolitic period and the beginning of the Chalcolithic period.

Butzer (1958) also concluded that the Near East between 7,000 and 4,400 B.P. was moister than that of today even though local temperatures were higher. When referring to a decrease in precipitation after 5600 B.P., Butzer (1958) gave evidences from Egypt and the Sahara and not from the Levant. Pollen diagrams from the eastern Mediterranean are characterized by an increase in oak pollen. In Lake Zeribar, Iran (van Zeist and Bottema, 1977), the increase in arboreal pollen started at *ca.* 6,000 B.P. and it reached its peak during the end of the Chalcolithic period and the Early Bronze Age.

Pollen assemblage from the Chab marsh in northwestern Syria (Niklewski and van Zeist, 1970, cited in: Wright, 1976) shows also a high percentage of oak during the Chalcolithic period. The increase in oak pollen started *ca.* 9,000 B.P., reached its peak until 7,000 B.P., and even though was high during the next 2,000 years it started to decrease between 7,000 and 4,000 B.P. It reached its low *ca.* 4,000 B.P. Wijmstra (1969) (cited: in Wright, 1976) collected pollen data from Tenage Philippon, Macedonia. The trend of increase in quercus pollen during the Chalcolithic period is shown also in this curve. The same trend is seen in the arboreal pollen of Lake Xinias, Greece (Bottema, 1979, cited in: van Zeist and Bottema, 1982).

The magnificent Chalcolithic culture, with its artistic tradition and technical knowledge, lasted just for several hundreds of years. It suddenly disappeared exactly in the same unknown way as it appeared. All the sites were abandoned without any signs of violence. There are several hypotheses concerning the disappearance of the Chalcolithic culture. Hennessy (1969) suggested that they had to leave their settlements because a migration of a new wave of people, who established the Early Bronze Age culture. But archaeological remains do not reveal any exchange of cultures in the same sites. Others looked for a circumstantial connection between the disappearance of the Chalcolithic population and the expansion of the first Egyptian Kingdom of Naarmer about 5,000 years B.P. But there are no indication of violence involvement in the abandonment of the settlements.

The present authors are of the opinion that another phase of warming up, which caused changes in the environmental conditions could be the best explanation for the collapse of the whole system of the Chalcolithic settlements. This opinion is based on the higher values of ^{18}O observed in the sediments of the Sea of Galilee starting at *ca.* 5,300 B.P. (Issar, 1990) and in Lake Van between *ca.* 5,500 and 5,000 B.P. (Schoell, 1978) and high level of the Mediterranean sea around 5,000 B.P. (Raban and Galilee, 1985).

The settlements in the area of Beer-Sheva and the Judean Desert were the last to be abandoned (Gonen, 1989). The reason for this apperently was the fact that the people of this semi arid areas established either pastoral way of life or agricultural settlements along river beds (Govrin, 1991) and developed the technology of shallow wells.

In summary, the Chalcolithic period was mainly a humid period in Israel and the areas bordering the Mediterranean, except at its end when it became warm and dry. The next chapter will deal with the continuation of the settlement in Israel during the Early Bronze Age, and the changes in the material culture in which they went through. After the disappearance of the Chalcolithic people the new comers brought in a new material culture.

The Early Bronze Period (*ca.* 5,200 - *ca.* 4,000 B.P.)

The Mediterranean shoreline receded by about 1 m from *ca.* 5,000 to 3,800 B.P. (Raban and Galili, 1985). Flemming (1989) discussing the tectonics and sea levels of the harbors of Caesarea Maritima gives estimates for the eustatic sea level:

Neolithic at 8,000 B.C. —	- 50 m
Chalcolithic at 5,000 B.C. —	- 10 m
Early Bronze Age at 3,000 B.C. —	- 1 m
After the Early Bronze Age B.C. —	± 1 m

Dead Sea sediments indicate a humid phase, by lisan-type laminated clay deposits, which has been approximately dated by Neev and Emery (1967) to about 5,500-4,300 B.P. A radiocarbon date of 4,410 +/- 320 B.P. from disseminated organic carbon found in the top of this clay bed, near the base of the uppermost salt tongue, has been taken to infer the end of this humid period. Later Neev and Hall (1977) suggested a dry period until 5,000 B.P. and a relatively high Dead Sea level (indicating a wet period) from *ca.* 4000 to 2,500 B.P. On the other hand the Dead Sea level observed by Frunkin *et al.* (1991) showed an increasing level from 5,200 B.P. reaching a peak at *ca.* 4,500 B.P. (Fig. 4-1). Carbon isotope analyses of organic matter from sediment cores from the Dead Sea shore was correlated with rainfall changes by Magaritz *et al.* (1991). They came to the conclusion that between 5,000 and 4,500 B.P. it was a wet period. Magaritz (1993) brought up parameters of water balance evaluated from the ratio of the amount of precipitation over evaporation, and came to a similar trend of fluctuations in in climate during the last 7000 years as Issar *et al.* (1989 and 1991).

Goodfriend (1988 and 1990) investigated the organic matter of land snail shells in the northern Negev. The study (Goodfriend, 1990) is based on 40 samples of snail shells (*Trochoidea seetzeni*), whose radiocarbon ages fall within the period 6,700-2,800 B.P. He concludes that the rainfall isohyets in the northern Negev during this period must have been located about 20 km south of their present positions, which implies roughly a twofold increase in the mean annual precipitation in the northern Negev (between the present-day 150 mm and 300 mm isohyets) relative to the present. Within this period from about 6,700 to 2,800 B.P., the last part (3,900-2,800 B.P.) may have been somewhat drier, since $\delta^{13}C$ values of organic matter of land snail shells appear to indicate a somewhat lower C3 plant presence in the southern part of the northern Negev (Goodfriend, 1990). The range of years (the period 6,700-2,800 B.P.) during which the Negev was wet according to the Good friend includes to the opinion of the present authors, two wet periods, namely the Chalcolithic and the Lower Bronze, a short dry spell in between them and the dry spell of the Middle Bronze from *ca.* 4,300 to 1,200 B.P. Goodfriend (1988) discusses his results in relation to other paleoclimatic data from the region, which seem to indicate a level of humidity not much above the present regime, although Horowitz (1979) shows a peak of relatively high arboreal pollen (mostly quercus) from the Hula Basin for this period. However, Tsukada (cited in: van Zeist and Bottema, 1982) shows variable but generally low arboreal pollen for this period.

The ^{18}O values of 41 speleothems from 10 caves in the northern part of Israel (Upper Galilee), (Geyh *et al.*, 1991, and Issar *et al.*, 1991) are depleted from *ca* 5,000 to 4,500 B.P., while in the core of the Sea of Galilee (Stiller *et al.*, 1984, Issar *et al.*, 1991) the depletion in ^{18}O values starts at *ca.* 4,400 B.P. (Fig.14-1). In the opinion of the present authors the differences are due to inaccuracy due to disparity in dating intervals. Geyh *et al.* (Manuscript) suggested a correction bias of +300 years for the speleothem time-scale compared to that of the Sea of Galilee. In general, one can see from Figure 1-14, that both curves, the one of the sediments of the sea of Galilee and that of the speleothem from the caves of Galilee have the same trend and there is a good correlation between them.

A palynological survey on Late Quaternary vegetation in Israel has been done by Horowitz (1973, 1974 and 1980) in Lake Hula. According to Horowitz, during the Holocene the climate resembled that of the present with a much more humid phase in the Atlantic period, from 7,000 to 4,500 B.P. This conclusion seems to the present authors to be due to a too big interval between ^{14}C datings of the section and because there were less than the average of 200 grains count per sample. Horowitz's conclusion marked the overall trend. Thus, although there are some minor fluctuation in the percentages of arboreal pollen from the Hula Valley, it is hard to make an accurate conclusion and to refer those fluctuation to an exact period of time.

Schoell (1978) reports on a depletion in the ^{18}O composition of the sediments of Lake Van. Hatzious *et al.* (1984) present a sedimentation rate diagram fom Kratilou plain in Greece. The low rates of sedimentation from *ca.* 6,000 to 4,500 B.P., present to the opinion of the present authors an evidence that the region was covered by trees, which prevented erosion. In coclusion it is suggested that during the Early Bronze age, the climate was cooler and the Mediterranean region enjoyed a more humid climate.

The Middle and Late Bronze Age (*ca.* 4,200 to 3,400 B.P.)

These periods, most probably were very warm and dry. This can be deduced from the following evidence. At *ca.* 4,200 B.P. there is a rise in the Mediterranean sea level of about 1 m. It reaches a peak at *ca.* 3,500 B.P. and then starts to recede (Raban and Galili, 1985). Raban (1987) reports also that this rise in sea level was accompanied by the filling up of the river inlets of Israel by sand. Other archaeological evidence indicates that Haifa Bay has dramatically filled in with sand (Sivan, 1982).

Supporting evidence to this severe climate change was found in the Coastal Plain of Israel. An archaeological investigation at Tel-Aviv (Ritter-Kaplan, 1984), located black clays deposited during the Early Bronze period, which contain a pollen assemblage showing characteristics for a humid climate. This climate terminated at the beginning of the Middle Bronze. Later sands without pollen were deposited. Lewy *et al.* (1986) report on a thick layer of dark swamp deposits containig artifacts of latest Neolithic period, earliest Chalcolithic period and Early Bronze Age. The layer is ovelain by friable calcarenite sandstone, which to the opinion of the present authers is an evidence to a transgressive due to eustatic reasons.

A heavier ^{18}O composition of the sediments of the Sea of Galilee from *ca.* 3,800 to 3,300 B.P. speaks also for a warm climate during this period. The same trend can be seen in the composition of the Galilee speleothem's curve, which reflects this event somewhat earlier (Fig. 4-1). However, this may be due to a bias in the calibrated radiocarbon time scale or insufficient correlation for the reservoir effect. Yet, the signature of a warmer phase is clearly seen in all curves and is also evidenced by the level of the Dead Sea which fell to such an extent level that its southern shallow part dried out altogether (Frumkin *et al.*, 1991). In the same period a total desertion of all the settlements in the Negev took place. The city of Arad was deserted at the end of the Early Bronze Age. During the Middle Bronze cap age the area is practically void of settlements. Ceramic evidence regarding the earlier age of these settlements, is brought by Cohen (1986). This continued almost all through the Late Bronze Age. During the Middle Bronze II only two camps, most probably military ones, were found.

The ^{18}O curve of Lake Van does reflects this warm phase by a heavy isotopic composition from *ca.* 4,700 to 4,000 B.P. and then a little lighter composition until *ca.* 3,300 B.P. The sedimentation curve in Central Greece (Hatzious *et al.*, 1984) shows a high rate of deposition from *ca.* 4,750 B.P. to *ca.* 3,300 B.P. To the opinion of the present authors this was due to the aridization and thus stronger soil erosion. Crown (1972) claims that the driest period of the Holocene, was from ca 4,500 to 3,500 B.P.

The Iron Age (*ca.* 3,400 to 2,600 B.P.)

At *ca.* 3,300 B.P. Raban and Galilee (1985) observed a short but distinct retreat of the sea. At the same time one can observe a rise in the levle of the Dead Sea (Frumkin *et al.*, 1991) and a depletion in the ^{18}O composition of the speleothems of the Galilee (Issar *et al.* 1991). The isotope curves of the Lake sea of Galilee (Issar, 1990) and Lake Van (Schoell, 1978) show also a depletion trend as well as a distinct rise in the level of Lake Van (Degens *et al.*, 1984) and a reduction of its salinity. There is a reduction also in the rate of sedimentation in cetral Greece (Hatzious *et al.*, 1984). On the basis of these observations one can conclude that following the dry period of the Late Bronze Age, a change into a colder and more humid climate in the Levant took place. This may correspond with a weaker monsoon expressed in the recorded low flood levels of the Nile which occurred from 3,180-3,130 B.P. (Butzer, 1980).

The Persian Hellenistic period (from *ca.* 2,600 to 2,300 B.P.)

An opposite direction in the trend of the environmental proxy-data one can observe at almost all the sites discussed in the previous paragraph. This brings to the conclusion that the climate of the Levant was warmer and drier during the period in which the Persians and the Hellens ruled this region.

The Roman and Byzantine period (*ca.* 2,300 to 1,400 B.P.)

At *ca.* 2,300 B.P. the trend reverses again. This continues with minor fluctuations to *ca.* 1,400 B.P., which brought Issar (1990) to suggest that the more benign climate in the deserts of Judea and Edom, promoted the flourishing of the Nabatean kingdom, which was established in the first century B.C. in the Negev by nomad tribes, apparently of Arabic origin. These tribes controlled the trade routes, especially that of the frankincense and myrrh, which were brought from southern Arabia to the Mediterranean ports. Their capital city, for some time, was in Petra situated in a canyon in the mountains of Edom. In the desert of the Negev, they built six beautiful cities, which flourished until the 7th century A.D. During these years one observs in the core in the Sea of Galilee an increase in the ratio of olive pollen and reduction in that of oak. (Stiller *et al.*, 1984). This is explained by Issar (1990) as a function of the cutting of the natural vegetation and planting of olives.

Liphschitz *et al.* (1981) investigated the assemblage of trees from the Roman siege ramp which was constructed for the siege of Masada, in the Judean Desert, at 70 A.D. The scarcity of Acacia wood and the abundance of Tamarix, relative to the present, indicate a change in climate to wetter conditions. Nevertheless, as only a few species have disappeared from this area during the last 2000 years and as the species composition appears to be fairly constant (Liphscchitz *et al.*, 1981), one may assume that apart from possible short fluctuations, the climate as a whole in the area has remained more or less unchanged.

Nir and Eldar (1987) investigated ancient wells in the Israel Mediterranean coast line region, for detecting young tectonical movements. The wells represent the period from the Late Bronze Age to the Crusader period. The curve shows to the present authors opinion the influence of change in precipitation, which during the Roman Byzantine time reached the maximum, and during moslem time receded.

Lev-Yadun *et al.* (1987), investigating annual tree rings as an index of climate changes intensity in the Levant, found that most suitable trees in this region for dendrochronology are : Red Juniper, Persian Pistacia and Cedar of Lebanon. The rings of Jerusalem Pine, for example, show that the range of changes in rainfall is about 400 to 900 mm per year. The changes are not linear and under 400 mm of rainfall there is no growth at all. Research in the rings of the Cedar of Lebanon and Red Juniper trees show that in the last 2,000 years there have been only slight changes of weather, some of which were regional and some only local. From 420 to 480 A.D. there were very narrow widths of of tree rings, which means drought.

The St. Catherine's Monastery in southern Sinai was established in the time of the Byzantine Emperor Justinian (525-565 A.D.). The widespread use of Populus, Phoenix dactylifera and Cupressus sempervirens logs as building material in St. Catherine's Monastery, from the 6th to 9th century A.D., suggests that these trees were common in the area in the past (Liphschitz and Waisel, 1976). In Lake Fucino, Italy, Giraudi (1989) observed a rise in the water table at *ca.* 2,000 B.P. This brought Claudius Ceasar to start a drainage project, which was concluded by Hadrian.

In the Iberian Peninsula Pons and Reille (1986) observed a high ratio of pinus and very low of quercus during the glacial periods, and vice versa. Garcia *et al.* (1986) show that towards the layer dated to be *ca.* 1,730 B.P., there is again an increase, in pinus and a decrease in quercus. At *ca.* 2,000 B.P. there is a layer of about 10 cm devoid of pollen. It is suggested by Issar, that this is equivalent to the period of high precipitation during the Early Roman period, in which the marshes turned into a flowing river, which caused the oxidation of the sediments. The warm Moslem period seems to be reflected by low Pinus ratio later.

The Arab Period (*ca.* 1,400 -1,000 B.P.)

During this period there is a rise in the Mediterranean Sea level (Raban and Galili 1985), the ^{18}O composition of the speleothemes of the Galilee and Sea of Galilee sediments become heavier (Issar, 1990). The level of the Dead Sea is low (Frumkin *et al.*

1991). In the same time the cities in the Negev are deserted. Kedar (1985) in his study on the Arab conquest and agriculture, in the seventh century A.D., maintains that the destruction was by no means the only repercussion of the Arab conquest on fertile crescent agriculture. The keen interest of the early caliphs in agricultural exploitation in the area between desert and sown is convincingly attested (Grabar, 1963). It is suggested by the present authors that the reason for the desertification was in the first place climatic. It is thus concluded that this period was warm and thus dry in the Mediterranean region.

The Crusaders Period (*ca.* 1,000 to 800 B.P.)

During this period one finds exactly the opposite trends, which were observed for the Moslem period and described in the preceding paragraph (See also: Fig. 4-1). It is suggested that this period was cool and humid.

Turkish Period (*ca.* 800 to 200 B.P.)

The data for this period is scarce in the Levant. The general impression from the curves presented in Figure 4-1, especially these from the Sea of Galilee and Lake Van, is that the warm period continued about 200 years. Later it became cooler most prbably in connection with the Little Ice Age.

The Industrial Period (*ca.* 200 B.P. to present)

Striem (1985) investigated the quantitative and qualitative aspects of the recent climatic fluctuations in Israel. The study was based on the systematic meteorological observations which have been made in Jerusalem since about 1850, and these constitute the longest record of climatic parameters in the Middle East. He found that winter (Dec. - March) temperatures rose by more than 1°C, while the summer (June-Sept.) temp. remained steady. The mean monthly barometric pressure rose by about 1 mb. The annual rainfall decreased from about 600 mm in the second half of the 19th century to about 500 mm in the first half of the 20th century, and there is at present (1985) a trend of increased rainfall. The pressure maximum used to occur before November (44% from 1861-1885), but only 17% from 1946-1970. Its occurrence in November became more frequent (an increase from 24% to 46%.) The retardation in the occurrence of the barometric pressure maximum is characteristic for drought years . A series of warm years occurred between 1920 to 1963. After 1930 the level of the Dead Sea dropped by about 2 meter from a level around -390 m, which had been maintained since 1893, to a level around -392 m.

Van Zeist and Bakker-Heeres (1979) report that in 1850 and in 1852 the lake of Aateibé near Damascus came to the edge of Tell Aswad the same as during the Early Neolithic times. In 1880 the southern part of the Lake had already receded to some extent and today only part of the north Lake of Aateibé remains. Giraudi (1989) reports that lake Fucino reached a very high level from 1861 to 1875 and had to be drained by a gallery.

4-4. CONCLUSIONS

From the data and discussion presented in this article, the following conclusions regarding climate changes during the Holocene were reached:

- A cold and humid period during the Pre Pottery Neolithic A period, i.e. *ca.* 10,000 and 9,500 B.P.
- A warm and humid period as a consequence of the monsoon influence during the Pre Pottery Neolithic B period, i.e. *ca.* 9,500-8,500 B.P.
- A cold and humid period during the Pottery Neolithic A period, i.e. *ca.* 8,500 and 7,500 B.P.
- A warm and dry period during the Pottery Neolithic B period, i.e. *ca.* 7,500 and 6,500 B.P.
- A humid most probably cold period during the Chalcolithic period (*ca.* 6,500-5,000 B.P.) except at the end of the period when the climate became warm and dry.
- A short warm period of the Upper Chalcolithic around 5,000 B.P.
- A cold and humid period during most of the Early Bronze Period, i.e from *ca.* 5,000 B.P. to 4,200 B.P.
- A warm period started at *ca.* 4,200 B.P. (Early Bronze VI) extending to *ca.* 3,400 B.P.
- From *ca.* 3,400 B.P. to *ca.* 2,600 B.P. (the Iron Age) there was another cold and humid period.
- From 2,600 to 2,300 B.P. (mainly Persian and Helenistic Empires) the climate was rather warm and dry.
- During the Roman period, i.e. from *ca.* 2,300 B.P. to about 1,400 B.P., (with a short interval lasting about 100 years, sometime at *ca.* 1,700 B.P.) a colder and more humid period prevailed.
- A warm and dry period started at *ca.* 1,400 B.P. (The Arab warm period) reached a peak of warmth and drought at *ca.* 1,200 B.P. and abated at *ca.* 1,000 B.P.
- From *ca.* 1,000 B.P. to *ca.* 800 B.P. the climate was rather cold and humid (The Crusaders cold period).
- At 800 B.P. a warm and dry period started which continued about 150 years (The Turkish warm period).
- From 650 B.P. to about 150 B.P. (the Industrial period) the climate was cooler and more humid, due to The Little Ice Age.
- From about 150 years ago the climate became more warm and dry.

REFERENCES

Avner U., Carmi I., and Segal D. (manuscript). Settlement in the Southern Negev from Neolithic to Bronze Age in light of radiocarbon dating.

Bar Yosef O. (1986a). The walls of Jericho: An alternative Interpretation. Current Anthropology, 27(2):157-162.

Bar Yosef O. (1986b). The land of Israel during the Neolithic period (in Hebrew). *In:* Ripel Y (Ed) The History of the Land of Israel, Vol. 1. Israeli Defense Ministry Publishing, Tel-Aviv, pp. 27-46.

Begin Z.B., Broecker W., Buchbinder B., Druckman Y., Kaufman A., Magaritz M., and Neev D. (1985). Dead Sea and Lake Lisan levels in the last 30,000 years : a preliminary report. Israel Geological Survey Report 29, 85:1-18.

Bottema S. (1987). The Late Glacial in the Eastern Mediterranean and the Near East. *In*: Brice W.C. (Ed.), The Environmental History of the Near and Middle East since the Last Ice Age. Academic Press, pp. 15-28

Bottema S. (1979). Pollen analytical investigation in Thessaly (Greece). Palaeohistoria, 21:19-40.

Bruins H.J. (1976). The Origin, Nature and Stratigraphy of Paleosols in the Loessial deposits of the N.W. Negev. M.Sc. Thesis, The Hebrew University, Jerusalem.

Bruins H.J. and Yaalon D.H. (1979). Stratigraphy of the Netivot section in the desert loess of the Negev (Israel). Acta Geologica Academiae Scientiarum Hungaricae, Tomus, 22(1-4):161-169.

Butzer K.W. (1958). Postpluvial variations of climate. *In*: Quaternary Stratigraphy of Climate. Johnson Reprint Corporation, New York, N.Y., pp. 103-128.

Butzer K.W. (1980). Pleistocene history of the Nile Valley in Egypt and Lower Nubia. *In* : Williams MAJ , Faure H. (Eds.), The Sahara and the Nile. Rotterdam : Balkema, pp. 253-280.

Crown A.D. (1972). Toward a reconstruction of the climate of Palestine 8000-0 BC. J. Near Eastern Studies, 31:312-330.

Degens E.T., Wong H.K, Kempe S., and Kurtman F. (1984). A geological study of Lake Van, Eastern Turkey. Geologische Rundschan, 73:701-734.

Flemming N.C. (1989). Tectonics and sea levels: *In:* Raban A (Ed.) (Oleson J.P. (general Ed) The Harbours of Caesarea Maritima; Results of the Caesarea Ancient Harbour Excavation Project, 1980-1985, Vol. I: The site and the Excavations. Center for maritime studies, University of Haifa, Publication no. 3. BAR international series, 491: 13-21.

Frumkin A., Magaritz M., Carmi I., and Zak I. (1991). The Holocene climatic record of the salt caves of Mount Sedom, Israel. The Holocene, 1(3):191-200.

Galili E. and Weinstein-Evron M. (1985). Prehistory and paleoenvironments of submerged sites along the Carmel coast of Israel. Paléorient, 11(1):37-52.

Galili E., Weinstein-Evion M., and Ronen A. (1988). Holocene changes based on submerged archaeological sites off the northern Carmel coast in Israel. Quaternary Research, 29:36-42.

Geyh M.A., Wakshkel E., and Franke H.W. (manuscript). Carbon - 14, carbon - 13 and oxygen - 18 data of speleothems from Upper Galilee.

Giraudi C. (1989). Lake levels and climate for the last 30.000 years in the Fucino area. Palaeogeography, Palaeoclimatology, Palaeoecology, 70: 249-260.

Goldberg P. (1977). Late Quaternary stratigraphy of Gebel Maghara. *In:* Bar Yosef O., Phillips J.L. (Eds), Prehistoric Investigations in Gebel Maghara, Northern Sinai, Qedem. Jerusalem: Institute of Archaeology, Hebrew University, no. 7, pp. 25-51.

Goldberg P. and Rosen A.M. (1987). Early Holocene palaeoenvironments of Israel. *In:* Levy TL (Ed), Shikmim I. Studies concerning 4th millennium societies in the northern Negev desert, Israel. British Arch. Reports Int. Series 356(i). Oxford: British Arch. (report).

Gonen R. (1989). Pastorals, farmers and copper workers - the Chalcolithic period. In: Introduction to the archaeology of the Land of Israel in the Biblical Times. The Open University, Tel-Aviv, pp. 112-181.

Govrin Y. (1991). The question of the origin of early Arad's population (in Hebrew). *In:* The Land of Israel. Researches in Geography and Archaeology, Vol 21, Ruth Amiran Book. The Society for Exploring the land of Israel and its Antiquities Publishing, pp. 107-110 (in Hebrew).

Goodfriend G.A. (1986). Holocene shifts in rainfall in the Negev Desert, as inferred from isotopic, morphological and faunal analysis of fossil land snails. *In:* Berger WH, Labeyrie LD (Eds.), Book of abstracts from the conference on abrupt climatic change. Scripps Institute Oceanography Ref. Ser., 86-8, pp. 122-124.

Goodfriend G.A. (1988). Mid-Holocene rainfall in the Negev desert from ^{13}C of land snail shell organic matter. Nature, 333:757-760.

Goodfriend G.A. (1990). Rainfall in the Negev Desert during the Middle Holocene based on organic matter in land snail shells. Quaternary Research, 34:186-197.

Goodfriend G.A. and Magaritz M. (1988). Palaeosols and Late Pleistocene rainfall fluctuation in the Negev Desert. Nature, 332:144-146.

Goodfriend G.A., Magaritz M., and Carmi I. (1986). A high stand of the Dead Sea at the end of the Neolithic period: Paleoclimate and archaeological implications. Climatic Change, 9:349-356

Gopher A. (1981). The prehistory of the Central Negev. The Society for Nature Protection, Sde Boker Field School, Israel (in Hebrew).

Grabar O. (1963). "Umayyad palace" and the "Abbasid revolution". Studia Islamica, 18: 5-18.

Garcia A.M., Morla J.C., Ruiz Z., Sainz O.H. (1986). Contribution al conocimiento del paisaje vegetal Holoceno en la submeseta Sur Ibérica: Análisis polénico de sedimentos higroturbosos en el Campo de Calatrava (Ciudad Real, España). *In*: Quaternary climate in western Mediterranean, Proceedings of the symposium on climate fluctuations during the Quaternary in the western Mediterranean regions. F. Lopez-Vera (Ed.). Universidad Autonoma de Madrid, pp. 189-204.

Hatziou M.E., Paepe R., and Suparan P. (1984). Landscape changes in Greece as a result of changing climate during the Quaternary. *In*: Desertification in Europe. Proceedings of the information symposium in the EEC programe on climatology. Held in Mytilene. Greece, 15-18 April.

Hennessy J.B. (1969). Preliminary report on the first season of excavations at Teleilat Ghassul. Levant, I: 1-24.

Herman Y. (1989). Late quaternary paleoceanography of the Eastern Mediterranean: The deep sea record. Marine Geology, 87: 1-4.

Horowitz A. (1973). Development of the Hula Basin, Israel. Israel Journal of Earth-Sciences, 22: 107-139

Horowitz A. (1974). Preliminary palynological indications as to the climate of Israel during the last 6000 years. Paléorient, 2(2):407-414.

Horowitz A. (1977). The climate and the settlement in the Negev mountain during the last Quaternary (in Hebrew). *In:* Sohar E. (Ed.) The Desert: past, present, future. Reshafin Publishing, Tel-Aviv, pp. 52-59.

Horowitz A. (1979). The Quaternary of Israel. New York: Academic Press.

Horowitz A. (1980). Palynology-climate and distribution of settlements in Israel. Kadmoniot 13(3-4):51-52. (in Hebrew).

Issar A. (1990). Water From the Rock. Springer Verlag, Heidelberg. pp 213.

Issar A. (1991). Climate changes in the Levant and the possibility of their mitigation. In: Paepe R., Fairbridge R.W., Jelgersma S. (Eds), Greenhouse Effect, Sea Level and Drought. Kluwer, Dordrecht, pp. 565-574.

Issar A. (1992). The impact of climate variations on water management systems and related socio-economic systems. UNESCO Project, International Hydrological Programme, Division of Water Sciences (unpublished report).

Issar A. and Eckstein Y. (1969). The lacustrine beds of Wadi Feran, Sinai: their origin and significance. Israel Journal of Earth Science, 18:21-27.

Issar A. and Bruins J. (1983). Special climatological conditions in the deserts of Sinai and the Negev during the Last Pleistocene. Palaeogeography, Palaeoclimatology, Palaeoecology, 43:63-72.

Issar A., Govrin Y., Geyh M.A., Wakshal E., and Wolf M. (1991). Climate changes during the Upper Holocene in Israel. Israel Journal of Earth Sciences, 24:179-185.

Issar A. Tzoar H., and Levin D. (1989). Climatic changes in Israel during historical times and their impact on hydrological, pedological and socio-economic systems, *In:* Leinen M., Sarnthein M. (Eds.), Paleoclimatology and Paleometeorology: Modern and Past Patterns of Global Atmospheric Transport. Kluwer Academic Publishers, pp. 525-541.

Kedar B.Z. (1985). The Arab conquest and agriculture, a seventh century apocalypse, satellite imagery and palynology. Asian and African Studies, 19: 1-15.

Leroi-Gourhan A. and Darmon F. (1987). Analyses palynologiques de sites archéologiques du pléistocene final dans la Vallée du Jourdain. Israel Journal of Earth Sciences, 36: 65-72.

Lev-Yadun S., Liphschitz N., and Waisel Y. (1987). Annual rings in trees as an index to climate changes intensity in our region in the past (in Hebrew, English abstract). Rotem, 22: 6-17.

Liphschitz N. and Waisel Y. (1976). Dendroarchaeological investigations in Israel (St. Catherine's Monastery in southern Sinai). Israel Exploration Journal, 26: 39-44.

Liphschitz N., Lev-Yadun S., and Waisel Y. (1981). Dendroarchaeological investigations in Israel (Masada). Israel Exploration Journal, 31: 230-234.

Lewy Z., Neev D., and Prausnitz M.W. (1986). Late Holocene tectonic movements at Akhziv, Mediterranean coastline of northern Israel. Quaternary Research, 25:177-188.

Luz B. (1991). Post-glacial paleo-oceanography of the Eastern Mediterranean. International workshop on regional implications of future climate change. Apr. 28-May 2, 1991. Ministry of Environment, Jerusalem, Israel.

Luz B. (1979). Palaeo-oceanography of the post-glacial eastern Mediterranean. Nature, 278(5707):847-848.

Luz B. and Perelis-Grossowicz L. (1980). Oxygen isotopes, biostratifigraphy and recent rates of sedimentation in the Eastern Mediterranean off Israel. Israel Journal of Earth Sciences, 29:140-146.

Magaritz M. (1993). Synchronized changes in regional water balance since the Mid-Holocene. Climatic Change, 24:179-185.

Magaritz M. and Goodfriend G.A. (1987). Movement of the desert boundary in the Levant from latest Pliestocene. *In:* Magaritz M., Goodfriend G.A. (Eds.) Abrupt Climatic Change - Evidence and Implications. D Reidel Publ. Co., Holland, pp. 173-183.

Magaritz M., Rahner S., and Yechieli Y. (1991). $^{13}C/^{12}C$ ratio in organic matter from the Dead Sea area: paleoclimatic interpretation. Naturwissenchaften 78: 453-455.

Neev D. and Emery K.O. (1967). The Dead Sea depositional processes and environments of deposition. Jerusalem Geological Survey, Israel Bull. 41, pp 147.

Neev D. and Hall J.K. (1977). Climatic fluctuations during the Holocene as reflected by the Dead Sea levels. Preprint represented at the International Conference on Terminal Lakes. Weber State College, Ogden, Utah, May 2-5, 1977.

Niklewiski J. and van Zeist W. (1970). A Late Quaternary polen diagram from north western Syria. Acta Botanica Neerlandica, 19:737-754.

Nir Y. and Eldar I. (1987). Ancient wells and their geoarchaeological significance in detecting tectonics of the Israel Mediterranean coast line region. Geology, 15: 3-6.

Pons A. Reille M. (1986). Nouvelles researches pollen analytiques a Padul (Granada): La fin du dernier glaciarire et l' Holocene. *In*: Quaternary climate in western Mediterranean, Proceedings of the symposium on climate fluctuations during the Quaternary in the western Mediterranean regions. F Lopez-Vera (Ed.). Universidad Autonoma de Madrid, pp. 405-422.

Raban A. (1987). Alternated river courses during the Bronze Age along the Israeli coastline. Colloque internationeaux C.N.R.S. deplacements des lignes de rivages en Mediterranee. (Ed). du C.N.R.S. Paris, pp. 173-189.

Raban A. and Galili E. (1985). Recent maritime archaeological research in Israel - a preliminary report. The International Journal of Nautical Archaeology and Underwater Exploration, 14(4):321-356.

Rasmussen L. (1991). Benthonic and planktonic foraminifera in relation to Early Holocene stagnation in the Ionian Basin, central Mediterranean. Boreas, 20: 357-375.

Roberts N. (1982). Lake level as an indicator of Near Eastern palaeoclimates: a preliminary appraisal. British Archaeology Report IS 133, pp. 235-267.

Rossignol-Strick M., Nesteroff W., Olive P. and Vergnaud-Grazzono C. (1982). After the Deluge: Mediterranean stagnation and sapropel formation. Nature, 295:105-110.

Schoell M. (1978). Oxygen isotope analysis on authigenic carbonates from Lake Van sediments and their possible bearing on the climate of the past 10,000 years. *In:* Degens ET, Kurtman (Eds.) The Geology of Lake Van. The Mineral Research and Exploration Institute of Turkey, Ankara, pp. 92-97.

Stiller M., Ehrlich A., Pollingher U., Baruch U., and Kaufman A. (1984). The Late Holocene sediments of Lake Kinneret (Israel) - multidisciplinary study of a 5 m core. *In:* Geological Survey of Israel, Ministry of Energy and Infrastrcture, Jerusalem, pp. 83-88.

Sivan D. (1982). Paleogeography of the Akko area in Holocene period. Third Annual Coastal Conference, Haifa, pp. 51-63.

Striem H.L. (1985). Quantitative and qualitative aspects of the recent climatic fluctuations. Israel Journal of Earth Sciences, 34: 47-48.

Tchernov E. (1982). Faunal responses to environmental changes in the Eastern Mediterranean during the last 20,000 years. *In:* Bintliff JL, van Zeist W (Eds), Palaeoclimates, Palaeoenvironments and Human Communities in the Eastern Mediterranean Region in later Prehistory. Oxford: B.A.R. International Series 133, pp. 105-127.

Tchernov E. and Bar-Yosef O. (1982) Animal exploitation in the pre-pottery Neolithic B period at Wadi Tbeik, southern Sinai. Paléorient, 8: 17-37.

Thunell R.C. and Williams D.F. (1983). Paleotemperature and paleosalinity history of the Eastern Mediterranean during the Late Quaternary. Palaeogeography, Palaeclimatology, Palaeoecology, 44: 23-39.

Thunell R.C. and Williams D.F. (1989). Glacial-Holocene salinity changes in the Mediterranean Sea : hydrographic and depositional effects. Nature, 338: 493-496.

Ussishkin D. (1986). The land of Israel during the Chalcolithic period. *In:* Ripel, Y (Ed.), The history of the Land of Israel, Vol. 1. Israel Defense Ministry Publishing, Tel Aviv, pp. 47-60 (in Hebrew).

van Zeist W. (1980). Vegetational and climatic interpretation of polynological data - Bat Sheva Seminar on approaches and methods in paleoclimatic research with emphasis on rich areas. Jerusalem and Rehovot, 20-30 Oct. 1980 volume of Abstracts, pp. 69-71.

van Zeist W. and Bottema S. (1977). Palynological investigations in western Iran. Palaeohistoria, 19:19-85.

van Zeist W. and Bottema S. (1982). Vegetational history of the eastern Mediterranean and the Near East during the last 20,000 years. *In:* Palaeoclimates, Palaeoenvironments and Human Communities in

Eastern Mediterranean Region in later Prehistory. Oxford: British Archaeological Reports, International Series 133, pp. 277-321.
van Zeist W. and Bakker-Heeres J.A.H. (1979). Some economic and ecological aspects of the plant husbandry of Tell Aswad. Paléorient, 5:161-169.
Williams D.F., Thunell R.C., and Kennett J.P. (1978). Periodic freshwater flooding and stagnation of the eastern Mediterranean Sea during the Late Quaternary. Science, 201:252-254.
Wijmstra T.A. (1969). Palynology of the first 30 meters of a 120 meter deep section in northern Greece. Acta Botanica Nederlandica, 18(4):511-528.
Wright H.E. Jr. (1976). The environmental setting for plant domestication in the Near East. Science , 194:385-389.

<u>CHAPTER 5</u>

QUATERNARY PERIODICITIES OF DROUGHT IN GREECE

ROLAND PAEPE[1], ILIAS N. MARIOLAKOS[2],
SOFIA S. NASSOPOULOU[2], ELFI VAN OVERLOOP[1],
AND NIKOS J. VOULOUMANOS[2]

[1] *Earth Technology Institute, Vrije Universiteit, Brussel-VUB, BELGIUM, and*
[2] *Applied Geology and Geodynamics, National Capodistrian University, Athens, GREECE.*

5-1. INTRODUCTION TO THE GREEK LANDSCAPE AND CLIMATE

The Greek landscape offers a large variety of plateau's, lake (subsidence) basins and coastal plains which have been shaped in the course of the Quaternary i.e. the timespan of roughly the last 3 Ma. million years. It is the kind of landscape which Budel (1982) described as the "Etesian Region of the Subtropical Zone of Mixed Relief Development". "Etesian" (from the Greek etos = year) points at the steady regularly reappearing south-bound drift of the lower atmosphere over the eastern mediterranean and adjacent lands, in the summer from mid-May through mid-September.

It dominates the Adriatic, Ionian and Aegean seas and the surrounding regions of the Balkan, Turkey, the Levant and the Sinai/Nile area, but also southern Italy, the Spanish Meseta and southern Iberian peninsula together with the Maghreb region of Morocco, Algeria and Tunesia. Similar etesian regions outside the Mediterranean are parts of California, Chile, south-western Africa and south-western Australia. It should be reminded that all these regions occupy a similar climato-geomorphological position namely to the north-west of a winter-cold and warm-arid zones. The mediterranean etesian belt is by and large the most extensive one lying east of the winter-cold Pamir and Himalayan arid zone and north of the Saharian-Saoudi warm arid zone. Similarly, lies the Californian etesian belt east of the winter-cold arid Nevada and warm-arid Mexican desert. In the southern hemisphere, the Chilean, south-western African and south-west Australia etesian belts lie due south of respectively the warm arid zones of Peru (Desierto de Sechura) and Northern Chile (Desierto de Atacama), of Namibia and of Middle Australia.

The "etesian regimen" is comparable to the monsoonal regimen, except that it is dry and that it is not replaced during the winter by a drift from the opposite direction. Indeed, areas of the "Monsoonal Region of the Subtropical Zone of Mixed Relief Development" are lying in the same latitudinal belts as the etesian regions but at the

NATO ASI Series, Vol. I 36
Diachronic Climatic Impacts on Water Resources
Edited by Andreas N. Angelakis and Arie S. Issar

eastern sides of the same aforementioned wintercold-dry and warm-dry regions of the continents: north-eastern China south-west of the Mongolian Desert, including the loess plateau within the rectangular bend of the Huang Ho; south-eastern United States from Arizona till Florida; south-eastern Australia; south-eastern Africa and finally the Gran Chaco and the Pampas areas of Argentina and immediate surroundings.

It is quite remarkable, although not entirely surprising, that both etesian and monsoonal subtropical regions provide comparable and most complete soil-stratigraphic sequences of the Quaternary. Etesian and monsoonal belts are indeed subordinated to periodically changing north-south wind systems which are not hampered by west-east bound mountain ranges so that they can operate freely consequently to seasonal changes.

5-2. LONG TERM, MIDDLE TERM AND SHORT TERM CYCLES

An extensive number of geological quaternary sections including number of archaeological sites (Paepe and Mariolakos,1984b), have been recorded by Paepe in the period 1967 through 1993 in different parts of Greece of which the most important ones were found on the East Coast of Attica (Marathon and Rafina), in the Kifissos Valley (Kratilou, Kalikleios), in the Peloponnesos (Koroni, Megalopolis, Sparta) and Crete (Kandanos, Retimnon, Ambelaki) (Fig. 5-1). From the field records it may be readily seen that not only seasonal local changes controlled these regions but also global ones pertaining to the identification of major palaeoclimatic variations of the last three million years. The combined study of these sections makes it possible to create a new quaternary stratotype region for the Eastern Mediterranean.

Palaeoclimatic variations in both etesian and monsoonal systems are remarkably well expressed in the long term, middle term and short term sequences of soil/sediment series (Paepe and Van Overloop, 1990a). In order to disentangle todays climatic situation, especially with regard to drought, it is absolutely mandatory to relate these geosoil/sediment sequences with variations of the climate during the long term Pleistocene Glacial/Interglacial Cycles of the last 3 Ma, the Middle term Stadial/Interstadial Cycles of the Last Glacial/Interglacial Stage of the last 130 Ka (thousand years) and the Short term cycles of the current interglacial like the present Flandrian Stage of the Holocene Series say the last 10,000 years B.P. (Before Present).

From all recorded evidences an attempt will be made to show that the Greek records for the long term, middle term and short term records may be regarded on basis of their correlation with world wide spread geological records, as the land stratotype sections for the Eastern Mediterranean Belt.

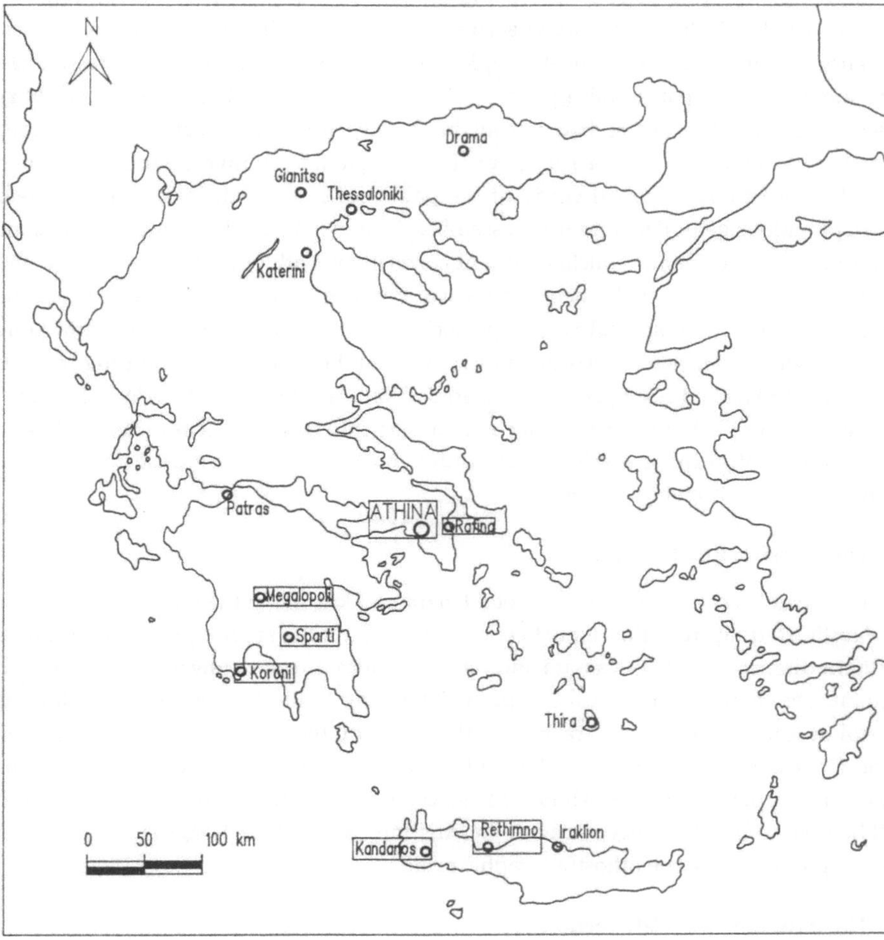

Figure 5-1. Location map of studied geological sections in this publication

5-3. THE LONG TERM GEOSOIL SEQUENCE OF THE SPARTA PLATEAU

The Land Stratotype of the Eastern Mediterranean

The Sparta Rift Basin of the Peloponnesos is flanked to the North by a plateau which actually is an uplifted former sedimentation basin in which the whole series of the Miocene, Pliocene and continental Quaternary sequences, is exposed. Its "continentality" is clearly outspoken and highly differentiated as to facies from the pure "lacustrine" sediments of the Megalopolis Basin and from the more distant "coastal marine" sediments of the Kokkino Limanaki section nearby Rafina in Attica. This continental sequence is characterised by a most complete sequence of the landrecord expressed in a continuous record of palaeosoils (geosoils) of at least the last 3.0 Ma years (see hereafter). Indeed, the geosoil record reveals to contain all oxygen isotopic stages of the deepsea record. N. Vouloumanos is carrying out the palaeomagnetic investigation combined with the micromorphological study of the palaeosoils. Nevertheless, from the field record of various geosoil stratigraphic assemblages, it may be readily concluded to some preliminary palaeoclimatic conclusions and correlations with other land sequences and not at least with those of the deepsea record. Because of its completeness, it will be regarded as the Stratigraphical Stratotype Section for the land record. The Sparta Long Term Geosoil Sequence was recorded in two different but closely neighbouring sections which will be labelled respectively as "Road Section" and "Afission East Wall Section". The two sections have been put together in vertical stratigraphical order to form a composite stratigraphic section. Hereunder, both parts will be described in one composite profile from top till bottom.

The Afission Road Section

The Road Section extends over a total length of 3 km, along the road from Sparta to the North, starting from the cement-concrete factory till the splitting of the road into a northern direction towards Vassara and into an eastern direction towards Chrissafa and Agriani. The total thickness of the deposits belonging to the Quaternary overlying the latosol of Tertiary age (T.S.), reaches a thickness of 160 m of which only the 50 m uppermost part of is believed to be continuous and provides good evidence of the Holocene, Upper and Upper Middle Pleistocene geosoil stratigraphic assemblages. It will be used to build up together with the Afission East Wall Section, the upper part of the composite Pleistocene lithostratigraphic profile.

The Afission East Wall Section

The Afission East Wall Section offers probably the most continuous and complete Lower Pleistocene sequence hitherto found in the Peloponnesos and in Greece as a whole. It shows furthermore good lithostratigraphic correlation with the overlying Lower Middle Pleistocene sequence of Affision Road Section as well and lies conformably on the Pliocene as will be described hereafter.

Description of the Composite Sparta/Affision Section

Lithostratigraphic description. The lithostratigraphic description is putting main emphasis on the geosoil stratigraphic assemblages in between gravel bed layers occurring at seemingly steady (time) intervals. References made to chronostratigraphic ages are based on comparison with the deepsea OIS-oxygen isotope stages and comparison with the magnetostratigraphic and susceptibility record of the Megalopolis Lake Basin section as will be shown hereafter (Fig. 5-2):

a) 160 - 148 m: HOLOCENE (0 - 10 Ka[3]) consisting of 12 holosoils (HS) interfering with gravel hard-banks, especially at the base the BASAL HARD BANK MEMBER, of the current interglacial or OIS 1 (oxygen isotopic stage 1).

b) 148 - 145 m : UPPER PLEISTOCENE (UP-estimated between 75 - 130 Ka) consisting of a (GS) geosoil assemblage series of **4 geosoils** (# S1 to S4) corresponding with OIS 5.

c) 145 - 108 m : MIDDLE PLEISTOCENE (UMP-Upper part estimated between 130 - 620 Ka). Gravel bed *A* (145 - 136 m) consisting of thick boulder deposits subdivided into three parts by 2 interstadial geosoils (# S5 and S6 at 150 Ka and 170 Ka) inside the gravel bed overlying a (GS) geosoil assemblage series of **8 geosoils** (geosoils # S7 till S14) occurs, starting with 2 interglacial geosoils of OIS 7 (# S7 & S8 at 200 Ka and 250 Ka), 2 interglacial geosoils of OIS 9 (# S9 and S10 at 290 Ka and 340 Ka), 1 interglacial geosoil of OIS 11 (# S11 at 405 Ka of the 390 Ka and 420 Ka interval), 1 interglacial geosoil of OIS 13 (# S12 at 480 Ka) and 2 interglacial geosoils of OIS 15 (# S13 and S14 at 530 Ka and 620 Ka interval).

d) 108 - 91 m : MIDDLE PLEISTOCENE (LMP-Lower part estimated between 620 - 910 Ka). Gravel bed *B* (108 - 100 m) consisting of thick coarse gravel beds at which base a (GS) geosoil assemblage series of 8 geosoils (geosoils # S15 till S22), grouped into two sets of each four geosoil assemblages all showing palaeosoils interfering with coarse gravel deposits.
The uppermost four geosoils are the weakest in development and most probably correspond to OIS 17 (# S15 and S16 at 670 Ka and 700 Ka interval) and OIS 19 (# S17 and S18 at 765 Ka and and 790 Ka interval); the lowermost four are the strongest developed geosoils corresponding to OIS 21 (# S19 and S20 at 810 Ka and 870 Ka interval) and OIS 23 (# S21 and S22 at 900 Ka and 910 Ka interval).

---

[3] Ka reads as "thousand years"

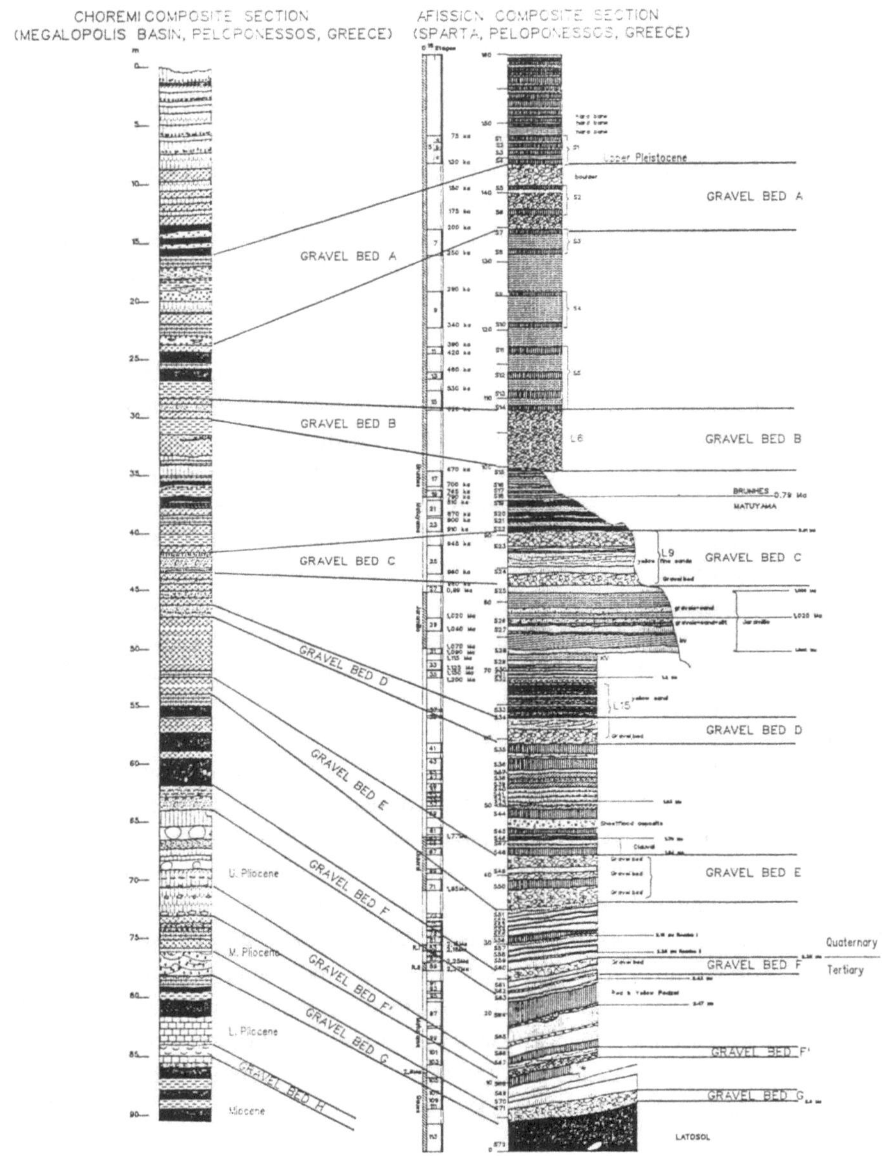

Figure 5-2. Geological sections of Choremi (Megalopolis) and Afission (Sparta) (Paepe *et al.*, 1993)

e) 91 - 64 m : LOWER PLEISTOCENE (ULP-Upper part between 910 Ka - 1.3 Ma[4]) Gravel bed *C* (91 - 83 m) consisting in the upper half of two thick boulder beds with in between a characteristic yellow fine sand bed, each of the boundaries being occupied by a well developed soil horizon most probably corresponding to respectively OIS 25 (# S23 and S24) and to OIS 27 (# S25); consisting in the lower half of alternating layers of gravels and sands/silts. At the base of the gravel and sands series follows a (GS) geosoil assemblage series of another 8 geosoils (geosoils # S26 to S34) four of which are alternating with pebble bands and four others (humic horizons) finally alternating again with yellow sands at the base, totalising 10 Brown Geosoils below the uppermost gravel bed.

f) 64 - 43 m : LOWER PLEISTOCENE (MLP-Middle part between 1.3 - 1.84 Ma) Gravel bed *D* (64 - 59 m) consisting of an upper coarse crossbedded sand layer and a lower boulder layer at which base a (GS) geosoil assemblage series of 13 Brown Yellow Geosoils (geosoils # S35 to S48) first alternating with sand layers corresponding with OIS 41 till 51 (# S35 till S40 in between 1.32 and 1.52 Ma) and gradually changing into an alternating soil and pebble band/gravel bed system corresponding with OIS 41 till 67 (# S41 till S48 in between 1.52 and 1.84 Ma).

g) 43 - 27 m : LOWER PLEISTOCENE (LLP-Lower part between 1.84 - 2.35 Ma) Gravel bed *E* (43 - 35 m) consisting of three gravel bed layers separated by 2 geosoils of OIS 69 and 71 (# S49 and S50 at 1.89 and 1.94 Ma) overlying a (GS) geosoil assemblage series of **10 Brownish Red Geosoils** of respectively OIS 73 till 91 (# S51 to S60 in between 1.99 and 2.35 Ma).

h) 27 - 06 m : UPPER PLIOCENE (UPL-Upper part between 2.35 - 2.6 Ma) Gravel bed *F* (27 - 25 m) consisting of an uppermost boulder bed overlying an alternating series of gravels and 3 geosoils of OIS 91, 93 and 95 (# S61, S62 and S63 in between 2.38 and 2.43 Ma) followed downwards by another **4 Red & Yellow Geosoils** of OIS 97 (basal part), 99, 101 and 103 (# S64, S65, S66 and S67 in between 2.43 and 2.6 Ma) and still below under a thinner gravel layer indicated as GRAVEL BED F'(14 - 12 m) still another series of **4 Red Geosoils** of OIS 105, 107, 109 and 111 (# S68, S69, S70 and S71 estimated to end up at 3.0 Ma).

---

[4] Ma reads as "million years"

i) 06 - 00 m : MIDDLE PLIOCENE (MPL-Upper part older than 3.0 Ma) Gravel bed G (06 - 04 m) consisting of thick valley scree overlying a thick (10 m observed) **Dark Red LATOSOL** of OIS 113 (# S 72) the bottom of which has not been reached.

The Lithostratigraphic Subdivision

The description above leads to the distinction of a number of lithostratigraphic Formations and Members in the litho-pedological record. The whole record belongs to the Eastern Mediterranean Group.

HOLOCENE: Formation I

(1) Light Brownish Yellow Holosoil Assemblage Member
(2) Basal hard bank member.

PLEISTOCENE: Formation II

(3) Yellow Brown Geosoil Assemblage Member (S01 - S04)
(4) Gravel bed A member (with 2 geosoils S05 - S06)
(5) Brownish Red Geosoil Assemblage Member (S07 - S14)

Formation III

(6) Gravel bed B member(with no geosoils)
(7) Reddish Brown Geosoil Assemblage Member(S15- S22)
(8) Gravel bed C member (with 3 geosoils S23 - S25)
(9) Brown Geosoil Assemblage Member (S26 - S34)

Formation IV

(10) Gravel bed D member (with no geosoils)
(11) Brown Yellow Geosoil Assemblage Member (S35 - S48)
(12) Gravel bed E member (with 2 geosoils S49 - S50)
(13) Brownish Red Geosoil Assemblage Member (S51 - S60)

Formation V

(14) Gravel bed F member (with 3 geosoils S61 - S63)
(16) Red and Yellow Geosoil Assemblage Member (S64 - S67)
(17) Gravel bed F' member (with no geosoils)
(18) Red Geosoil Assemblage Member (S68 - S71)

Formation VI

(19) Gravel bed G member (with no geosoils)
(20) Red Latosoil Assemblage Member (S72)

Chronostratigraphy of the Sparta/Afission Composite Section

Thanks to the lithostratigraphic reference layers Gravel bed C and Gravel bed B together with their underlying geosoil stratigraphic assemblages, correlation and geometrical connection of the Afission Road section with the Afission East Wall section, resulting in the composite profile, could easily be established in the field. Thus, a continuous section of 160 m thickness composed of geosoil stratigraphic assemblages of the complete Pleistocene Series of geosoils interfering with important gravel beds, was reconstructed. Quite obviously the sequence remained totally avoid from erosional hiatuses from the Holocene Series at the top until the contact with the Pliocene Series at its base. It is the first time that such a continuous land sequence has been found in the European terrestrial sequences.

A Sixfold Cyclothem Sequence

The lithostratigraphic subdivision of the Pleistocene record revealing the afore described six Gravel beds (A through F) all linked up at their base with the six associated geosoil stratigraphic assemblages are totalising 60 geosoils. Below, the weathering of the geosoils becomes considerably stronger which may point to dramatic changing climatic conditions as well. Taking the facies of the sediment layers and the similar nature of the soil development into account, it leads to a sixfold cyclothem subdivision of the Pleistocene Series within the reach of the first 60 geosoils. In referring to the classical stratigraphical subdivision of the Quaternary these gravel/soil assemblages read as at least 60 major climatic changes of alternating humid/warm (geosoils) and dry/cool (clastic sediments) phases and 6 major periods of longlasting droughts (gravel beds).

Correlation of the Sparta/Afission Gravel/Geosoils Assemblages with the Deepsea ODP 677

The lithostratigraphic sequence is now projected against the deepsea oxygen isotope record of ODP 677 in which each of the odd stages are representing interglacial stages (Fig. 5-3). As stated above, they are supposed to coincide in the land record with the development of an interglacial geosoil. Actually, isotopic atmospheric changes are registered via the vegetation cover in each of the geosoils the latter representing the weathering interface between atmosphere, biosphere and lithosphere namely the pedosphere (Paepe and Van Overloop, 1990a). During phases of drought (cold/dry conditions) the vegetation cover disappears from the landsurface whereas during phases of prevailing warm/wet conditions, an abundant vegetation cover invades the former dry lands again. Interaction between lithosphere and atmosphere resulting in the pedosphere/geosoil development will greatly depend on the presence or not of the vegetation cover as well as on the type of vegetation assemblage that is dealt with. As a result registration of the stable isotope record will totally differ between various phases

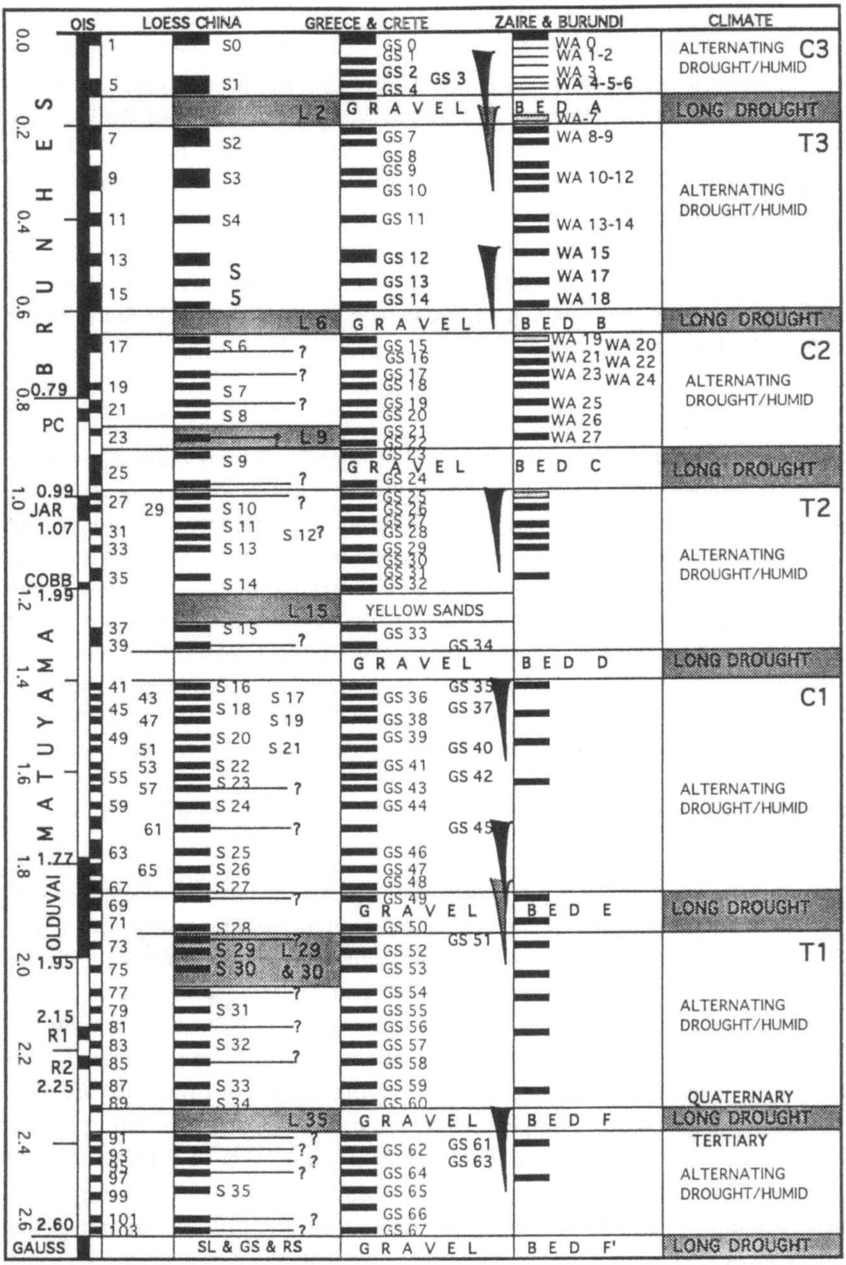

Figure 5-3. Correlation of Sparta/Afission geosoils with deepsea and other land records

of vegetation cover and certainly with those phases without any vegetation cover at all. The changes in vegetation cover then play a similar role as the planktonic/benthonic variations in the deepsea sediments. For this reason it becomes quite possible to correlate geosoil levels and eventually their related oxygen isotopic changes, with those of the deepsea oxygen isotopic stages (OIS).

Moreover, the geosoil diversity is a direct consequence of the "time-stability hypothesis of the geological/ecological diversity". Indeed under sufficiently long periods of stable climatic conditions fauna and flora could remain unchanged under specific environmental conditions. This leads to typical ecological conditions under specifically prevailing global atmospheric conditions leading to characteristic vegetational assemblages on land and planktonic/benthonic assemblages in the ocean occurring within specific intervals of time. It leads also to the "time stratigraphic stability concept of a single interglacial geosoil" as was advocated by some of the authors earlier (Paepe and Van Overloop, 1990a). The latter concept points to the fact that interglacial geosoils in the long term sequence, occur at the standard deviation level of most probably 0.1%. Indeed as it will be shown hereafter with the study of the Holocene soil sequence, a single interglacial soil occupies maximum 10 % of the 10.000 years timespan of the current interglacial stage. This is when it occurs with a periodicity of 1000 years as is the case for the current (Holocene) interglacial. Within the scope of the middle term 100,000 years of the Last Interglacial - Glacial scale, 10,000 years is again 10% of these very 100,000 long lasting interglacial years from which it may be readily deduced that the timestability of an interglacial geosoil which is maximum 10% of each very interglacial, becomes 10 times smaller in the actual 100,000 years i.e. 1% and equally another 10 times smaller within the 1,000,000 years time interval, i.e. 0.1%.

This reasoning leads to the extraordinary conclusion that interglacial geosoils in the long term range of say 1.0 Ma do not shift stratigraphic position more than 1 year per 1000 years, this means not more than 100 years/100,000 years. If one is dealing with geosoils of the 100,000 periodicity in the long term sequence of 1 million and more, its absolute position may be located at the precision at 100 years and less. There is no other fossil which is showing such a "time stratigraphic stability" with regard to its last or first appearance position (LAD[5] or FAD[6]) in the deep sea records.

As a consequence, each geosoil level of the land record (which is actually representing a total landscape environment), reappears and disappears within well fixed limits of time of the Pleistocene (say Quaternary) record with a maximum error of 0.1 %. This is a very stable periodic cycle of extremely great impact on Planet Earth's surface. Correlation of geosoils with deepsea OIS is therefore quite feasible. Indeed, all 103 odd isotopic stages of the deepsea record down till the Matuyama/Gauss boundary at 2.6 Ma

5 LAD reads as last appearance datum

6 FAD reads as first appearance datum

years, must reveal the same time stability of occurrence as they are directly a result of the same changing geochemistry of the atmosphere which in turn is controlling the occurrence of decay of the vegetation cover i.e. the corresponding geosoil.

One arrives at exactly 72 interglacial geosoils of the land Sparta/Afission stratotype section within the 113 OIS and 67 geosoils within the 104 OIS reach,taking into account that 11 of the odd isotopic stages seem to contain each a doublet of geosoils totalising 22 geosoils to be added to eighter 50 or 45 single geosoil/odd isotopic stages. The number of 67 geosoils represents almost the double amount of the number of geosoils observed in the stratotype section of Jiacun of the Chinese Loess Plateau. Indeed, in the latter section half of the duly occurring geosoils have not been registered as single value soils which may explain the above mentioned discrepancy.

OIS 104 falls in the middle of the of Gravel bed F' in between Geosoil S67 and S68, i.e. in between the Red and Yellow Geosoil Assemblage and Red Geosoil Assemblage Member. Both Geosoil Assemblages represent fossil soils which are more strongly developed than is usually the case with Pleistocene palaeosoils of the former assemblages above.

This why it is believed that the geosoils of Pleistocene tradition starts only above the Gravel bed F which upper boundary is to correspond with 2.35 Ma. This time boundary of the Plio-Pleistocene Limit is in accordance with formerly expressed estimates of the beginning of the Quaternary at about 2.4 Ma. The time boundaries used in the lithostratigraphic description for delimitation of the Pleistocene Formations and Members above are based on the same principle using the changing degree and type of geosoil development for their correlation with the OIS levels.

5-4. THE LONG TERM GEOSOIL SEQUENCE OF THE MEGALOPOLIS LAKE BASIN

The Lake Basin Stratotype of the Eastern Mediterreanean

N. Nassopoulou has undertaken parallely to the study of the Sparta/Afission section the study of the Quaternary section in the lignite exploitation pit of Megalopolis Lake Basin. The lithostratigraphic sequence on which a series of laboratory researches are carried out shows good comparison with the Afission geosoil section of Sparta (Fig. 5-2). The Megalopolis lignite studies are furthermore completed by an additional borehole study carried out by S. Nassopoulou and E. Van Overloop in 1991 and an open pit stratigraphic section described by S. Nassopoulou and R. Paepe in 1992.

The Megalopolis/Choremi Section

In order to make a possible subdivision of the Quaternary, a lithostratigraphic correlation of the Megalopolis Lake Basin with the Sparta Loess Plateau sequences seems most appropriate. Actually, comparison of two totally different sedimentation environments should lead to the distinction of the climato-environmental changes which occurred during the Quaternary.

Lithostratigraphic Description

a) 00.0 - 12.5 m: HOLOCENE (0 - 10 Ka) consisting of 8 holosoils (HS) (5 uppermost Yellowish Brown silt clays and 3 lowermost Gray brown humic clays) interfering with clayey loesslike sediments [as originally interpreted by Lttig and Vinken (1968)]; at first sight they seem to correspond to the Holocene soil sequence at the Afission Road Section of Sparta but further study should be carried out with regard to their true age.

b) 12.5 - 15.8 m: UPPER PLEISTOCENE (UP-estimated between 75 - 130 Ka) consisting of a (GS) geosoil assemblage series of 4 geosoils (1 Dark Gray Brown humic clay soil at the top and 3 Lignite layers interfering with iron concretions layers) of the Last Interglacial.

c) 15.8 - 33.2 m: MIDDLE PLEISTOCENE (UMP-Upper part estimated between 130 - 620 Ka) Gravel bed A (15.8 - 23.8 m) consisting of gravels, clays and sand layers containing 1 humic claysilt geosoil (probably 2 interstadial geosoils in the borehole) and overlying a (GS) geosoil assemblage of 8 geosoils (1 humic gray clay, 2 strong black lignites, 2 humic gray silty clays, 1 dark gray peaty clay, 2 black lignites) comparable to those of the Afission Road Section.

d) 33.2 - 41.7 m: MIDDLE PLEISTOCENE (LMP-Lower part estimated between 620 - 910 Ka) Gravel bed B (33.2 - 35,6 m) consisting essentially of clays with gravelly layers and loess nodules interfering with 1 Green Gray clay loam geosoil overlying a (GS) geosoil assemblage of 6 geosoils (1 humic clay silt, 2 Black lignites, 3 Dark Greenish Gray humic clay silts) comparable to those of the Afission Road Section.

e) 41.7 - 45.9 m: LOWER PLEISTOCENE (ULP-Upper part between 910 Ka - 1.30 Ma). Gravel bed C (41.7 - 43.3 m) consisting of coarse sand layers and clayey silts with loess nodules (N) at the base (43.3 m) overlying a (GS) geosoil assemblage of 3 geosoils (3 Dark Greenish Gray humic clay silts).

f) 45.9 - 52.8 m: LOWER PLEISTOCENE (MLP-Middle part between 1.30 - 1.84 Ma). Gravel bed D (45.9 - 47.3 m) consisting of coarse sandy silt with a double row of loess nodules (N) at the base (47.3 m) containing 1 dark greenish gray strong humic clay overlying a (GS) geosoil assemblage of 7 geosoils (Dark Greenish Gray silty clay) interfering with silts and loess nodules (N).

g) 52.8 - 61.9 m: LOWER PLEISTOCENE (LLP-Lower part between 1.84 - 2.35 Ma). Gravel bed E (52.8 - 54.0 m) consisting of coarse sandy silt overlying a (GS) geosoil assemblage of 10 geosoils (1 lignite, 2 lignitic clays, 2 lignites, 1 strong humic clay) and 4 lignites with Unio bivalves and yellow sand phosphate balls at the base).

h) 61.9 - 70.5 m: PLIOCENE (UPL-Upper part between 2.35 - 2.58 Ma). Gravel bed F (61.9 - 64.2 m) consisting of coarse sandy silt with Septaria levels overlying a (GS) geosoil assemblage of 6 geosoils (Gray sandy silts and Septaria levels).

i) 70.5 - 76.3 m: PLIOCENE (MPL-Middle part between 2.58 - 3.00 Ma). Gravel bed F' (70.5 - 72.9 m) consisting of coarse sandy silt with septaria overlying a (GS) geosoil assemblage of 4 geosoils (Gray sandy silt).

ia) 76.3 - 83.9 m: PLIOCENE (LPL-Lower part between 3.00 - 3.40 Ma). Gravel bed G (76.3 - 78.0 m) consisting of coarse crossbedded sandy silts overlying a (GS) geosoil assemblage of 3 black lignite geosoils interfering with clay-silt layers on compact basal peat.

ib) 83.9 - 99.0 m: MIOCENE (UM-Upper part between 3.40 - 3.82 Ma). Gravel bed H (83,9 - 84.8.0 m) consisting of coarse crossbedded sandy silts overlying a (GS) geosoil assemblage of 3 lignite levels.

Correlation between Megalopolis/Choremi and Sparta/Afission Sections

The stratigraphic correlation between the Choremi/Megalopolis Basin and the Afission/Sparta Basin offers a comparison between a subsided sequence of almost pure lake deposits and an uplifted continental plateau sequence of fluviatile, eolian and occasionally waterlaid deposits. The correlation is based once more on the Gravel bed series and their interfering geosoil stratigraphic assemblages. Furthermore, the Choremi borehole section which was studied in the laboratory by Nassopoulou with regard to changes in magnetic polarity, susceptibility, stable isotope geochemistry and $CaCO_3$ content, offers a further and more direct possibility for correlation with the deepsea record.

Thus thanks to the lithostratigraphic correlation between the sections of Sparta/Afission and Megalopolis/Choremi, the "geosoil stratigraphic assemblages" of the first could also be projected on these laboratory results as well. Evidently this lead to further precision of their location in the time record independently from the afore described correlation with the oxygen isotope stages. Indeed from this exercise it is clearly shown that the Brunhes/Matuyama Polarity Boundary which is located in the triple series of gray clayey silts above Gravel bed C in the Megalopolis/Choremi section correlates perfectly with its position in the Reddish Brown Geosoil Assemblage Member above the Gravel bed C member of the Sparta/Afission section. It is also the boundary between high and lower susceptibility as well of the boundary in the oxygen isotope record and the high and low $CaCO_3$ values. Similarly, the Quaternary/Tertiary boundary occurs in the Megalopolis/Choremi section at the position of Gravel bed member F in between the triple series of lignite layers with sand phosphate balls above and the septaria layers below which tally respectively with the Brownish Red Geosoil Assemblage Member above and the Red and Yellow Geosoil Assemblage Member below. The Matuyama/Gauss Polarity Boundary in the Megalopolis/Choremi section is located at the next Gravel bed member F' occurrence which corresponds to its esteemed position from the correlation with the deepsea OIS in the Sparta/Afission section at the top of the corresponding Gravel bed member F'. This limit furthermore is characterised in the Megalopolis/Choremi section by the lower High Susceptibility and upper Low Susceptibility boundary, the oxygen isotopic change and the high and low $CaCO_3$ values. With other words the correlation of the Gravel bed members and the Geosoil Assemblage Members of the Sparta/Afission section with the deepsea OIS of ODP 677 finds confirmation in the correlation between both the Peloponessos sections.

5-5. MAGNETIC AND CHRONOSTRATIGRAPHIC BOUNDARIES

The Brunhes/Matuyama Boundary

In the framework of the above combined lithostratigraphic and chronostratigraphic subdivision, the position of the Brunhes-Matuyama boundary which occurs in the deep sea record at the base of OIS 19 is likely to coincide with geosoil S18 i.e. in between the series of Gravel bed B and C. As a consequence the Middle Pleistocene/Lower Pleistocene boundary should then be located at the next outspoken change in facies hereunder, indicating a dramatic climatic change as is probably characterised by the boundary between geosoil S22 and Gravel bed C corresponding with the base of OIS 23. It is the best approximation so far obtained for the Middle/Lower Pleistocene boundary which is moving in time from the previously arbitrarily choosen boundary at 800 Ka to a more realistic figure as 910 Ka, not to forget that this is expressed in the new Lower Pleistocene time scale as recently established by Shackleton *et al.* (1993) on odp Site 677.

The Plio-Pleistocene Boundary

As previously described a series of 8 deeply weathered red geosoils (S64 through S71) occurs under Gravel bed F which itself is containing three weakly developed geosoils (S61 through S63). All 8 red geosoils below are separated by gravel lags and a major gravel bed is found to separate the four upper red ones from the four lower and even stronger developed, dark red coloured ones of the series. S67 is tallying with OIS #103 below which the next (cold stage) Gravel bed F′ occurs containing the Matuyama/Gauss boundary located at 2.6 Ma according to the new time scale.

The uppermost geosoil of the series #S64 is a Red and Yellow Podzolic type of soil and is encompassing OIS 97 so that the top of the geosoil may be located somewhere around 2.43 Ma and its base at about 2.47 Ma. The dramatic change in soil development which occurs in these geosoil series infers that the Plio-Pleistocene transition took place certainly at the level of the Red and Yellow Podzolic geosoil. It is quite likely that the three red geosoils above the gravel at 2.6 Ma and the four geosoils below this Matuyama/Gauss boundary definitely belong to the Pliocene. In this line of thought it is believed that Gravel bed G and the Latosol below it already belong to the Lower Pliocene.

N. Vouloumanos is carrying the palaeomagnetic and the micromorphological studies on this section in order to disentangle the location of the Plio-Pleistocene boundary precisely. Anyhow it is possible to correlate as of now this section with the lake basin section of Megalopolis studied by S. Nassopoulou.

5-6. LONG TERM PERIODS OF DROUGHT IN PELOPONNESOS

The afore proves that the climatic changes of the land record in both sections described in the Peloponnesos have a global climatic signature. A final attempt will be made hereafter to reconstruct the climatic evolution during the whole of the Pleistocene Epoch especially with emphasis on the phases of drought. This reconstruction of the drought cycles will essentially be based on the analysis of the Gravel bed Series and the interfering geosoil stratigraphic assemblages (Figs. 5-2 and 5-3).

The Gravel Bed Members

Both sequences of Megalopolis/Choremi and Sparta/Afission have a series of seven Gravel bed member A through G (including F′), in common on which basis time stable lithostratigraphic correlation for the whole of the Pleistocene Series and the Pliocene Series was possible. Eventually when considering Gravel bed member H the upper part of the Miocene might be included as well. Actually, as of Gravel bed member F the Pliocene Series are reached in both sections. In both facies changes occur below: strong red soil development in Afission versus gray sandy silts with Septaria lags in Choremi.

Furthermore, Gravel bed member F′ and Gravel bed member G separates in both sites the Pliocene Series into an Upper, a Middle and Lower Pliocene Subseries.

Finally, Gravel bed member H most probably tallies with the topmost part of the Miocene Series as a new dramatic change in facies occurs hereunder showing strongly developed compact gray clay beds interfering with compact lignites layers. It is quite evident that the above subdivision based on correlation of the lithostratigraphic Gravel bed member sequence in both Sparta and Megalopolis basins are but a first approximation. A number of laboratory studies must still follow such as palaeomagnetism, susceptibility, oxygen isotope and micromorphology especially in the Sparta Plateau sequence. However, after comparison with the 103 OIS of ODP 677, they show already together with the intercalated geosoil stratigraphic assemblages, a recurrent periodicity of circa 400 k-years independently in each of the sections. Environmentally the gravel beds which delimiting each of the 400 k-years cycles, are attributed to severe cold arid stages during which periods a high amount of coarse angular clastic deposits was build up over a timespan of 70 to 100 k-years. This leaves an average of 300 k-years for the building up of the geosoil stratigraphic assemblages series which occur in between.

Geosoil Thermomer and Cryomer Cyclothems

The Pleistocene Long Term Geosoil Stratigraphic Assemblages. The 400 k-year cycle is well known from the deep sea and from the land record as well. Paepe and Van Overloop (1990a) made an attempt to subdivide the Pleistocene on basis of this 400 k-years periodicity in Super Stages which were called Thermomers (T=with warmer interglacials) and Cryomers (C=with colder interglacials) (Fig. 5-3). The sections in the Sparta (Afission) and Megalopolis (Choremi) Basins confirm former statements. In taking 2.35 Ma as the start of the Pleistocene a subdivision into six cyclothems, three Thermomers and three Cryomers, seems quite feasible:

a) **Thermomer 1** starts right after Gravel bed member F about 2.35 Ma, encompassing both Reunion Events in the reach of odd stages 89 through 73 tallying with GS 60 through GS 51 i.e. the Brownish Red Soil Assemblage Member. Thus it may be concluded that the Pleistocene begins with a phase characterised by warm interglacials of which the red strongly developed geosoils in Afission and the lignite levels at 62m, 57m and 55m are testifying in Choremi. It furthermore points at the gradual transition from the Pliocene onwards into the Pleistocene resulting in a continuous cooling off as is attested by the first five stronger and the last five weaker developments of the 10 geosoils. As the timespan is of maximum 300 k-years these geosoils and interfering with thin gravel lags attesting short severe cold periods in between, pointing at a periodicity of a 23,000 Ka recurrence cycle or precession cycle. Finally it is ending up with a threefold arid phase of almost 100 Ka long corresponding with the Gravel bed member E development which is interrupted by the two milder periods of weaker geosoil developments most probably because one is dealing with the warmer Thermomer.

Again the precession periodicity is likely to control this long cold phase as well.
b) **Cryomer 1** starts right after Gravel bed member E with four strong red geosoil levels encompassing the Olduvai Event around 1.84 - 1.79 Ma (GS 48 - 45) followed by ten weaker geosoil developments (GS 44 - 35) after the short cold sheetflood interruption of about 1.7 Ma in the Afission section and an equal amount of humic gray silts and clays in Choremi. As it ends up about 1.54 Ma pointing at a 300 Ka for the whole timespan in which geosoils occurred, a precession periodicity is most likely to have again controlled the geosoil sequence which above was labelled as Brown Yellow Geosoil Assembalge Member. Gravel bed member D does not contain any supplementary geosoil development most probably as a result of the prevailing Cryomer conditions although lasting in time till about 1.45 Ma.
c) **Thermomer 2** starts right above Gravel bed member D with the series of yellow sands which facies and stratigraphic assemblage position tallies with the aquiferous loess sand bed labelled L15 of the loess sequence in China. It withholds a number of humic horizons of which the two lowermost are labelled GS 34 and 33 whereas an uppermost one is labelled GS 32. The latter encompasses with the base of OIS 35 located at 1.2 Ma which reminds the chronostratigraphic position of the upper part of L15 again. The facies changes into a series of gravelly clays and pebble bands in which the 7 geosoils of rather weak development occur (GS 31 - 26). A soil sterile gravel and sand deposit closes up this sequence which complies with the upper part of the Jararamillo Event between 1.00 Ma and 1.02 Ma. In Choremi the sequence is rather reduced with only three light gray humic clay levels however, imbedded rather thin layers of gravels, sands and silts. The latter facies tallies with a similar facies interfering with geosoils (S34 - S26). The sequence in both sites is sealed off by Gravel bed member C in which milder interruptions with geosoil development (GS 25 - 23) occurred. Thus the gravel is composed of an intermediate yellow sand body seperating an upper and lower gravel deposits. The whole is encompassing the time interval of 70 Ka in between 980 Ka and 910 Ka which is fairly good corresponding with layer L9 of the Chinese sequence.
d) **Cryomer 2** starts right after Gravel bed member C with four strong geosoil developments (GS 22 - 19) followed immediately by four weaker developments (GS 18 - 15) the whole labelled as Reddish Brown Geosoil Assemblage Member. It is overlain by a soil sterile gravel Gravel bed member B at the top developed within the timespan of 670 - 620 Ka. The Brunhes/Matuyama Boundary occurs then at the treshhold of the strong and weak developments which in the new timescale is located at 790 Ka. Gravel bed member B occupies the position of layer L6 in the Chinese sequence. The soil series in Afission perfectly correspond to the gray humic clayey silt of Choremi developing upwards into the series of lignites and olive green silty clays which are affected by faulting.
e) **Thermomer 3** starts right after the cold phase of Gravel bed member B with three extremely well developed geosoils (GS 14 - 12) corresponding and ressembling perfectly

to the famous deeply weathered S5 soil complex in China. The threefold geosoil complex shows at Afission and elsewhere in a plateau position in Greece, always the same geosoil stratigraphic assemblage and is thus representing a real benchmark in the correlation. It is followed upwards in a loessoid matrix by successively geosoils GS 11 (corresponding to S4 in China), 10 and 9 (corresponding to doubled S3 in China), 8 and 7 (corresponding to the doubled S2 in China). The whole labelled Brownish Red Geosoil Assemblage Member is capped by Gravel bed member A which is subdivided into three gravel members due to the presence of two weakly developed geosoils (S6 at 175 Ka; S5 at 150 Ka) of which no equivalent seems to have been recognised in China so far. Probably one is dealing here with interstadial soil developments. The latter two geosoils correspond to a couple of humic clay levels in the lake basin facies of Choremi in the middle of a hughe thick gravel deposit overlying two strong lignite layers most probably to be correlated with the strong geosoil development dating from the beginning of this cyclothem in Afission. Quite clearly there seems to exist a serious sedimentation gap again in the lake basin sequence as far as the correlation with the geosoils is concerned. Such discrepancy was also observed in the lowerlying lithostratigraphic sequences when it came to comparison of the plateaubound geosoils and lake levels.

This cyclothem shows a definite change in the sedimentation pattern with the foregoing cyclothems inasmuch that the interfering sediment facies in between the geosoil levels are definitely biased by more wind blown components and thus pointing at more arid conditions, whereas nicely spaced into four groups of interglacials (comparable hereafter with the Chinese record namely S5, S4, S3, S2). It may therefore be concluded that a new climato-sedimentation pattern starts about 600 Ka with a thermomer (Thermomer 3). This type of sedimentation is in sharp contrast with the previous one which started at about 1.2 Ma i.e. another 600 Ka before and composed of two cyclothems starting with Thermomer 2 and followed by Cryomer 2. Furthermore, it is preceeded by another doublet of cyclothems Thermomer 1 and Cryomer 1 which encompasses the timespan of 2.35 Ma till 1.2 Ma i.e. which reads as another 1.2 Ma.

f) **Cryomer 3** starts right after the cold phase of Gravel bed member A with a complex of three well developed geosoils (GS 4 - 2) in Afission which find their equivalents in the three strong lignitic peat layers of the Choremi pit. From the ongoing pollen study (Qiu Hong Rong and E. Van Overloop), the susceptibility/palaeomagnetic study (S. Nassopoulou and J. Hus) and the stable isotopic study (S. Nassopoulou and E. Keppens) one is surely dealing here with the Last Interglacial Stage. Geosoil S1 may still belong to this Last Interglacial although it might already belong to one of the Last Glacial interstadial soil levels. The whole of the geosoil sequence is labelled as Yellow Brown Geosoil Assemblage Member which latter is covered by the basal hard bank member of the Holocene.

It is followed upwards by the series of 10 weak geosoil levels interfering with hard banks in Afission and 12 weak geosoil levels interfering with silt and iron concretions in

Choremi. The latter were already labelled Holocene by Lüttig in 1966 (personal communication) although further work should be carried out in order to make sure they are not dealing with soil developments of the Last Glacial Stage which also contain 12 geosoil levels.

Anyhow, a Holocene Age is most acceptable as will be advocated hereafter. Comparing with foregoing sequences of gravel bed and geosoil assemblage members, one may conclude that with the 12 Holocene Soils or Holosoils (HS) the second interglacial stage in Cryomer 3 has been reached. Some 10 others still have to come before the next longlasting Gravel bed member with extreme arid conditions, will be developed i.e. by about 250,000 A.D. Indeed the Holocene Stage is coming to an end within 3 Ka followed 75 Ka of the next Glacial Stage and twice the timespan of 130 Ka for the two new Interglacial-Glacial Cycles Thermomer 4 and Cryomer 4 to come hereafter : in total this adds up to approximately 3+75+200= 278 Ka. This total added up to the start of the Last Interglacial Stage 127 Ka yields approximately a 400 Ka timespan until the next Gravel bed member will be formed following on Gravel bed member A.

5-6. THE LONG TERM GEOSOIL SEQUENCE OF KANDANOS (CRETE)

The roadcut nearby the Village of Kandanos, south of Chania in the Western part of Crete and just West of the Lefka Ori range (2453 m), reveals a complete Pleistocene geosoil section similar to the afore described ones, as well. The section has been recorded by Paepe and Mariolakos (1990) and studied from the stratigraphic and neotectonic point of view. It revealed the presence of 15 reddish geosoils alternating with gravel deposits. At the base of eight of the gravel beds hughe wedges have been developed. These wedges show absolutely no resemblance to fault lines (I. Mariolakos) or earthquake induced cracks. Quite instead they recall frost-wedges (R. Paepe) related with extreme cold periglacial periods. Their position at the base of eight of the gravel beds concurs to the assumption that the gravel beds were phases of longlasting severe cold phases. Sometimes they are found associated with cryoturbatic types of structures as well. Both phenomena are strengthening the idea of periodically prevailing severe cold conditions on the island of Crete during the Quaternary (Fig. 5-4).

In the section of Kandanos halfway the west coast and Lefka Ori (2453 m) the highest mountain in Western Crete, number of such periglacial phases occurred and have stratigraphically been recorded. In comparing the Kandanos Roadcut section with the Afission Section one observes the striking evidence that all Gravel bed member A through F occur in the sequence. Most of the eight frost wedge levels seem directly asssociated with gravel beds. Five occur respectively together with Gravel bed member A, C, D, E, and F; three occur in the first half of C3, T3 and C1 each time right after important phases of geosoil development like respectively the important Yellow Brown Geosoil Assemblage Member corresponding to OIS 5, the lower part of the Brownish Red Geosoil assemblage

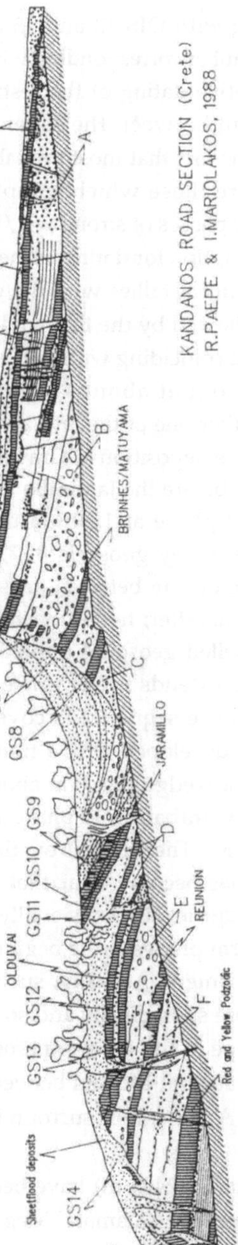

Figure 5-4. Quaternary Geosoil and Gravel Bed Sequence

Member corresponding with OIS 13 and 15 and the lower part of the Brown Yellow Geosoil Assemblage Member corresponding with OIS 65, 63 and 61.

An attempt for relative dating of the restricted number of warm humid phases of geosoil development in between the seven cold gravel beds is also possible. The reasoning is based on the fact that most probably the geosoils which are building up the restricted sequences are those which comply with the deepest weatherings coree-sponding to the climatic phases of strong wet/humid conditions.

The following assumption for dating of the geosoil is made hereby: GS 14 in between Gravel bed member F and E tallies with the geosoil development around the Reunion Event (2.15-2.25 Ma) followed by the first cold at about 2.0 Ma at the base of Gravel bed member E; GS 13 and 12 coinciding with the Olduvai Event (1.84-1.79 Ma) is followed by an extremely strong cold at about 1.73 Ma at the beginning of the sheet flood sedimentation and another one at the end about 1.69 Ma; the next severe cold of about 1.33 Ma occurs just before deposition of Gravel bed member D; GS 11, 10 and 9 represent the geosoil series of just before the Jaramillo Event say in between 1.09 and 1.02 Ma and is followed by the cold phase at 1.00 Ma the whole covered by a thick Gravel bed member C. It is followed by geosoils 8, 7, 6 and 5 with the Brunhes/Matuyama boundary in the middle of it in between geosoils 7 and 6 and a gradual transition into Gravel bed member B herafter; hereupon follows the complex of at least two strong developed geosoils labelled geosoil 4 immediately followed by a weak cold phase at about 450 Ka. The cold extends in the gravel lag above it and is overlain by another strong geosoil 3; the entire sequence is covered by Gravel bed member A in which interstadial geosoil 2 is developed after a transitional cryoturbatic phase and seconded by an extreme cold frost wedge phase at about 150 Ka; finally geosoil 1 represents the same situation with a cryoturbatic cold phase at the bottom and a frost wedge generating cold at the top about 20 Ka. The whole is overlain by modern Holocene deposits.

From the afore it has become clear that the cold phases are well represented by gravels and by frost wedge levels, occasionally by cryoturbatic levels wheras the geosoils represent the initial warm phases at the beginning of each cyclothem alike whether they are thermomers or cryomers. Actually, such frost related structures have been also observed, however, more sporadically and so far never in stratigraphic sequential order by Mariolakos and Paepe in the Peloponnesos namely in the sheetflood deposits in the surroundings of Kalamata and Messini between 1986 and 1992. All occurred at the base of Gravel bed member A in the hills surrounding the Megalopolis Basin. This leaves no doubt that :

a) The Taiyetos Mountains (2405 m) have been glaciated provoking a local periglacial environment in the region of Kalamata, Megalopolis and Sparta of which they are the surrounding mountains.

b) The cold phases corresponding to the Gravel bed member were acting simultaneously in the Peloponnesos and in Crete.

Most probably, the same periglacial activity was also duly operating elsewhere in the Greek mainland as cryoturbatic structures and frost wedges have been observed Mariolakos and Paepe (1988) in Beotia, by Livaditis and Paepe(1986) nearby Kaparelli, and by P. Anastasiadis and R. Paepe (1981) north of Drama in Macedonia. These unpublished observations point at periodically prevailing periglacial climatic conditions due to local glaciated areas in the whole of Greek peninsula. It was probably this geographical aspect together with the high mountainous topography which enhanced local glaciation in Greece easily.

It is not surprising although quite striking that the Kandanos section provides a complete record of the gravel bed sequence as encountered in the Afission section whereas each of the geosoil series is reduced to only 15 out of the 60. This is a most important statement because it proves that on basis of gravel bed sequences a complete sequence of the Quaternary may be detected. It may be questioned though which and why of the single geosoils in between these gravel beds are in fact telescoped together into a compact version of two or three soils. Most probably it deals with problems of differential erosion in areas of non-subsidence or subsidence which can only be disentangled by comparative stratigraphy of several sections in different geographical entities.

5-8. PERIODS OF EXTREME LONG, SEVERE SHORT AND INTERFERING TERM DROUGHTS IN GREECE

A summarising approach of the afore leads to the recognition of the following major longterm droughts during the Pleistocene beginning at 2.35 Ma ago: Gravel bed member a: 130 Ka to 200 Ka, Gravel bed member B: 620 Ka to 670 Ka, Gravel bed member C: 910 Ka to 980 Ka, Gravel bed member D: 1.25 Ma to 1.32 Ma, Gravel bed member E: 1.84 Ma to 2.00 Ma, Gravel bed member F: 2.35 Ma to (Top).

Shorter cycles occur in between these major droughts alternating with the geosoils as a result of the astronomical forcing and their periodicity : Eccentricity 100,000 years, Obliquity 41,000 years and Precession 23,000 years.

Within the timespan of 2.35 Ma which has been considered as the time boundary between the Tertiary and the Quaternary, 60 shorter variations of dry and humid phases occur tallying with the number of 60 geosoils and interfering dry loess or cold sand-gravel phases. It was furthermore quite possible according to the global intercontinental correlation to detect at least 11 major levels frostwedges of extremely cold but short phases interfering with the Gravel bed members : Frost wedge level 1 at 18,000 B.P., Frost wedge level 2 at 150,000 B.P., Frost wedge level 3 at 270,000 B.P., Frost wedge level 4 at 350,000 B.P., Frost wedge level 5 at 460,000 B.P., Frost wedge level 6 at 980,000 B.P., Frost wedge level 7 at 1,140,000 B.P., Frost wedge level 8 at 1,410,000 B.P., Frost wedge level 9 at 1,680,000 B.P., Frost wedge level 10 at 1,880,000 B.P. and Frost wedge level 11 at 2,320,000 B.P.

Summarising the droughts of the Long Term Range on basis of the threefold recognition of cold phases there are: Six long lasting cold phases occurring at 400 Ka time intervals. Eleven extremely severe cold but short phases grouped in the last 500 Ka and in between 1.0 Ma and 2.0 Ma. Sixty medium cold phases interfering with the geosoils in 100 Ka, 41 Ka and 23 Ka cycles.

5-9. MIDDLE TERM GEOSOIL SEQUENCE OF KORONI

The Middle Term Last Interglacial - Glacial Cycle

The most complete Last Interglacial - Glacial section was found in the southern part of the Peloponnesos namely at Koroni. It is an important one because it reveals the geosoil details of a middle term cycle of 100 Ka. It shows more light on what kind of geosoil cycle may be expected for each the 100 Ka intervals of the long term cycle where at the same time it clearifies the climatic evolution in between two interglacial stage and the position of interstadials and stadials. It shows a rather complete series in between the Last Interglacial Geosoil and the Current Interglacial Geosoil of 10 interstadial geosoils (Fig. 5-5).

Lithostratigraphic Description of the Koroni Section

It could be summarized as follows:

a) SML-2.4 m	Neogene marine deposits.
b) 02.4-3.8 m	127 Ka and 115 Ka. Pedocomplex or PK 1 Last Interglacial geosoil corresponding to OIS 5e on top of lagunal deposits.
c) 03.8-4.5 m	Fine whitish homogeneous sands with frostwedge at the top corresponding to OIS 5d.
d) 03.8-5.8 m	115 Ka till 73 Ka. Dark Brown interstadial geosoil assemblage of warm/wet woodland interstadials interfering with shorter severe cold stages in between : IS[7] 1 and IS 2 of OIS 5c and IS 3 of OIS 5a with cold stage OIS 5b in between.
e) 05.8-06.5 m	Fine whitish homogeneous sands with frostwedge at the top. At 73 Ka the first severe cold strikes and deserts form.
f) 06.5-07.3m	Brown interstadial geosoil assemblage of relatively mild/moist interstadials of park steppe and tundra geosoils and climate tallying with the full (pleni) glacial stage: IS 4-6 73 Ka-55 Ka of OIS 4 Early Glacial
g) 07.3-10.7 m	Pale Brown interstadial geosoil assemblage of cool/dry interstadials of

7 IS reads as Interstadial Soil

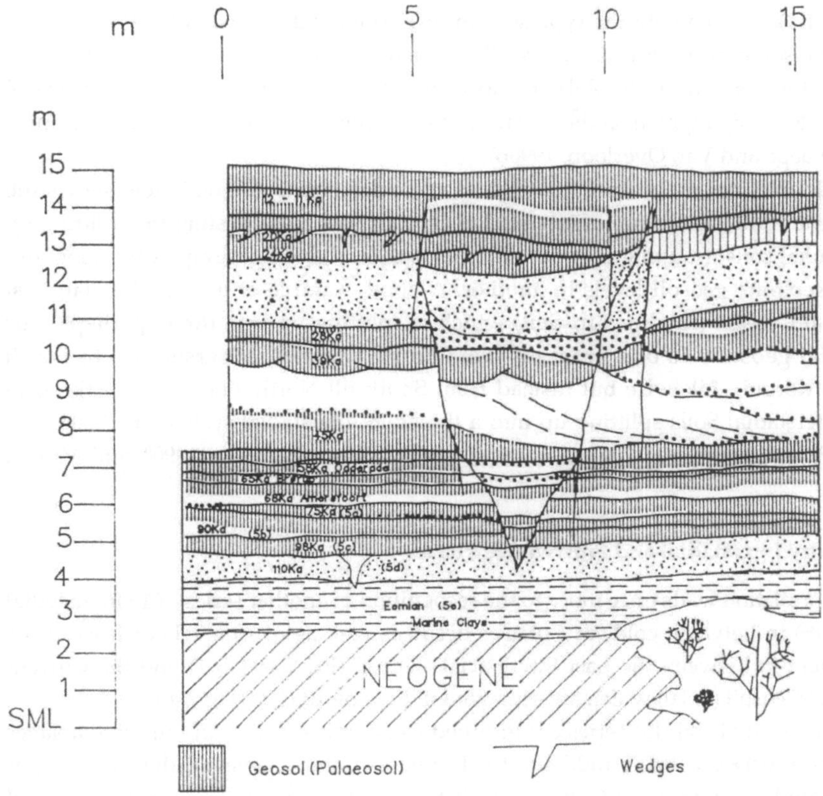

Figure 5-5. Koroni: Upper Pleistocene deposits with approximate ages (Paepe *et al.*, 1984)

	tundra geosoils: IS 7-9 45 Ka-28 Ka of OIS 3 Middle Glacial
h) 10.7-13.2 m	Gravelly whitish sands becoming finer at the top with pebble band and frostwedge row. At 18 Ka the second and even more severe cold strikes and deserts form again of maximum cold of OIS 2.
i) 13.2-13.5 m	Dark Brown interstadial geosoil assemblage of steppe soil of the Late Glacial. Only one single level is represented here: IS 10 15 Ka-10 Ka of OIS 2 Late Glacial
ia) 13.5-15.0 m	Whitish sands with Modern Holocene Soil at top.

Correlation of Koroni with North Sea Basin and West African Rift

The correlation of the Koroni Section with the North Sea Basin was carried out step by step from its very position in the Eastern Mediterranean through Sardinia and the loess of France and Belgium to the North Sea Basin Coversand Region of the Belgian/Dutch Border namely the type section of Zelzate and Moershoofd (Paepe, *et al.*, 1990b). The same exercise was succesfully carried out towards no less than the West African Rift in comparing the Zelzate and Koroni Sections with the Last Interglacial / Glacial Cycle of Gigunghwe on the Burundi/Zare Border along the River Ruzizi (Ilunga, 1984 and Paepe and Van Overloop, 1990b).

This double correlation both continental and intercontinental, definitely shows the global event of the occurrence of interstadial geosoils. It is also attesting the splitting up of interstadial soils grouping into one complex soil in a given geographical region and showing separate geosoil levels in another. A good example of it is the Belgian Last Interglacial Geosoil (the Rocourt Soil) which is splitting up into three, perhaps four interglacial geosoils as one proceeds from North (Zelzate/Moershoofd) to South (Sardinia/Koroni). Likewise but instead from South till North, one observes the Late Glacial Interstadial Soils splitting up into a threefold interstadial system at 17,000 B.P., 12,000 B.P. and 11,000 B.P. (Ilunga, 1984; Paepe and Van Overloop, 1990b and Musisi, 1991).

Cold/dry Interstadial and Frostwedge Phases

This correlation leads once more to the recognition of another series of 13 interstadial geosoils (IS) in between cold/dry phases occur as well as of 10 levels of frostwedge development in between the Last Interglacial Geosoil (PK 1 or GS 1) and the Current Interglacial Geosoil (GS 0) which are separated in time by 100 Ka (Fig. 5-6):

The interstadial soil (IS) series are grouped as described before in four assemblages of geosoils in between many mild short cold phases occur. However, the assemblages are separated by longer cold phases with frostwedge levels attesting severe cold conditions at approximately time levels as described hereunder: Frost wedge level at 18 Ka, Frost wedge level at 28 Ka, Frost wedge level at 45 Ka, Frost wedge level at 73 Ka, Frost wedge level at 90 Ka, Frost wedge level at 98 Ka, Frost wedge level at 105 Ka, Frost wedge level at 115 Ka.

It proves that in the timespan of 100 Ka a complex series of at least 13 cold cycles dominated by the strong cold phases of the 23 Ka and 41 Ka periodicity are prevailing during the Last Interglacial-Glacial Cycle. These are supplementary cycles to the already described long term cycles. One must understand that each of the 100 Ka time interval are to be subdivided by an even number of interfering cycles which renders the climatic evolution of the Pleistocene already quite complex.

The Timespan of the Last Interglacial

Maximal cold stages are recorded in between, at 50 Ka and 18 Ka, and cooler dry periods in between the geosoil. Periodicities of 23,000 and 41,000 years are prevailing.

The Last Interglacial is lasting no longer than the current one, namely 12,000 years,

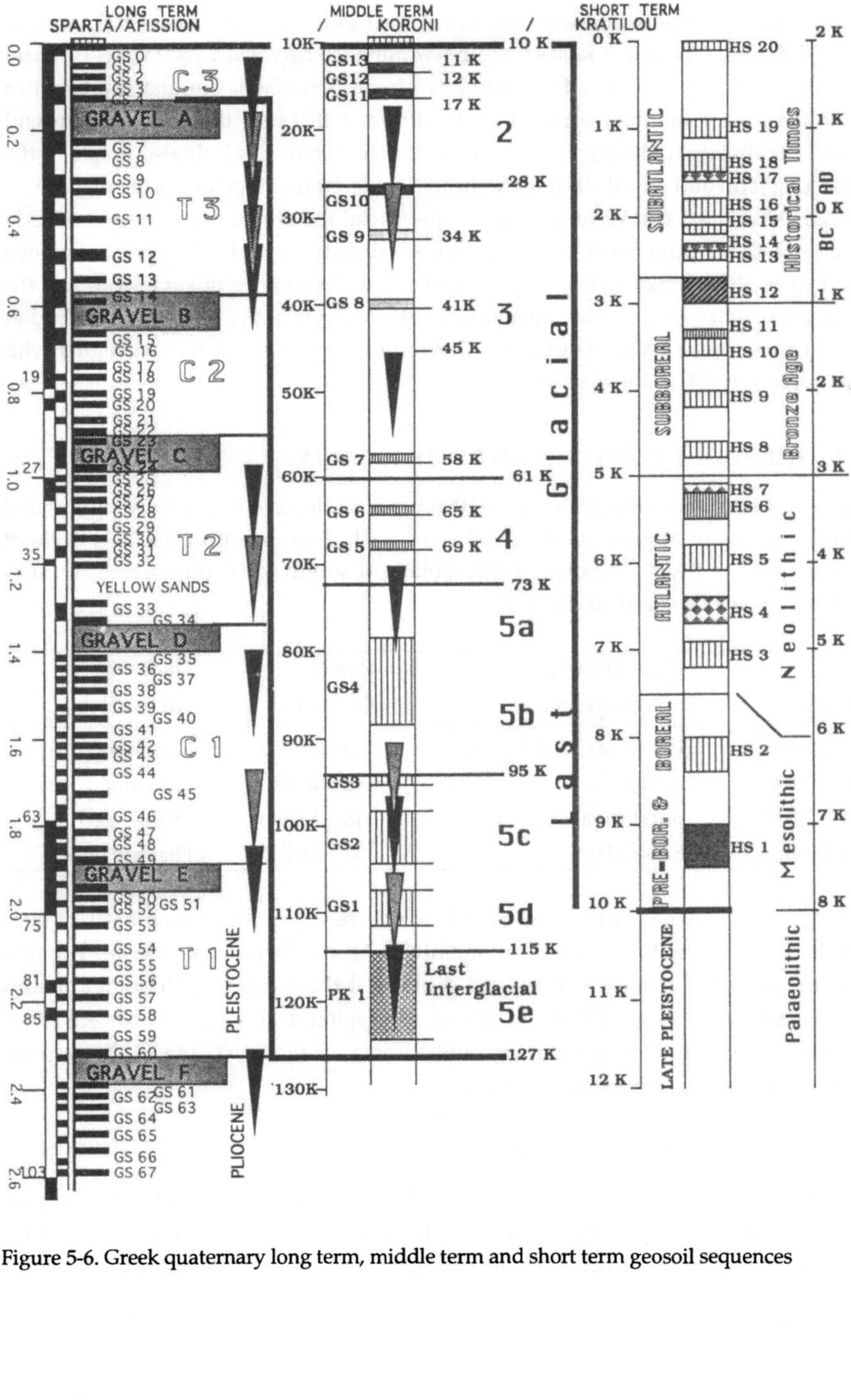

Figure 5-6. Greek quaternary long term, middle term and short term geosoil sequences

followed by a rather long transition period until 73 Ka. Hereafter the climate is rapidly decaying although the maximum cold is reached only at 18,000 years B.P. (Before Present). The Late Glacial witnesses a rapid warming up between 13,000 years B.P. and 10,000 years B.P. as the current Holocene Interglacial is reached. Comparison between the Last Interglacial and the Current Interglacial shows that now the latter nears its end and is on the decline to the next glacial, the second of Cryomer C_3. It also points to the fact that long term and middle term periodicities are operating together at the present.

Ice cores revealed that during the Last Interglacial the amount of atmospheric CO_2 was about 270-280 ppm whereas today figures already as high 353 ppm have been reached. It is also known that interglacial of the older long term cycle were on the average 2.5°C warmer than the current one which, eventually, points to still higher amounts of atmospheric CO_2. During the Cretaceous they were 4 to 8 times higher with temperatures 5° to 10° C above today's.

5-10. THE SHORT TERM CURRENT INTERGLACIAL

The short term sequence of the last 10,000 years (Holocene) is generally subdivided into four sections of approximately 2,500 years each. It is based on the CO_2 cycle of SUESS and on the Blytt/Sernander palaeobotanical subdivision from which the sub-stage names have been derived (Fig. 5-7) :

Suess CO_2-Cycle	Blytt/Sernander Sub-Stages names	Paepe Geosolis	Hadziotis Archaeology
10 Ka-7,5 Ka	Pre-Boreal and Boreal	HS 1 and 2	Melolithic
7,5 Ka-5,0 Ka	Atlantic	HS 3 till 9	Neolithic
5,0 Ka-2,7 Ka	Sub-Boreal	HS 10 till 14	Bronze age
2,7 Ka-0,0 Ka	Sub-Atlantic	HS 15 till 20	Historic times

The geosoil sequence of the Holocene has been set up by one us on basis of geo-archaeological investigations carried out mainly in Attica (Paepe *et al.*, 1984a). The first region comprises the Marathon Plain and south of it the series of archaeological sites along the East Attica coast till Vraonas. The second region deals with the archaeology of the Kifissos Valley in between Athens and Pireefs on the west side of the Attican peninsula debouching into the Saronic Gulf.

The Marathon Plain and the East Coast of Attica

The site of the Athenian / Persian Battle of Marathon (490 B.C.) offers a real opportunity to to establish the geosoil sequence of the Holocene as many well dated historical constructions and levels of ceramics intermingled with sediments and geosoils occur

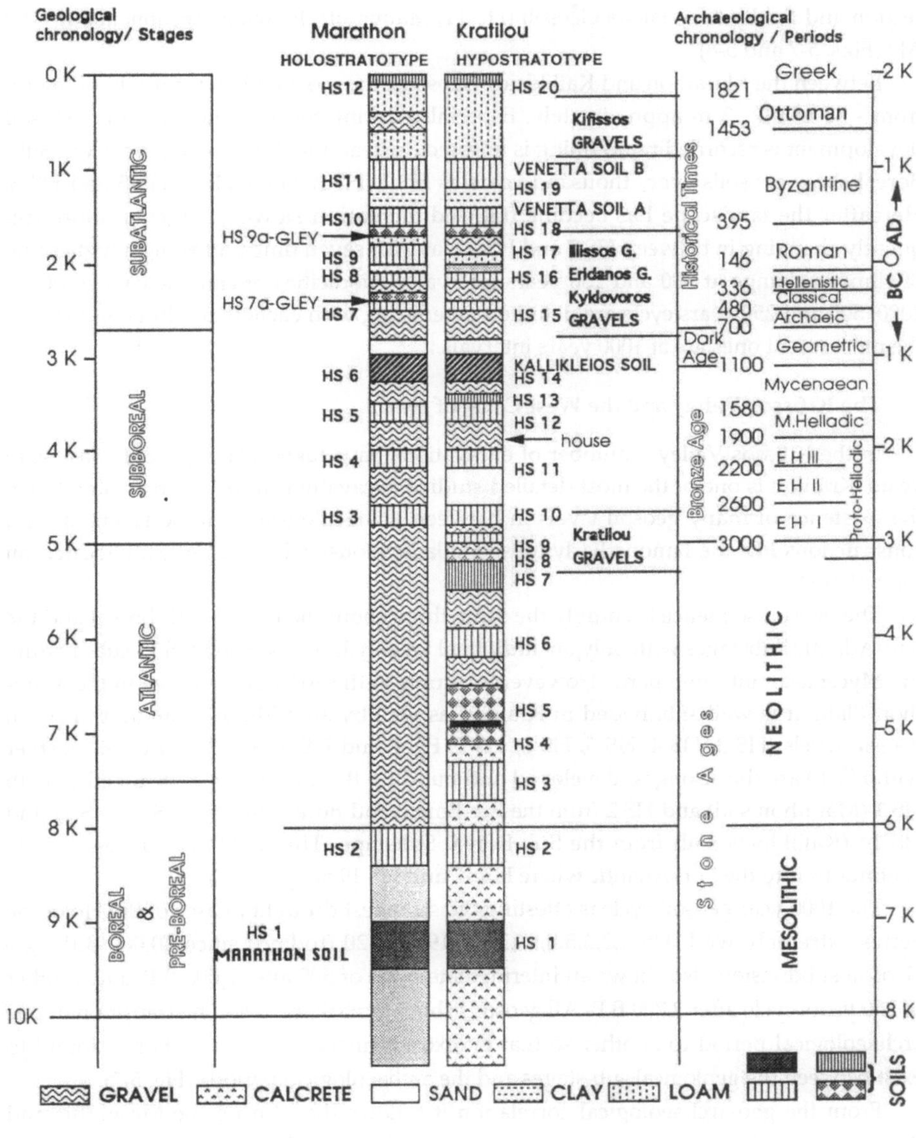

Figure 5-7. Holocene geosoils and alternating sediments in Greece

scattered all over this battlefield. The section shows a remarkable twofold subdivision marked by three well developed geosoils namely the Marathon Geosoil (HS 1) about 7000 B.C. at the base, the Kallikleios Geosoil (HS 6) about 1100 B.C. in the middle of the section and finally the Venetta Geosoil (HS 11) sealing off the whole section, about 1100 AD (Figs. 5-7 and 5-8).

Between the Marathon and Kallikleios Geosoils, the coastal plain is rapidly filling up from - 40 m till - 3 m approximately. Especially during the Atlantic where no geosoil development is recorded rapid filling is inferred. Instead the Pre-Boreal, Boreal and Sub-Boreal show geosoils every thousand years as HS 1, HS 2, HS 3, HS 4, HS 5 and HS 6. Hereafter the landscape has become flat, sedimentation slower, however, more frequently changing in between HS 7 and HS 11, at least seven times, attesting a higher rate of climatic change at 500 and 250 years interval. Periodicities of climatic change of the 1000, 500 and 250 years cycle are definitely interfering with eachother. HS 11 and HS 12, the most recent ones are at 1000 years intervals.

The Kifissos Valley and the West Coast of Attica

In the Kifissos Valley a number of excavations have taken place since the last sixty years. Kratilou is one of the most detailed studied excavation since 1934 revealing today the existence of many geosoil levels in between archaeological ceramic layers or/and constructions like the famous Early Meso-Helladic House which is perfectly located on top of HS 11.

The geosoil sequence is entirely the same throughout the Boreal, Sub-Boreal and the Sub-Atlantic Substages with only an additional weakly humic geosoil HS 13 subdividing the Mycenaean into two parts. However, contrary to the Atlantic Substage in the Marathon Plain, it is well subdivided in Kratilou as well by an additional amount of seven geosoil levels (HS 3, HS 4, HS 5, HS 6, HS 7, HS 8 and HS 9). HS 3 and HS 6 together with HS 10 are the strongest developed and build up the 1000 years cycle together with HS 1 (Marathon soil) and HS 2 from the Pre-Boreal and Boreal and the HS 10, HS 11 and HS 14 (Kallikleios soil) from the Sub-Boreal Substage. This 1000 year drought cycle continuous into the Sub-Atlantic where HS 17 and HS 19 are part of it.

The 1000 year geosoil cycle is attesting the strongest drought phases of the Holocene Series extremely well [HS 1,2,3,5,8,9,12,16, 19 and 20 (today)] since 10,000 B.P. ago. Geosoil subdivision also shows an intermediate cycle of 500 after 7,000 B.P. and another of 250 years cycle after 2,700 B.P. All geosoils show remarkably well the transition of one archaeological period to another so that an extraordinary good correlation is found to exist between the geological sub-stages and the archaeological periods (Fig. 5-7).

From the geo-archaeological correlation it follows that during the Mesolithic and Bronze Age periods prevailing climate was generally dry whereas during the Neolithic and Historical Times climate was periodically moister. In the former the Kratilou Gravels were built up whereas in the latter successively the Kyklovoros, Eridanos and

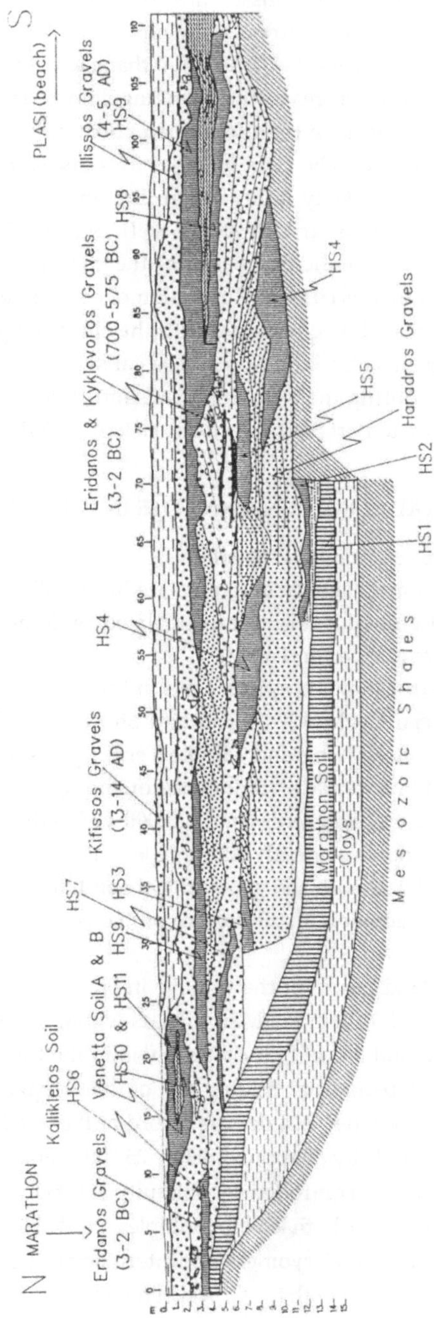

Figure 5-8. Marathon plain

Illisso Gravels and finally the Kifissos Gravels were laid down interfering with phases of geosoil development as the figure is showing.

This leads to the remarkable conclusion that each of the single archaeological phases is expressed successive stages of the changing climate since almost the last 5000 years. Mankind and Society have therefore been constantly adapting themselves to these changing environments. The current Sahel drought is most probably to be associated with the 1000 years cyclicity following in line with droughts of Byzantine/Medieval, Roman and Early Geometric times as well. Sea level has been rising and falling accordingly, so that periodic occupance of the coastal fringe and adjacent valleys was greatly influenced by this climatic fluctuation of drought and flood. The Biblical Flood is situated at about 5000 B.P. coinciding with the climatic optimum when sea level was at its highest. It explains also why Roman coastal sites are today under water since it was dryer than today during the Roman (expansion) Period. Cleopatra could not finish the canal of Alexandria to Port Said because of lack of water!

5-11. GENERAL CONCLUSIONS : THE INTERFERENCE OF PERIODICITIES

These could be summarized as follows:

a) The etesian climatic changes in Greece are like in all monsoonal belts, well expressed in the display of geosoil levels and intercalated sediments like gravels and windborne deposits (loess, coversands and alike).

b) The geosoil series show complete series in the Long Term (3 Ma), Middle Term (Last Interglacial/Glacial Cycle, 127-10 Ka) and Short Term Cycle (Current Interglacial or Holocene, 10 Ka). Both Long and Middle Term Cycles are controlled by specific periodicities enhanced by external astronomical forcing such as the eccentricity (~100,000 yrs), obliquity (~41,000 yrs) en precession (~23,000 yrs) induced by the earth's rotation. The Short Term cycle shows evidently shorter cycles than the afore mentioned ones which in most cases are related with the geochemical condition of the atmosphere and with the effect of the sun's radiation on the surface of Planet Earth (2500 yrs, 1000 yrs, 500 yrs and shorter ones).

c) Especially the land record of the geosoils in Sparta (Greece) reveals the existence of 67 geosoils i.e. equally 67 minor cold/dry phases within the entire Brunhes/Matuyama time interval of 2.6 Ma and 60 cold/dry phases within the timespan of the Pleistocene of 2.35 Ma. Both point at a periodicity nearing that of the obliquity say 40,000 years. If compared with the 103 OIS of the deepsea drilling record of the last 2.6 Ma or 89 OIS of the last 2.35 Ma, an average periodicity of approximately 25,000 years is obtained. Land signals seem at first sight more closely controlled by obliquity (41,000 yrs), ocean signals by precession (23,000 yrs). Furthermore, both lead to multiples like 400,000 and 100,000 of themajor above mentioned Thermomers/Cryomers andInterglacial/Glacial cycles as well as the long lasting droughts observed with the Gravel bed members (70,000 to 80,000 yrs).

d) As the Interglacial/Glacial cycle is also controlled by the very same external astronomical forces obliquity and precession, it very much looks as if there is a permanent simultaneous interference of periodicities controlled by the eccentricity (100,000 yrs), the obliquity (41,000 yrs) and the precession (23,000 yrs) throughout the Long Term and Middle Term cycles. The interstadial geosoil of the Middle Term are therefore interfering, just like the frostwedge levels, with the 100,000 yrs cyclicity of the very interglacial levels of the Long Term record as well.
e) The 20 geosoil levels of the Short Term cycle of the last 10,000 yrs or Holocene, show the predominance of the 500 yrs periodicity and its multiple the 1000 yrs. It reflects the situation of all other interglacial levels revealed in the Long Term and Middle Term Cycles. With regard to the latter, it has now become quite evident that it lasted from 127 Ka through 115 Ka. This could infer that climatic changes of the Current Interglacial started already in the Late Glacial with the milder interstadials at 11,800 yrs (Bvlling) and 12,000 yrs (Allervd) and even including earlier interstadials of 15-16 Ka. It means that climatic variations of the Holocene have now come to an end and that the changing climate which is expressed now is located on the treshhold of the Current Interglacial to the next Glacial which is inevitable. The next extreme cold peaks are foreseen for 5,000 A.D., 23,000 A.D. and 63,000 A.D.
f) As an ultimate final conclusion is that Short Term Cyclicities certainly played their role as well in all former interglacial and most probably during the glacial phases as well. This may be at the origin of the variety of geosoils encountered in the record. The astronomical forcing is a long lasting system in which today's climate has been predicted by a big jumble of periodicity treads which all unrafle systematically as time proceeds. It is an urgent matter to disentangle this in order to identify the real effects of the man made Greenhouse Effect in order not to confuse it with eventually a natural warming of the present climate which in my modest opinion would be even more desastrous.

REFERENCES

Büdel J. (1982). Climatic Geomorphology. Princeton University Press, Princeton, N.J., pp 443.

Ilunga P. L. (1984). Le Quaternaire de la plaine de la Ruzizi. PhD Thesis Earth Technology Institute (ETI), Free University Brussels (VUB), pp. 353.

Musisi J.H. (1991). The Neogene-Quaternary Geology of the Lake George-Edward Basin, Uganda, PhD thesis Earth Technology Institute (ETI), Free University Brussels (VUB), pp. 299 .

Paepe R., Van Overloop E., Hadziotis M.E., and Thorez J. (1984a). Desertification cycles in Historical Greece. Progress in Biometeorology, 3 55-64.

Paepe R. and Mariolakos I. (1990). Paleoclimatic reconstruction in Belgium and in Greece based on Quaternary lithostratigraphic sequences. Current Issues in Climate Research. Proceedings of the E.C. Climatology Programme Symposium, Sophia Antipolis, pp. 113-130.

Paepe R. and van Overloop E., (1990a). River and soil cyclicities interfering with sea level changes, *In* : Paepe R., R.W. Fairbridge and S. Jelgersma, (Eds.), Greenhouse Effect, Sea Level and Mitigation of Drought, Kluwer Academic Publishers, Boston, pp. 253-280.

Paepe R., Mariolakos I., Van Overloop E., and Keppens E. (1990b). Last Interglacial- Glacial north-south geosoil traverse, Quaternary International, 5: 57-70.

Shackleton N.J., Berger A., and R. Peltier W. (1990). An alternative astronomical calibration of the Lower Pleistocene time scale based on ODP Site 677. Transactions of the Royal Society of Edinburgh : Earth Sciences, 81: 251-261.

CHAPTER 6

CYCLIC CLIMATIC CHANGES IN CYPRUS AS EVIDENCED FROM HISTORIC DOCUMENTS AND ONE CENTURY'S RAINFALL RECORDS

DEDALOS C. KYPRIS

Head, Water Resources Division,
Water Development Department, CYPRUS.

6-1. INTRODUCTION

Climatic changes, as related to the average weather conditions experienced over a period of 30 years, have been observed in Cyprus in recent decades. This is testified to continuous reduction of recorded rainfall. The most dramatic impact of such change, relates on the insufficient output of the waterworks of the Country, as compared to that for which they were designed for. Because river flow records of only a short timespan were available at the time of the design of the waterworks, simulated flows derived from much longer rainfall records have been used instead for most catchments in Cyprus, since 1916. The mathematical models used to derive the simulated flows, from the historic rainfall, did not take climatic changes into consideration, which up to that time have not been studied in Cyprus. As such, historic high rainfall values from the beginning of the century, produced high simulated flows, which have been used for the design of the waterworks.

Any climatic change or trend, which may be foreseen, would have a decisive influence on the future water resources management policies to be followed. This consideration let to this research on past climatic changes in Cyprus, with the objective of recognizing any such trend or recurrence pattern. It must be noted that this research was recently undertaken and with this presentation the work done may not be considered concluded. Previous studies, carried out for the past and present climate of Cyprus, revealed that the air temperature recorded at Nicosia since 1896, show an increase of about 0.5 °C until 1985 (Hadjioannou, 1987). The starting point of the gradual increase of the Nicosia average annual temperature is located somewhere around the early twenties (Fig. 6-1). In the same study the rainfall over Cyprus reveals a decreasing trend with average yearly precipitation values for the period 1916-1950 503 mm, for the period 1941-1970 489 mm and for the period 1951-1980 477 mm (Hadjioannou, 1987).

In this study the evidence which was looked for in Cyprus and presented, refers to

NATO ASI Series, Vol. I 36
Diachronic Climatic Impacts on Water Resources
Edited by Andreas N. Angelakis and Arie S. Issar

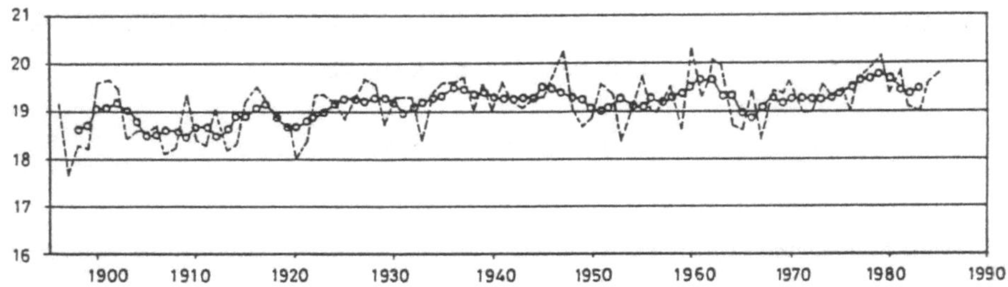

Figure 6-1. Annual (dashed) and five year moving average (dots) of Nicosia air temperature, in centigrades, as from 1896 to 1985 (Hadjioannou, 1987)

rainfall only and no attempt was made to correlate this with temperature changes at this stage. Rainfall over Cyprus, with which this presentation deals, is the only primary source of water. The output of all Island aquifers, springs and rivers feeding the waterworks, depend entirely on rainfall and rainwater. Therefore climatic changes affecting the amount of rainfall that falls on the Island, are of utmost importance for the people of Cyprus.

An answer to the question, as to whether the decrease in the average rainfall that Cyprus is experiencing during this century is a permanent trend or this is part of a cyclic climatic change, is from the practical point of view extremely important. The aim of this paper is to contribute towards such an answer. The problem will be approached by: (a) Analyzing the instrumental records available in Cyprus, which cover the present century and the last part of the previous one, as far as rainfall is concerned, (b) presenting the findings from the study of the widths of the growth rings (no other dendrochronological parameters have been studied as yet) of a pine tree found in Cyprus, to be over 1000 years of age and (c) examining critically documents drawn out of the rich historic past of Cyprus.

6-2. SETTING AND RAINFALL

Cyprus, with an area of 9,251 km^2 is the third largest Mediterranean Island after Sicily and Sardinia. The Island is located at 35° north and 33° east in the eastern end of Mediterranean sea. Its shape is an elongated one, with maximum length (Cape Apostolos Andreas to Paphos coast) of about 223 km and a maximum width (Cape Kormakiti to Cape Gata) of about 93 km (Fig. 6-2). The climate of Cyprus is typically Mediterranean, with mild rainy winters and dry hot summers. Rainfall by 90% occurs from November to March, while the remaining 10% occurs from April to September. The geographic distribution of rainfall is controlled mainly by the relief of the country. In the

Mesaoria Plain and the flat eastern areas of the Island, yearly rainfall averages about 300 mm. On the high ground of Kyrenia Range (1000 m), along the northern coast, the average rainfall rises to about 500 mm. On the summit (1950 m) of Troodos Mountain is 1100 mm (Fig. 6-2). The dominant role which the Troodos mountains play in the

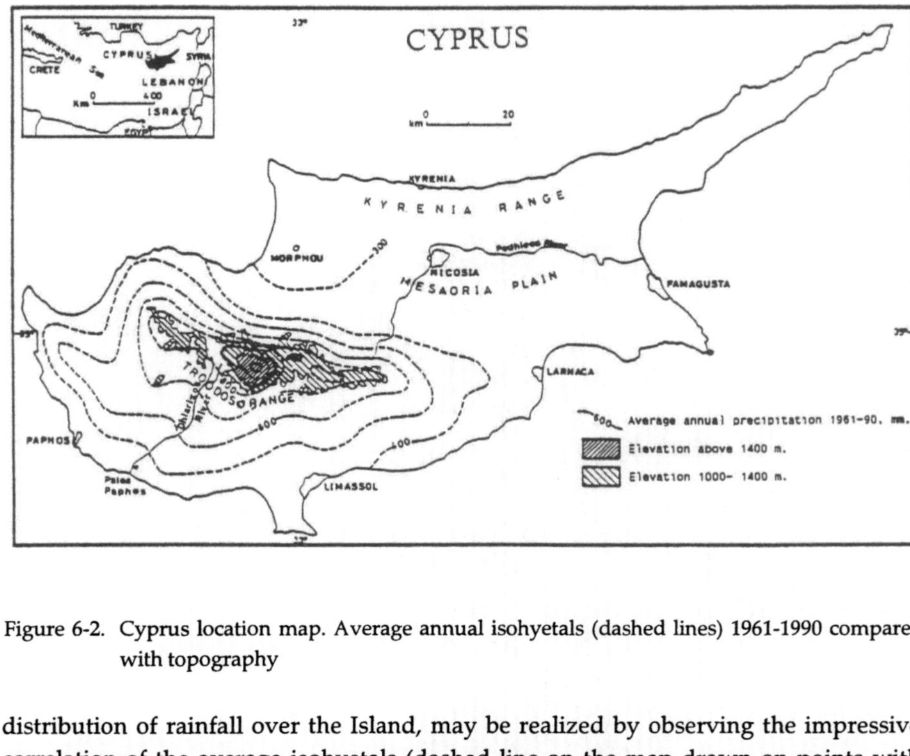

Figure 6-2. Cyprus location map. Average annual isohyetals (dashed lines) 1961-1990 compared with topography

distribution of rainfall over the Island, may be realized by observing the impressive correlation of the average isohyetals (dashed line on the map drawn on points with equal rainfall), to the topographic contours (Fig. 6-2). However it is worthwhile to note that the northern slopes of the mountain receive less rainfall at the same elevation than the southern ones.

6-3. EVIDENCE OF RAINFALL DECREASE FROM METEOROLOGICAL DATA

The present "normal" rainfall value, 1961 to 1990, is 503 mm for the southern part of

Cyprus, which is not under the occupation of the Turkish troops and meteorological data are available at the Meteorological Service of Cyprus. If extrapolated to include the areas where recent meteorological data are not available, the present rainfall normal value, for the Island, is 462 mm. Comparing the average rainfall over Cyprus from 1916

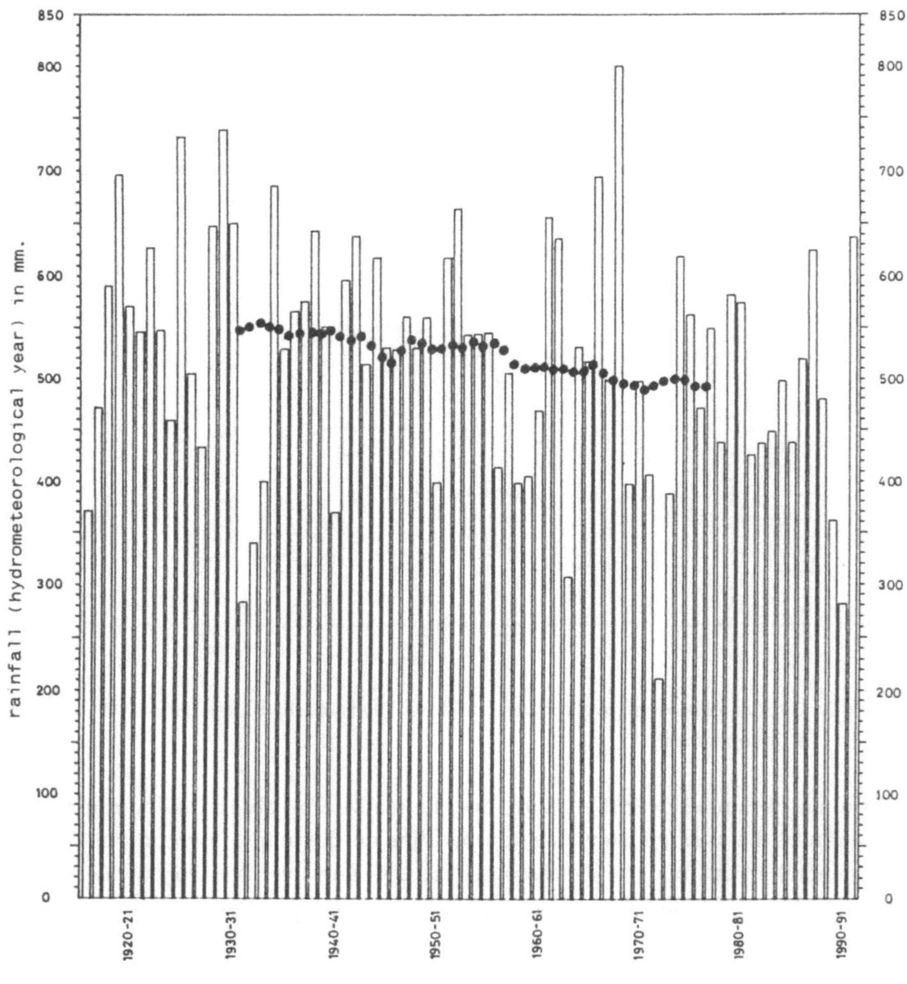

Figure 6-3. Annual average rainfall in the non-occupied area of Cyprus, where data are available until present. The dotted line is the thirty year moving average (Rainfall data from the Meteorological Service of Cyprus)

to 1950 (503 mm) with the three last decades (462 mm), we observe that rainfall at the beginning of the century was 9% higher. This rainfall reduction does not appear as a step change, but follows an almost steady trend as revealed by the thirty years moving average of the mean rainfall of Cyprus from 1916 to present (Fig. 6-3). This situation is an alarming one for the well being of the people of Cyprus, which spent a considerable amount of money on waterworks, to provide storage of winter river flows for multiannual utilization. The situation becomes even more serious by the fact that the runoff reduction is not only 9% but much greater.

6-4. PROJECTION OF RAINFALL REDUCTION TO RUNOFF

For the purpose of a gross evaluation of the percentage of water lost for Cyprus due to rainfall decrease, a general Runoff to Rainfall Deviation Diagram has been prepared for Cyprus (Fig. 6-4). The deviation from normal of the yearly flows as recorded at flow gauging stations for six important rivers of Cyprus, have been plotted against the deviation from normal of rainfalls for each respective catchment. The yearly normal values used for rainfall have been derived in the same way as the yearly normal values

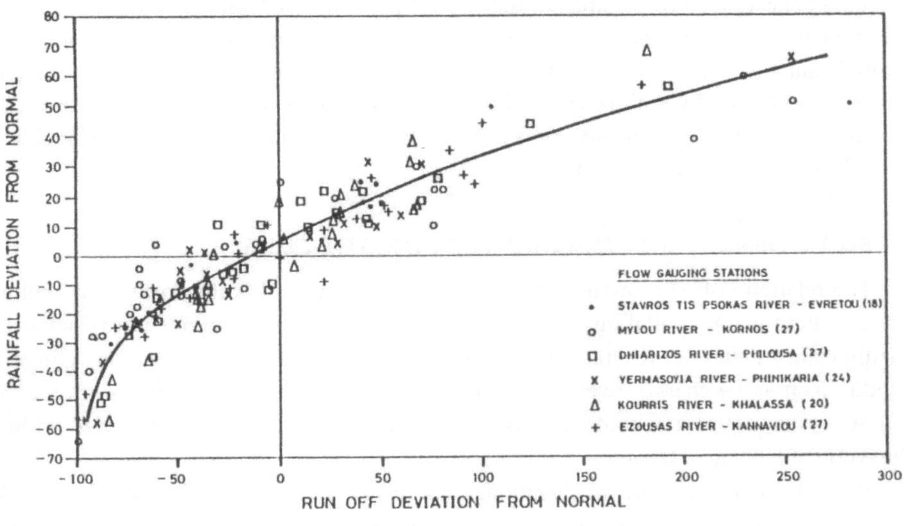

Figure 6-4. Generalized rainfall to runoff deviation diagram. The numbers in parenthesis refer to years of flow records

for each river flow, as the arithmetic mean of the rain fall values over the catchment, obtained for the same years for which flow records were available. It has to be noted that although 50.7% of rainfall are above normal values, the number of flows above normal are only 42.7%. This is attributed to the fact that runoff is not in linear relation to rainfall. As a result, a normal rainfall gives less than normal runoff by about 14%. Making a correction for this, we find out that at the beginning of the century the water resources of Cyprus were about 27% higher than at present, which is a great loss to our generation and creates anxieties for the generations to come.

6-5. EVIDENCE OF RAINFALL INCREASE AT THE BEGINNING OF THE CENTURY

However a message of hope is being received from the records (1881 to the present) of Nicosia rainfall station. The hopeful message relates to a rainfall increase as from the beginning of the records until the early years of this century (Fig. 6.5). The downward trend developed later follows the same pattern as that for Cyprus average (Fig. 3). Specifically the Nicosia rainfall average 1901-1930 is by about 4% higher than the respective figure for 1882-1911. Such behavior gives us the right to consider the decrease of rainfall in Cyprus during our century, as part of a cyclic climatic change or disturbance and thus hope for better days to come and a reverse of the present trend.

It is possible that an argument be brought up, that the slightly lower rainfall in Nicosia at the end of last century was just a disturbance within a general trend of rainfall decrease persisting since the ancient times until present. Such an argument can not be accepted, because if this century's rate of rainfall change is applied to the historical times, rainfalls must have been then several times higher than to day. Such high rainfalls have not been evidenced by the historic and other information available. On the contrary all information, available to us, indicate a more or less stable climate, but with wet and dry periods of a not well defined duration.

6-6. EVIDENCE FROM THE GROWTH RINGS OF TREES

The relation of tree growth to the weather conditions and particularly to rainfall at places where water availability to the tree is limited, is well known. Cyprus enjoys a Mediterranean climate, where rainfall is not abundant and at low elevations climatic conditions may be characterized from this aspect as semiarid. Thus dendrochronology might prove a useful tool in identifying any rainfall or climatic changes recorded within the wood of long aged trees.

Recently, at a place with an elevation of about 1500 m and an average yearly rainfall of about 900 mm, a pine tree of over 1000 years of age was found in Cyprus. The tree found on site uprooted, of the species *Pinus nigra,* had a stem diameter of 1.5 m and a height of about 14 m. A first effort has just started to investigate the possibility that any rainfall changes with time, or of any other climatic parameters, could be identified on the

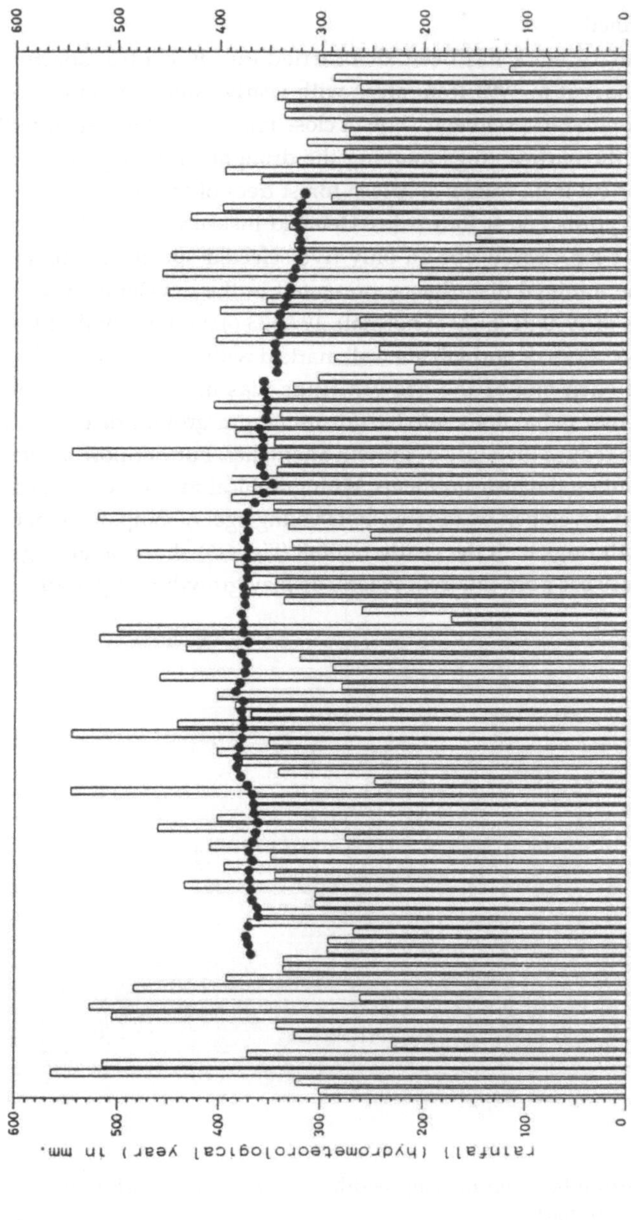

Figure 6-5. Nicosia annual rainfall. The dotted line is the thirty years moving average (Rainfall data from the Meteorological Service of Cyprus)

growth rate of the tree. The work is in an entirely preliminary stage, the only parameters measured being the widths of the growth rings. Still the results, which will be presented later on, are considered worth of mentioning. The year when the tree died has not, at the moment, been defined.

Some preliminary work has been also carried out on control specimens, of the species Pinus brutia, cut in 1993 from areas with yearly rainfall of about 550 mm. The work carried out on these specimens, show a close reliance of the tree growth with the observed rainfall, the narrow rings marking the droughts. Investigations were carried out on fifteen slabs cut from trunks of fifteen forest trees of the species Pinus brutia and provided by the Cyprus Forest Industries. Detailed measurements of the width of the growth rings has been carried out on only five selected specimens, because of their roundness and the minimal disturbance presented in the development of those rings. The drought years of 1990-91 (56% of normal), 1972-73 (41% of normal), 1963-64 (60% of normal) and 1931-32 (53% of normal) are well marked with very narrow rings.

The width of each one of the growth rings was measured, on these selected specimens, along two radial lines selected for minimum growth disturbance and their values averaged to reduce the noise of growth anomalies. Furthermore some adjustment has been carried out on the measurements, with graphical means, to compensate for the wider growth rings developed by the tree in its young age. A comparison of the adjusted width of the growth rings with the yearly rainfall (Fig. 6-6) shows a fairly good relation between them. Trees cut from different places, display growth ring patterns comparable

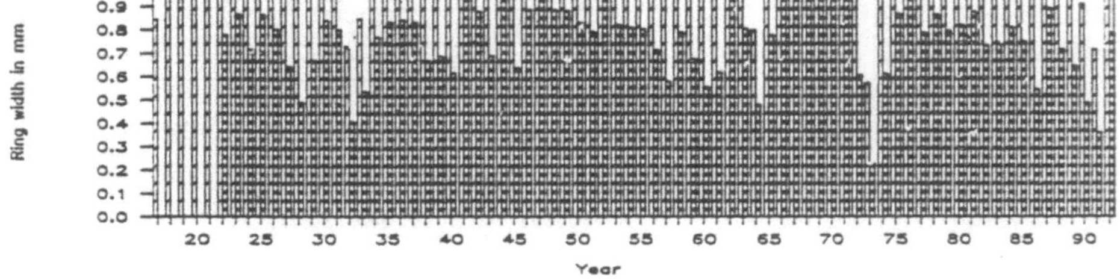

Figure 6-6. Comparison between the ring widths of *Pinus brutia* (dark shaded), with yearly rainfall 1916-1992

to each other, the highest agreement between them presented among the rings formed in dry years, when the bad weather conditions were influencing the whole Island (Fig. 6-7).

As from the limited work carried out so far on specimens from Cyprus on this subject, we conclude that it is possible to identify with confidence the dry years of the past. The analysis of a good number of century aged trees of suitable species will eliminate or reduce the influence (or the noise) on the pattern of the growth rings that special conditions could have on individual trees (diseases for instance). By averaging the results, the confidence on the conclusions about any climatic change would be increased.

Although long aged trees existing for centuries are numerous in Cyprus, they have not been sampled for dendrochronological climatic studies. Such trees are considered in Cyprus real monuments of nature and they are protected. One of those trees of the species Pinus nigra, as referred to previously, was found dead and made available for study. This tree lived for many centuries on a steep slope of the serpentinitic rocks of the Troodos Mountains. No spring was found near the tree and no moisture appeared on the ground surface either. The difficult conditions under which this tree found itself from the beginning of its life, may be judged from its low growth rate. Its stem diameter at the age of 50 years was 4.5 cm and at the age of 100 years was only 11 cm. It took 350 years for this tree to reach the height of 9.2 m. Under such circumstances the tree is believed to have relied on the limited moisture supplied by the rainfall and stored into the weathered rock, its yearly growth depending closely on the availability of such moisture.

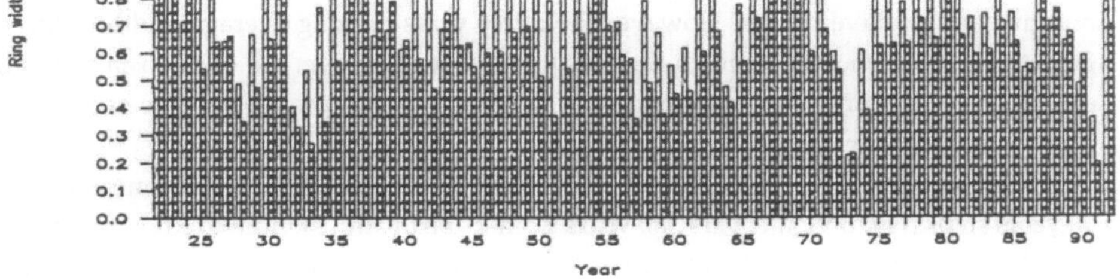

Figure 6-7. Comparison between the ring widths of *Pinus brutia* trees from different localities for the years 1922-1992

It is unfortunate that the trunk of this tree got rotten in the inside, so that a proper slice could not be obtained. Work has been carried out on two pieces of wood obtained at one meter above ground. The one with 380 growth rings is from the central wood while the other with 570 rings from the periphery. There is a gap between the two because of the rotten wood at this place. Additional samples taken later from other parts of the trunk at higher levels, have not been studied yet.

It is interesting to note from the diagram (Fig. 6-8) of the widths of the outer 120 growth rings, that some very narrow ones appear every few years. It is unfortunate that

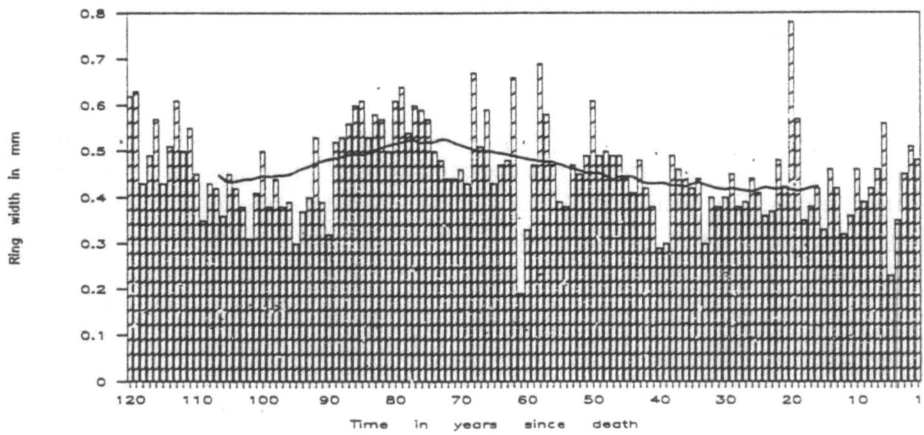

Figure 6-8. Ring widths of the long aged *Pinus nigra*. The line is the 30 years moving average width of growth rings

there is no rainfall station in the area the tree was found and the year of the tree's death is also unknown. An attempt made to match these narrow rings with the low values of the average yearly rainfall for Cyprus or the pattern of growth rings of other trees, did not give any positive results. However the thirty years moving average width of these growth rings when compared to the thirty years moving average of Nicosia rainfall, exhibit some similarity such as a part with increasing trend, probably at the end of last century and another part of this century, with a decreasing trend (Fig. 6-5).

Another very interesting diagram is the thirty year's growth of the tree measured at 10 years steps (Fig. 6-9). It covers 380 years from its birth and, after an unknown interval because of missing wood, 570 years until its death. In this diagram a gradually increasing rate of growth and then a decreasing one are repeated in a cyclic way at intervals between 140 to 180 years. It is our belief that such gradual changes in the growth rate in this tree, both in the increase and the decrease direction, coming back

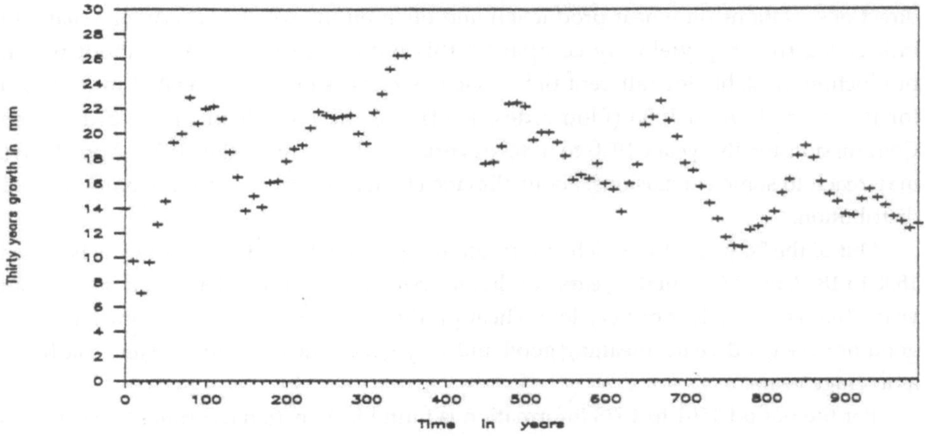

Figure 6-9. Thirty year growths of the long aged *Pinus nigra* taken every 10 years. The gap is for unknown years

every 140 to 180 years, may not be attributed to cyclic local changes that affected this individual tree in this way. Although there is no direct evidence linking these changes of growth rates of this particular tree with rainfall, we believe that rainfall changes was the main cause for this, without excluding other climatic parameters, such as temperature changes, playing a substantial role. However this is evidence provided by a single tree and as explained before a number of long aged trees must be studied in order that concrete conclusions are reached.

6-7. HISTORIC EVIDENCE FROM WHEAT PRODUCTION IN CYPRUS

Irrigated agriculture is practiced nowadays in about 13% of the arable land of Cyprus. In the past irrigation was much more limited, agriculture being practically rainfed. The main agricultural product was wheat, the level of production of which was characterizing the agricultural years as good, medium or bad years. Wheat in Cyprus is a rainfed crop with yields depending on adequate rainfall and distribution in time (Hadjichristodoulou, 1982). Information provided by chroniclers, or other sources, although limited, if critically examined and quantified, may allow the comparison of rainfalls for different periods the various documents refer.

A certain chronicle, covering the years 1800 to 1878, was written by an unknown author, from the village of Lysi (Kyriazis, 1931). This village lies in an important wheat producing area of Cyprus, the Mesaoria Plain (Fig. 6-2). The chronicler among other historic information for Cyprus, gives also an account of wheat production. Each year he refers to is characterized as to whether it was a good one, a bad one or an average, with direct observations on wheat production and often on the weather conditions that had influenced the crop yield. By comparing this with other information about wheat production available for different times, such as records providing related information for the years 1501 to 1533 (Ploumides, 1974) and the records of the Cyprus Grain Commission for the years 1950 to 1969 (Cyprus Grain Commission, 1970, App. 1), we may reach to some conclusions, about climatic changes relating to rainfall amounts and distribution.

Out of the 76 years, for which information is given in the Chronicle of Lysi, between 1800 to 1878, the 27.6% of the years are characterized by the chronicler as bad or very bad years, that is with a low or very low wheat production, the 53.9% were characterized as good or very good years, meaning good and very good wheat production and the 18.4% as average years.

For the period 1501 to 1533 information is found in Venetian Documents, relative to areas occupied at that time by Venice, referring to wheat production in Cyprus. We are informed from those, that out of the 11 years with information about wheat production, 36% were bad, 54% good and 10% medium harvest years.

Wheat production in Cyprus has been estimated by the Cyprus Grain Commission for the years 1950 to 1969 in metric tons. To make this information comparable with that from the Chronicle of Lysi and the Venetian Documents, these years have been classified in the same way as good, bad and average years. So it has been judged that low harvest has occurred for 20% of the years, high harvest for the 65% and medium harvest for the 15% of the years.

The first observation is that both in the 16nth and 19nth centuries, years of drought were more frequent than for the period 1950 to 1969. Wheat productivity in an average year was the same for all the three periods, as supported by information that the yields per one bushel of seed were on the average seven bushels for the 16nth century A.D., in the times of the Turkish ruling of Cyprus (1571-1878) and the nineteen fifties (Christodoulou, 1959). This drives to the conclusion that, there are no big differences between the above said periods, as far as the weather characteristics influencing wheat production are concerned, but there are differences in the frequency of dry years.

6-8. HISTORIC EVIDENCE FROM COTTON PRODUCTION

Although information on the production of cotton in Cyprus in the past can not be quantified in the same way as the information about wheat production, critical examination of the evidences available, may lead to certain conclusions relating to the general climatic conditions prevailing at the different times the information refers to.

According to Giovanni Mariti (an official of the Imperial and Tuscan Consulates, in Cyprus 1760-1767), cotton in the past was one of the chief products of Cyprus (Cobham,

1971). Until the middle of this century it was extensively cultivated, grown on rainfall only, rainfall and springtime irrigation, or perennial irrigation. At about the middle of this century the proportions depending on rainfall to springtime to perennial irrigation was 3:5:2, with 80% of cotton cultivations depending heavily on rainfall (Christodoulou, 1959). In the older days, due to the limited waterworks, the percentage grown on rainfall was much higher.

Mariti compares cotton production of his times with that of the beginning of 18nth century and that of Venetian rule in Cyprus. According to information he collected from old persons of his time, cotton production (in units used at that time) was 8000 bales at the beginning of the eighteenth century and up to 30000 bales during the Venetian rule (sixteenth century). In his time the production dropped to only 3000 to 5000 bales, which he attributes to two natural causes, want of rain and the strong hot winds of July (Cobham, 1971). At another place of his book, Mariti refers to the village of Citti (near Larnaca, now called Kiti), saying "Much cotton was grown here; now the land is more used for wheat, barley and mulberry trees" (Cobham, 1971). Kiti is in the flood plain of a river, very sensitive to rainfall, having nowadays a spring flow only in very high rainfall years. The replacement of cotton plantations, most probably in fields flooded with spring water diverted from the river, was perhaps a necessity when the flow of the river was reduced because of lower rainfalls that persisted for decades.

A critical examination of this information, about the change in the production of cotton in Cyprus and specifically in the Larnaca area, may lead to a conclusion about a possible climatic change during the eighteenth century, much alike the one we are experiencing now in the twentieth century, with the beginning of the century being wet turning gradually more dry, but still wet enough to allow some reduced production of the cotton plantations.

6-9. HISTORIC INFORMATION ABOUT RIVERS AND GROUNDWATER

Direct information concerning the water resources situation in Cyprus in the past, although limited is important, because of their direct dependance on rainfall. Such information is available for the 16nth century and can be drawn out from a report with the title "Report about Cyprus" which has been prepared by a certain A. Savorgnaro, sent in 1562 from the rulers of Venice to study the defense matters for Cyprus, in the expectation of an attack by the Turks to conquer the island (Kyriazis, 1936). There it is reported: "I am sorry not to be able to talk about rivers, except for torrents with little running water. There are everywhere wells from where they raise the water with buckets on wheels (alakatia). This place very often suffers from lack of water" (Kyriazis, 1936). Remnants with such alakatia (local name of what is known as persian wheels) may still be seen all over Cyprus, one of which at least was still operating a few years ago in a village near Nicosia. In another place, Savorgnaro refers to Nicosia town as follows: "...but has nothing for a river but a torrent. She has springs and waterwells and very good soil... (Kyriazis, 1936).

The condition at which Savorgnaro found the river of Nicosia, is a little different to that described 200 years later by Mariti (in Cyprus 1760-1767), saying "But in our days it has no water, the little which it collects in the Winter months being spread over the neighboring country" (Cobham, 1971). Mariti by using the "But in our days" he is contrasting with situations in the past when apparently he had information that Pedhieos river had more water "discharging it near Famagusta" as he says further down.

About 200 years earlier than Savorgnaro, Martoni (a traveller), who visited Nicosia in the year 1394 on the 10th of December, speaks about "...a brook through the middle of Nicosia which a man can cross on stones when it does not rain. When it is rainy weather a great quantity of water runs down..." (Cobham, 1969). The "brook" is Pedhieos river at its natural course. Giovanni Mariti gives the information that the river "Pedicus" passing in the old times through the middle of the ancient town of Nicosia, was diverted outside in 1567 (Cobham, 1971). The wording "when it does not rain" that Martoni used, is interpreted as referring to the dry season and that Pedhieos had then a certain baseflow.

The river about which Savorgnaro speaks of "torrent", Mariti as "Pedicus" and Martoni as "Brook" is the Pedhieos river (Fig. 6-2). The situation at which Martoni and Savorgnaro found Pedhieos river is very close to the one that prevailed also early this century. At about 50 years ago, there were still springs in Pedhieos valley, upstream of Nicosia, irrigating a few orchards, sustaining also, in some years, a low summer river flow. In Mariti's time it seems that there was no summer flow in Pedhieos river to reach Nicosia, but only winter flow. The most dramatic description of the situation of the water resources of the Island, is given for the end of the third and the beginning of the fourth century A.D., by Leontios Makhairas, the chronicler of Cyprus of the medieval times (15th century A.D.). In his "Recital concerning the Sweet Land of Cyprus entitled 'Chronicle'", Makhairas describes the situation about the condition of the water resources of Cyprus in those times as follows: "Constantine the Great, after he was baptized, saw that our land of Cyprus has been without inhabitants for thirty six years, because there was a great famine from lack of rain and all the tillage was spoiled. And the famine became great and all the waters of the springs failed, and men were going from place to place with their flocks to find water that they might live they and their flocks. And everything dried up both cisterns and springs; and they deserted our most admirable Cyprus and crossed over to this place and to that wheresoever each one of them found a resting place. And the island remained without inhabitants for thirty six years" (Dawkins, 1932).

Such dry weather conditions, persisting for such a long time (longer than the 30 years period used to establish normal conditions), that Cyprus has experienced, according to Makhairas, at the end of the third and the beginning of the fourth centuries A.D., may very well be described as a short duration climatic change.

6-10. CONTINUITY BREAKS IN ANCIENT SETTLEMENTS, GODDESS OF RAIN

Such weather changes leading to the drying up of springs and rivers, as that described by Makhairas, most probably occurred during the historic and prehistoric times of Cyprus.

As it happens to day that people pray for rain in the periods of drought, it was also happening in the past. A statue of a Goddess found in Ancient Salamis (Hadjioannou, 1980), with the name "Omvrios", which in Greek means "of the rain", is attributed to the attempt of the people of Salamis, an ancient town near Famagusta (Fig. 6-2), to influence the supper powers to bring rains and terminate a long drought.

Long periods of low rainfall, with the result of sensitive springs and rivers drying up, most probably occurred during prehistoric times. Suitable places near rivers or springs with perennial flows were the appropriate sites, which ancient Cypriots selected for their settlements. Many of them were continuously occupied for centuries until deserted for some obvious reason such as earthquake or destruction by enemies. But there are in Cyprus Neolithic settlements where there occupancy has an unexplained gap (Khirokitia, Kalavassos-Tenta, Cape Andreas-Kastros and elsewhere). Their Neolithic I culture, according to the evaluations of archaeologists (Karageorgis, 1982), did not end as the result of destruction. The settlements were reoccupied later (Neolithic II culture, 1500 years later at Khirokitia). Considering the sensitivity of the rivers in this area to drought, the gap observed in the occupancy of these prehistoric settlements may well be attributed to climatic changes and a prolonged drought. Drying up of the water sources of those settlements, may have driven the occupants to other places of Cyprus, where they could still find some water, or even forced most of the inhabitants of the Island outside the country, as it happened in the third and fourth centuries A.D., about which we have historic information that it happened.

6-11. INFORMATION FROM AN ANCIENT GREEK TRAGEDY WRITER

Cyprus seems to have also experienced periods of higher rainfall than at least during the last decades. There is a short description relating to the irrigation of the land of Ancient Paphos (Fig. 6-2), given by Euripides, the reknown ancient tragedy writer. In the choral of the tragedy "Bacchae" he speaks about the river named "Varvaros" or according to some scholars "Vokaros", that makes Paphos land fertile without rain. The river crossing the land of ancient Paphos is what we call to day "Dhiarizos River" (Hadjioannou, 1983, p.168-169). In this choral of "Bacchae", Euripides says as follows, in a free translation: "Hundreds of diversions of Varvaros (Vokaros) river make Paphos fertile without rain". With the phrase "hundreds of diversions" Euripides gives us the extent at which the land of Ancient Paphos was irrigated from the Water of Dhiarizos river. With the statement that the diversions of Vokaros river make Paphos fertile without rain, that Vokaros, or Dhiarizos river had a summer flow, in the times of

Euripides (fifth century B.C.), used to irrigate the gardens of the town. The ancient Paphos, locally known as Palea Paphos, about which Euripides speaks, is located a few kilometers to the east of the Paphos town of to day, on the southwestern coast of Cyprus (Fig. 6-2). For the Dhiarizos River we have flow records since 1965 both at Kouklia village, which has developed at about the same place where ancient Paphos was, and higher up on the mountains at Philousa village.

Statistics from the Kouklia records, show that for only three years out of the 27 Dhiarizos river has still some flow during the dry season. This situation is much different from what Euripides describes. It is true that since the ancient times much agricultural development occurred uphill and considerable quantities of river water are withdrawn upstream of this area. However statistics even from Philousa records show that for only 10 years out of the 27 there was enough base flow of the river to reach the site of ancient Paphos, which again does not justify the description of Euripides. This leads us to the conclusion that at the time of Euripides rainfall in Cyprus was more than in the period of our flow records from 1965 until present, resulting then in higher runoff of Dhiarizos river. However if we imagine the situation of Dhiarizos flows at the beginning of this century, with rainfalls higher than today at about 9% and an approximate 27% more runoff, the situations at Euripides time and the beginning of our century may be comparable.

6-12. EVIDENCE FROM CHANGES IN THE FOREST COVER

Cyprus in ancient times was famous as much about the copper mines as for the thick forests, which gave to the Island, among other names, that of "Dasoessa", which in Greek means "forest covered". The forests were extensively exploited, providing the wood for the construction of what ancient historians describe as "great fleets" and for smelting the copper minerals mined in the Island. The copper produced was exported either as copper plates or as manufactured copper articles.

The amount of wood consumed for the smelting alone, estimated from the slag heaps in Cyprus, would require the forests of the Island to have been cut sixteen times (Constantinou G., Director of the Geological Survey Department of Cyprus, personal communication). This means that the regeneration capacity of the Cyprus forests would have been much higher than to day, particularly if we consider that most of the slag heaps are located at elevations around 200 m, with yearly rainfalls around 350 mm. With nowadays rainfalls and soils, forests are productive at elevations higher than 150 m, while at lower elevations they do not in general regenerate (Peonides L., Director of Forests Department of Cyprus, personal communication). It is logical to suppose that the timber used in smelting the copper minerals and in shipbuilding, would have been taken from the vicinity of the works, that is from the lowlands, today without forest regeneration capacity and the hilly ground near by, which lies in the marginal zone. This is leading us to the conclusion that the climatic conditions, particularly as relates to rainfall, would have been more favorable for forest regeneration, when the copper industry was flourishing in Cyprus (3rd millennium B.C. to Roman times).

An interesting information, coming in support of the above, is given by Strabo, the geographer of the beginning of the Christian times, speaking about the forests of Cyprus. Strabo by obviously contrasting the poor forest cover of Cyprus at his time (first century A.D.) with the situation that Eratosthenes (third century B.C.) described to prevail before his time, says "Eratosthenes talks of the plains as being formerly full of wood run to riot, choked in fact with undergrowth and uncultivated. The mines were here of some little service, the trees being cut down for the melting of copper and silver; and of further help was shipbuilding, when men sailed over the sea without fear and with large fleets. But when even so they were not got under, leave was given to those who would and could cut them down to keep the land they had cleared in full possession and free of taxes" (Cobham , 1969).

The composition of the Cyprus forests would have been also different at that time. The plains were also thickly forested. A sample from a borehole drilled near Larnaca town, dated with radiocarbon to be 1456 plus or minus 120 B.C., contained significantly more oak (Quercus) pollen than any other sample from this area dated to be later in time. This is an evidence for oak stands of a lowland forest close to the area of sampling (Gifford, 1978). Oaks nowadays are found on the igneous rocks of the Troodos massif, many miles away.

The most probable explanation about the changes in the forest cover of Cyprus, which became most apparent at about the beginning of the Christian times, is a gradual climatic change resulting in lower rainfall.

6-13. CONCLUSIONS

From the evidence presented so far, so much from the instrumental records, the historic evidence and the one thousand years information preserved in the wood of a long aged pine tree, it is concluded that Cyprus did experience, during the last few thousand years, cyclic climatic changes. However the evidence does not support that the changes have gone beyond the limits of what we know to day as the Mediterranean Climate of Cyprus, which the Island seems to have enjoyed at least during the historic period.

REFERENCES

Christodoulou D. (1959). The evolution of rural land use pattern in Cyprus. *In:* The World Land Use Survey, L Dudley Stamp 2. Geographical Publications Limited, London, pp. 123-137.

Cobham C.D. (1969). Excerpta Cypria. The Library, Nicosia-Cyprus. Photo-offset reproduction of Cambridge University Press, Edition 1908, pp. 3-26.

Cobham C.D. (1971). Giovanni Mariti-Travels in the island of Cyprus. Zeno booksellers and publishers,

London. Reprint of the Cambridge University Press, Edition 1909, pp. 40-112.

Cyprus Grain Commission. (1970). Annual report for the year 1969-70. Nicosia-Cyprus.

Dawkins R. M. (1932). Leontios Makhairas-Recital concerning the sweet land of Cyprus entitled "Chronicle", Vol I. Oxford at the Clarendon Press, Oxford, pp. 3-5.

Gifford J. A. (1978). Paleogeography of archaeological sites of the Larnaca lowlands, Southern Cyprus. A thesis submitted to the Faculty of the Graduate School of the University of Minnesota, pp. 166-168.

Hadjichristodoulou A. (1982). The effects of annual precipitation and its distribution on grain yield of dry land cereals. J. Agric. Sci. Camb., 99:261-270.

Hadjioannou K. (1980). The ancient Cyprus in ancient Greek sources, Vol IV/2. Holy Archbishopric Cyprus, Nicosia, pp. 110-112 (in Greek).

Hadjioannou K. (1983). The ancient Cyprus in ancient Greek sources, Vol V. Holy Archbishopric Cyprus, Nicosia, pp. 168-169 (in Greek).

Hadjioannou K. (1985). The ancient Cyprus in ancient Greek sources , 2nd Edn,vol II. Nicosia(in Greek).

Hadjioannou L. (1987). The climate of Cyprus: Past and present. *In*: Meteorological note series, Meteorological Service 3. Nicosia-Cyprus, pp. 4-5.

Karageorgis V. (1982). Cyprus from the stone age to the Romans. Thames and Hudson Ltd, London, pp. 25-27.

Kyriazis N. (1931). Chronicle note. Cyprus Annals year 8 part B:81-95 Kyriazis N. (1936). Report about Cyprus by A. Savorgnaro. Cyprus Annals year 12 part A:85-103 (in Greek).

Ploumides G. (1974). The Greek lands under Venice rule between the second and third war among Venice and Turkey. Ioannina-Greece, pp. 98-100 (in Greek).

CHAPTER 7

AVOIDING WATER SHORTAGES
SOME ANCIENT GREEK SOLUTIONS

DORA P. CROUCH

Professor Emeritus of Architectural History, School of Architecture, Rensselaer Polytechnic Institute Troy, New York, U.S.A.[1]

7-1. INTRODUCTION

One of the most important features of water provision in ancient Greek cities is that as many sources of water as possible were utilized, thus reducing the reliance on any one source. That is, if the enemy cut off the aqueduct, or if an earthquake interrupted the flow of the spring, there was cistern water to fall back on. If the cistern went dry, there was the fountain. If the rains were delayed, there was still the well or the spring. This meant that the failure of any one source had less potential to devastate the community. Use of multiple sources of water was an evolutionary advance. The survival of the community was assured by the deliberate redundency of the system.

This system had relative advantages as well as absolute ones. With a redundant system, shortages were less likely. With a major third of the system -the domestic cistern- maintained by each family, the independence and security of the family were enhanced. With some water available in every courtyard, the advantages of urban living were most equally shared, even by women and slaves. Further, public fountains and baths were not concentrated in one area of the city, but somewhat scattered so that the greatest number could benefit from them.

7-2. CLIMATIC AND GEOLOGICAL CONSTRAINTS

Before water shortages can be avoided or managed, they must be recognized. Avoiding water shortages in ancient Greek lands was an annual problem, as the Mediterranean climate has rainfall mainly confined to the winter months whereas the need for water is highest for both crops and humans in the midst of the heat of summer. Therefore the Greeks had strong incentives to notice and utilize every possible source of water, and to conserve it for re-use. Already in ancient times, the writings of Plato and

[1] Correspondence address: 739 Yale St. # 6B, Santa Paula, CA 93060, USA

NATO ASI Series, Vol. I 36
Diachronic Climatic Impacts on Water Resources
Edited by Andreas N. Angelakis and Arie S. Issar

others in the fourth century B.C. reflect a venerable tradition as to the causes of water shortages. Plato writes of deforestation in the hills of Greece in his *Critias* iii D-F,

> In consequeunce of the successive violent deluges . . . there has been a constant movement of the soil away from high elevations; and, owing to the relief of the coast, this soil, instead of laying down alluvium as it does elsewhere, has been perpetually deposited in the deep sea around the periphery of the country, or, in other words, lost.... All the rich, soft soil has melted away, leaving a country of skin and bones [so that rainfall] is allowed to flow over the denuded serface [directly] into the sea.

Plato joined *Herodotus* (sixth century) and Aristotle (fourth century) in citing overgrazing as the agent of destruction.

Modern analysis, however, has revealed a much more complicated situation, with the goat as only one factor, and probably not the most significant. After discussing the different theories about alluvial deposits in river valleys of the Mediterranean, Renfrew and Wagstaff (1981) continue,

> Man-induced soil erosion is normally put forward as an alternative to a climatic explanation for late Holocene aggredation in the Mediterranean region. It is well recognized that man's interference with natural vegetation and soil systems in semi-arid environments can have considerable repercussions in terms of soil erosion (e.g. Butzer 1974). In Greece the present degraded vegetation and soil cover has been attributed to deforestation and poor management by several writers
>
> . . . The available paleobotanical evidence from Greece has been summarized by Grieg and Turner (1974) who identify a contrast in the vegetational history of northern and southern Greece. In the north, thick oak forest had developed by about 7000 BC. It persisted up until and probably through the classical period into the mediaeval period when degeneration commenced. In southern Greece the pattern indicates that a previous forest cover was very degraded, probably by the bronze age, *ca*. 3500 B.C., if not earlier. The vegetation present today in southern Greece is comparable to that which appeared then and has persisted since that time. Greig and Turner emphasize that the prime cause of forest decline in north and south Greece was deforestation by man. Concurrent with the decline in tree pollen, the spread of agriculture can be observed from the increasing frequency of cultivated species in the pollen spectra. There is no evidence from the palaeontological record to suggest climatic change in Greece since 7000 B.C. Butzer . . . lists pollen studies from elsewhere in the Mediterranean region which agree with this conclusion. It appears, therefore, that deforestation by early peoples could be responsible, at least in part, for the present degraded soil and vegetation cover in Greece. . . Man-induced soil erosion would not necessarily start simultaneously everywhere. . . A number of

> elements may be important in valley and coastal aggradation, and these are not mutually exclusive. . . A change in climate may have occurred in Greece, but at present the available evidence does not necessarily implicate climatic change as an explanation for holocene aggradation. . . The partial cause and effect relationship between man and soils in the bronze age is postualted largely on the basis of the apparent synchroneity between the spread of bronze age activities and the dates for the initiation of erosion.

That shortages did occur is evident from the archaeological evidence. The drought of the fourth century (Camp, 1982) was signaled at Athens by the filling of wells and the opening of cisterns, as the water table receded. The drought lasted for some 25 years. Athens, which had not founded any new colonies for nearly 300 years, was forced to send out her last colonists. It is now thought that a century-long cold spell, possibly triggered either by volcanic eruptions or by natural variation in the cycle of average annual temperatures, or both, brought on the drought.

Climate change has often been postulated to explain deforestation and other changes in Greek lands during the classical period and since then. The papers collected in di Castri and Mooney (Mediterranean Type Ecosystems, 1973) may stand as representatives of the discussions that have been going on during the last forty years. I have read a great many of these position papers, pro and con of such a change, and concluded that the fact of a drastic change in Greek times is "not proven." Normal fluctuation of the climate is recognizable but nothing as severe as a new Ice Age. However, volcanologists have evidence for a cool period during the last half of the last millenium B.C. (Climates of Change, 1963).

Many writers have commented on the tendency of karst terranes to self-destruct. From the extensive reading I have done on this matter as it concerns Greek deforestation during the fifth to first centuries B.C., I have concluded that in areas where there are active karst processes, the actions of human beings can amelliorate or exaccerbate the results. As an example, where careful terracing is constructed, agriculture can extend onto the upper hillsides without too much topsoil being carried away in torrential rains. Even if the terraces are abandoned and allowed to grow up with shrubs and trees, the results are not too bad. But let sheep and goats be turned loose to climb over the terraces and break them down, and the negative results are striking in terms of alluviation of the valleys. Apparently such a negative process was going on during the fourth century B.C., running parallel with the shift from independent city states to the Hellenistic empire of Phillip of Macedonia and Alexander the Great. At this stage of the research, we do not know whether political instability caused agricultural degredation, or vice versa.

It is tempting to wonder whether drought and agricultural collapse contributed to the willingness of men to enlist under Alexander the Great, and their determination to conquer the Persian lands. A similar drought in the eighth century B.C. also documented by Camp (1979), had earlier spurred the colonization process. Groups from the Greek cities in Ionia, the Aegean islands, and mainland Greece had colonized northeastward into the Black Sea area and westward to southern Italy, Sicily, southern France, eastern Spain, and southward to North Africa (Fig. 7-1).

Figure 7-1. Map of Greek Lands

My recent book, Water Management in Ancient Greek Cities (Oxford, 1993), has shown how careful was the selection of sites of these eighth century settlements, based on a strong tradition of selecting geologically suitable terrain and on swiftly developing technology of water management. With collegues who are geologists and hydraulic engineers, I am now studying the dependence of Greek settlement on its geological base, which also will published by Oxford as Geology and Settlement: Greco-Roman Patterns. One example of this process, is our comparative investigation of Corinth and Syracuse to determine whether the resemblance of terrane between them is conceptual only, or whether in some important ways the geology of the mother city Corinth resembled the geology of her colony Syracuse. If we find similarities in how the water behaves in the geology of the two places, this will suggest that the choice of a colony 's site was not accidental.

Gradually the Greeks mastered the water potential of their environment, to be discussed below. Already in the seventh or sixth century B.C. they were building underground aqueducts to supply fresh drinking water to their cities. This technology was imported from the Persians, who in turn learned it from the Armenians (Los Angeles Times, Sept. 29, 1971). A water-bearing layer in a karst terrane was tapped or milked, and the water carried as far as one hundred miles, in natural karst channels, in man-made channels, or in a combination of the two (Wulff, 1968).

First, a careful survey of the terrain is made by a water expert. Frequently the expert can locate a seam of water in an alluvial fan by noting traces of seepage or changes of vegetation. A trial well can be dug by two diggers in the ground and two at ground level to operate the windlass for pulling up the debris. The debris is piled around the mouth of the shaft, to prevent surface waters from infiltrating the well. Once they reach water level, they clear out the shaft to the impermeable clay floor, and for the next few days monitor the rate of flow into the well. It it is more than a trickle, they have indeed struck an aquifer. They may sink additional shafts nearby to verify that they have found an aquifer of sufficient flow for the purpose in mind. The surveyor next plots the path of the qanat downhill, He plans a slight slope of one foot in 500 to 1500 feet, so that the mild flow of the water will not wash out the channel. Additional shafts are dug along the selected path, with the difference in elevation between the highest and lowest shaft bottm being allocated evenly to each of the shafts along the path. That is, if the final level of the water is to be 100 feet lower than its origin, and there are 50 shafts, then each must be two feet deeper than the one uphill from it. The shafts are located 50 to 100 feet apart.

Once the shafts are in place, they proceed to dig the connecting conduit, beginning at the mouth end. Since ancient times, the diggers have lined up three widely spaced lamps in a row in order to keep such a tunnel to a straight line. The tunnel may be reinforced for the lowest 10 to 15 feet to protect it from damage by stormwater. As they approach the head well at the top of the qanat, the diggers work slowly and carefully, hoping to guard themselves from a sudden in-rush of water that could mean death. At the uphill end of the tunnel, small auxiliary tunnels may branch out from the main tunnel, like fingers in an outstretched hand. These "milk" the hill of additional water and thus supplement the supply of the city downstream. Depending on the local terrain, the builders may construct additional qanats paralled to the first but several hunderd meters away, to convey more of the mountain waters to users in the populated areas. In 1968 there were 22,000 qanats in Iran, forming 170,000 miles of underground conduits delivering 19,500 cubic feet of water per second, an amount equivalent to 75% of the discharge of the Euphrates river.

The earliest qanat system we know of in Greek lands was on the island of Lemnos, built by the engineer Theodorus of Samos in the year 575 B.C. (Feldhaus, 1931, p. 122). Failure of this system was recorded just after World War II, when L. Picard (1956-58, 1964) noted that the natural water drainage on Lemnos was producing a swamp on one side of the island, while the major city on the other side of the hills was desperate for water. The ancients had learned to intervene in such natural wastage, diverting the water to where it was needed, but this information has since been forgotten.

7-3. WASTE

Waste and misuse of water are major causes of water shortages. As settlements grew in size and complexity, from hamlet to village to town to city to metropolis, waste and misuse became less and less feasible. The demands of the climate and the constraints imposed by the geology necessitated close attention to the management of water.

Travels in many Mediterranean countries where sinkholes and dolines are found indicates a nearly universal custom of filling up such holes with any kind of waste product, solid or liquid. As recently as 1978, certain sinkholes on the Tripoli plain of the Peolopnesos were thought, as a result of tracer experiments, to have no connection at all with the karstic springs of the areas near Argos (Marinos, 1978). Therefore, it was allowable to funnel municipal waste waters into two of the sinkholes, which were assumed not to be connected with the network of channels to the springs below. Unfortunately, reality turned out to be a great deal more complicated, as was realized already by 1986 (Morefis *et al.*, 1986):

> The bad hygenic condition of the water of the springs [is] connected with one or more sinkholes especially when the wastes of town or villages are collected into these sinkholes . . . due to quick circulation of the underground water within the karstic system.

It is possible that the temptation to use karst shafts for dumping municipal waste waters goes back to at least Greek times, for at the site of Priene in Turkey, I found evidence that the drains of the main street of the town may empty into a shaft just outside the western gate of the city. My team of geologists and hydraulic engineers is currently investigating this possibility. At Priene the contamination of the local water table may have been less of a problem than between Tripoli and Argos, as Priene was built on a terrace immediately above the ancient shore line, and the postulated drainage shaft emptied directly into the sea.

In other places, municipal waste waters may have been dumped into irrigation supplies; this alternative seems likely at the metropolis of Athens and at the small provincial town of Morgantina (Sicily). At Athens, the Great Drain of the Agora collected the water from the Acropolis area and carried it out of the Dipylon Gate where this "man-made river" was used for irrigation. In the same way, drains at Morgantina empty out of the gates onto the south slope of the hill the town sits on. It would have been logical to use these waters to grow trees for fuel and building materials, just as today the overflow of the spring on this slope waters the vines and olives.

Re-use of these waters as irrigation-with-fertilizer solves problems of municipal drainage, but causes other problems unless carefully managed. The chief problem in re-using waste water is proliferation of disease-bearing entities. This is particularly serious if the wastes from one species are re-used in growing food for the same species, without being treated. Modern hydraulic engineers have the rule of thumb, "People's wastes to grow animal food, animal wastes to grow human food." This technique could have been ascertained by the ancient peoples by careful observation. Until recently, most of our modern municipal waste water was dumped with little or no treatment into the nearest body of flowing water. While most of the waste so dumped was organic, this was a reasonably good solution for small and medium-sized cities. The Greek lands of the Mediterranean, however, lack perenniel rivers to use as dumping grounds, since most rivers here are periodic responses to storm runoff. Thus polluting the rivers was not a viable option.

Still largely unexplored are the relationships between municipal water arrangements, such as the wells, cisterns, fountains, and drainage channels found at Delos and Pompeii, and the irrigation of farm land in the hinterland of a great city like Corinth or a small town like Morgantina. Unfortunately, nothing like Doolittle's Canal Irrigation in Prehistoric Mexico (1990) or Adams' Heartland of Cities (1981) for Mesopotamia has been written for Greek irrigation. Yet it was certainly possible that storm runoff and domestic waste waters were utilized to grow a steady crop of trees around Greek cities, to use as building materials and as firewood. Needs for both firewood and building materials were constant, and a viable city needed a long-range plan to supply them. When the excess from one activity - here, waste water - could supply the deficit of another activity - here, irrigation - then the interactive solution was sustainable over many generations.

In addition, modern studies have shown that how waste water is disposed of directly affects the ground water table. In some areas of Southern California, surplus water is slowly soaked into the ground to replenish near-surface aquifers and to act as a barrier against salt water intrusion. Over-pumping of aquifers for "cheap" water has allowed the problem of salt-water intrusion to occur. Thus water disposal and replenishment are directly related to the long-range viability of an area.

7-4. WATER SKILLS, SOCIAL SKILLS

A brief existence and squalor were the lot of cities which ignored resource constraints; long duration and beauty the lot of cities that made of the necessity for water and drainage a social focus and an excuse for ornamenting the public realm with fountains and baths. Indeed, one might argue that the cooperation and ingennuity necessary for water management had as their by-products the skills that made possible many aspects of Greek culture such as self-governance, building temples and producing dramas.

I have argued (Crouch, 1993) that by expending maximum thought on their water arrangements, the Greeks could expend minimum human energy and economic resources on them. It is noteworthy that the time and effort of the women and girls who fetched the drinking water for the family was considered valuable to the society, and worth conserving. To this end, the family cistern captured water from the roofs, requiring only annual maintenance once it was in place. In the public areas of the city, placement of fountains and other water elements was strongly correlated with placement of the buildings and spaces for public life. Adult male citizens were involved with water use and management in their roles as farmers, athletes, civic officials, architect-engineers, potters, fullers, and so on. The skills of obtaining and managing water supply (Lindh, 1987 and Lindh and Berthelot, 1979) carried over into the organizing of dramatic preformances, religious ritual, democratic elections, and warfare. Managers of water in an ancient Greek city had to factor in the local hydrogeology, the preferred urban form, the cultural ways of using water, the spaces and buildings adapted for water use. To maximize water use and to avoid shortage, they had to recognize both the maximum and minimum amounts of water available, and the range of human behaviors that were likely.

7-5. MULTIPLE SOURCES

Multiple sources of water are clearly manifest in the physical arrangements of these ancient settlements, as recaptured by archaeology. To see an extensive rain catchment system that still survives from Greco-Roman times, we have to go far into the desert, to Petra in Jordan, where the mountains were scarped by the Nabataeans into gigantic shallow "dishes" to collect rainwater, which was then channeled to large municipal cisterns (Crouch, 1969). An upper terrace on a mountain was cleared of dirt and plants and deepened in the center to the shape of a shallow dish. Then a channel was carved from the dish down the side of the mountain to a very large cistern or a small underground reservoir. This had been hewed out of the rock, and waterproofed inside with special hydraulic cement. Usually the opening was kept rather small, and preferrably shielded by an overhang, to cut down on evaporation. Similar "dishes" to catch rainwater may be seen today in areas as widely separated and southwest Turkey and southern California - where the annual rainfall is sparse and the demand for water high.

Because the collection and storage elements were of stone and the site long deserted, we can see at Petra many details of systems which undoubtedly existed in centers closer to the Greco-Roman mainstream but which have perished through later alteration of the site or from being made of less permanent materials. Still other storage systems have not yet been recognized because no one was looking for them, just as the elements of water wheels and of pumps have only recently begun to be recognized.

Water wheels of Roman times and before have been studied by Thorkild Schioler (1973) who concluded that sophisticated arrangements like the paternoster chain (a continuously moving chain with doorless compartments) were already in use in small towns like Pompeii by 100 B.C. He pointed out that Archimedes who lived in the first half of the third century B.C. and invented the screw that bears his name, may also have invented the *saqiya* which combines the paternoster chain, a right angle gear, and a circular track for the animals that powered the lifting device. If the third century is too early for this device, then the *saqiya* could have been developed after the right angle gear was invented in Alexandria in the first century B.C. I incline to the earlier date for the gear and the *saqiya*, as Vitruvius (who lived at the end of the first century B.C. and was a compiler not an innovator) provides the technical details of a mill with gears of this kind (*De Architectura*, Bk.X, Ch.5). The water wheel that turned horizontally on its vertical axis was probably invented in the second century B.C. in some mountainous area, and had the advantage of generating twice as much power because it was more efficient (Arnold, 1969). Already by the early second century A.D., the *saqiya* was used on a large scale for municipal water works. A measure of its sophistication is that the *saqiya* had 200 parts, whereas the simpler device called the *saduf* or sweep has only ten.

More recently, Schiøler has been studying pumps, but as yet I have not seen any of his results. We do know that in the third century B.C., Ctisibus of Alexandria experimented with air compression, and invented the force pump. Hero of that city described it in the first century as "a machine which expels water to a great height." A series of practical developments made pumps available first for expelling water from mines and later for other uses such as municipal water works and fire fighting. I have republished a diagram of a Roman pump found at Bolsena, now in the British Museum (Crouch, 1993). Many other pumps may await recognition in museum store rooms.

7-6. KARST

Ubiquitous in all the Greek areas that I have studied, springs were central to the provision of water. Karst geology is found everywhere in these lands. Water that falls on carbonate rocks finds its way into the rock through fissures, gradually enlarging them and creating a series of shafts and channels. This is fortunate for human beings, since it is a natural mechanism for capturing rainfall on mountain slopes, filtering it through soil and rock, and producing potable water downhill at places convenieasty becaause of the calcium carbonate it carries in solution.

A marked advantage of the karst is that water falling on mountainsides and entering the rock through fissures and shafts takes weeks, months, even years to reappear in springs at the foot. This dampened flow can provide water in the midst of the annual summer drought. Without karst geology, few sites in Greece would be habitable. Water from karst formations has the additional advantage of being fresh and tasty becaause of the calcium carbonate it carries in solution (Fig. 7-2).

As an example of a karst water supply, we will discuss the springs of Corinth (Fig. 7-3). The ancient city was built on a series of nine terraces above the Gulf of Corinth. Layers of poros, conglomerate, and impermeable clay/marl formed these terraces, with springs appearing at every stepping back of the relief. Within the ancient core of settlement there are three famous springs, Peirene, Glauke and the Sacred Spring, all

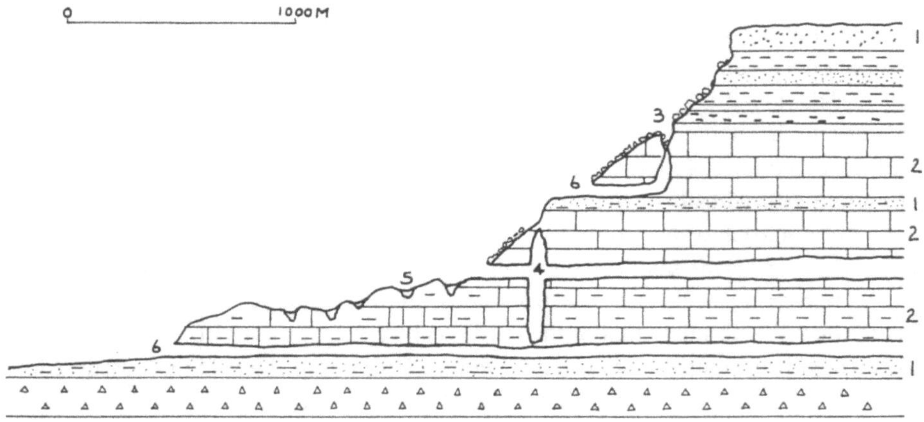

Figure 7-2. Diagram of karst system based on work of Pohl and Crawford: (1) Caprock/ impermeable strata, (2) Limestones, (3) Sinkhole where water enters rock, (4) Shaft, (5) Collapse sink, one of a series, (6) Resurgence

evidently karstic in origin and developed architecturally by the archaic, Hellenistic, and Roman builders of the city (Hill, 1964).

Peirenne and the Sacred Spring are tucked under the edge of the terrace that the forum sits on, flanking the main road between the sea and the mountaintop. The presence of the springs made a natural stopping place in the long climb, and led to the development of the business district of the city at this point. The Glauke fountain is both later and less natural, being a large hunk of rock left standing when the ridge was quarried for stone to build the seventh century Temple of Apollo that crowns the ridge. As the area around the temple developed, the trenches of the quarry became reservoirs, and the hunk of rock was supplied with piped water, making a useful fountain at the entrance to the temple precinct.

To the west, in the potters' quarter of the seventh century, an additional extensive aqueduct and reservoir system has been discovered (Robinson, 1965). The irregularity of its channels, like the similarly irregular channels under the agora and under the great Aesklepion/gymnasium complex (Roebuck, 1951) at the north edge of the city, speak to the possible karstic origin of the supply channels, with subsequent human intervention. Our team includes speleologists who are investigating these tunnels. Other exxperts are also working to understand water arrangements at Corinth: Dr. David Romano at the University Museum in Philadelphia is producing a detailed topographic map of the site, and Mark Landon of the University of California at Berkeley is writing a doctoral

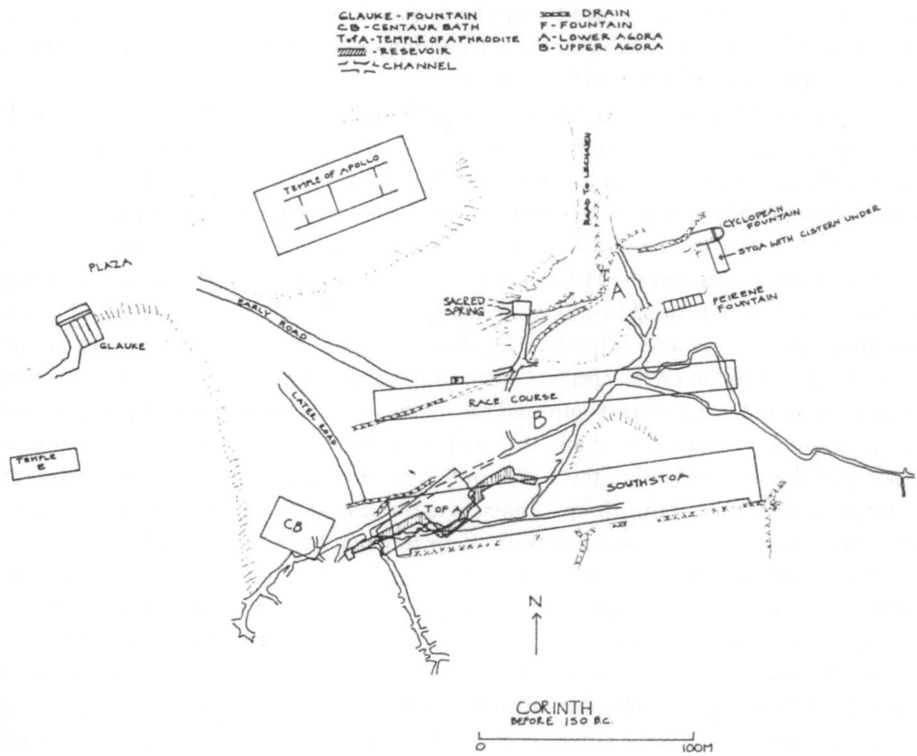

Figure 7-3. Karst channels underlying the central area of Corinth

dissertation on the water system elements of the site. Their work complements our own study of the relation between the underlying geology and the development of urban form at Corinth.

7-7. RIVERS AND WELLS

In later times, the Romans made use of river water as an ingredient in their water systems, but this option was not usually available to the Greeks, whose rivers were ordinarily dry in summer. However, at least one site in the Peloponesos did utilize a spring-fed river. This is a sanctuary of Apollo and its adjacent small town at a place called Gortys. The Gortion River, fed by karst spings in the mountains to the east, flows so abundantly that, in June, I observed it to be at least twelve feel wide and five feet deep. The sparkling cold water was led to supply the elaborate baths (Ginouves, 1959, Figures 46-53) whose ruins are today the main attraction of the site. The contrast between the brown stony aridity of the surrounding hills and the lush trees and green of the immediate vicinity of the river is striking.

At nearby Olympia, also, there were baths, gymnasiums, and lodgings for travelers, all lined up near the river. By the fifth century B.C., west of the ceremonial center, a bath building had been erected next to a swimming pool that probably tapped the river for its water. Immediately east of these stood a round building originally used as a *sudatorium* (sweat bath), later remodeled as a Heroon (shrine to a hero). The water buildings continued in use through the Hellenistic period, with a palestra and gymnasium being added northwest of them. The baths were enlarged in 300 B.C. and again in 100 B.C., when sweating rooms were added (perhaps at this time the former *sudatorium* became the Heroon). In Roman times, the ancient baths were abandoned in favor of the new ones that were were added to the north, south, and east of the ceremonial center, but the swimming pool disappeared earlier. Perhaps the pool was swept away in the same flood that destroyed the western end of the southern hall of the gymnasium. Drinking water for Olympia came from a spring three kilometers to the east, and was distributed throughout the sanctuary by a complex series of pipes, many fragments of which are still visible.

Much more common in Greek lands than the use of rivers, however, was the utilization of wells. For instance, at the Harapan city of Mohenjo-Daro near the Indus River, many wells are known from periods in the third millenium B.C. Since Mohenjo-Daro was in active commercial relations with the Middle East, I think we may safely infer that knowledge of wells was already widely disseminated in settled areas during the third millenium. Shafts common in karst terrane seem to have served as natural models for wells (Crouch 1993, Chapter 10). I have argued that natural models were imitated by the ancients when they constructed wells, reservoirs, underground water channels, etc. We have evidence of wells already during the eighth centhry B.C. on Cypress (Diakaios, 1961). Many wells of the archaic period were found in the agora at Athens (Fig. 7-3), and in the agora at Corinth.

At the provincial hilltown of Morgantina in central Sicily, with a mixed culture of Greek and native Sikel, we find already in the sixth century B.C. a well that was more than forty-five feet deep (Crouch, 1984). It is evident that the Greeks not only knew how to construct wells of sixty or even a hundred and twenty foot depth, but also that they were acquainted with the principle of draw down (Fig. 7-4): The laws of Solon (a sixth century Athenian law-giver) state that one may not dig a well within half a mile of a public well, so that dewatering of the public well was prevented. That use of water could pertain to the survival of the group is suggested by his law that if one had dug to sixty feet without finding water, one's neighbor had to allow a five-gallon jug of water to be filled twice a day at his well, since "Solon thought it well to help a man in need, but not to encourage laziness" (*Plutarch*, XXIII,6).

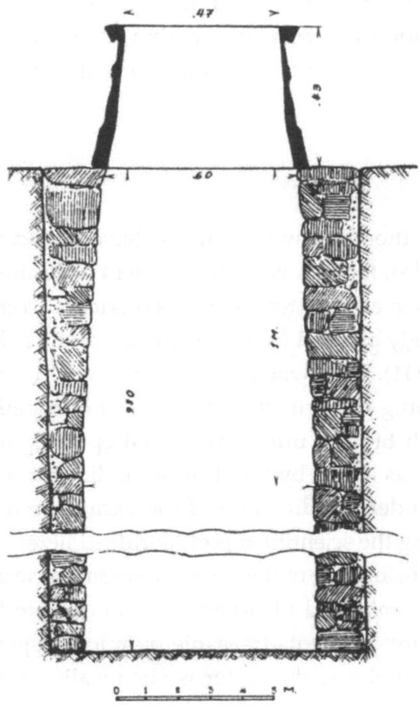

Figure 7-4. Well section (sixth century B.C.) from the Athenian Agora. Reprinted by permission from the American School of Classical Studies, Athens

7-8. CISTERNS

In most settlements, well water and spring water were preferred for drinking, but houses and many public buildings usually had cisterns as well. At first I could not understand why the people would want both fountains and cisterns, but gradually came to realize that to carry every drop of water from a spring to the house for bathing, washing, watering the donkey, cleaning the house, etc. would be an enormous task. The domestic cistern, usually placed under the courtyard and filled by downspouts from the roof, provided enough sub-potable water for these purposes for the entire dry season. Cistern water was technically potable for drinking in spite of its flat taste, as long as the cistern was cleaned every fall before the rains started. Thus cistern water could safely be used for cooking as well as other domestic purposes. Yet the fresh flowing water from a spring was more delicious and was preferred for drinking.

A good example of cistern use is in the city of Pergamon in Ionia. Near the row of palaces on the eastern side of the citadel are two very large, deep cisterns. Both were cylindrical in shape. One still has the central column that supported the stone roof, now partially missing. The amount of green vegetation growing now within the palaces suggests that additional cisterns were placed as usual under the courtyards (Brinker, 1990, p. 100, plus at least 17 Figures).

7-9. WATER QUALITY

Studies of water quality at different places, though few in number, have already indicated that "circumstances alter cases." At Delos, for instance, some ancient wells are still producing clear, ice-cold water in mid-summer, as I can testify from experience. Yet at Pompeii the wells tap into a water table strongly affected by the chemicals produced by Mt. Vesuvius (Crouch, 1993, and Maiuri, 1931). The water from such a well in a residential courtyard was convenient for washing etc., but the residents of Pompeii preferred to drink their aqueduct water which tapped uncontaminated springs in mountains to the east. The water of Pompeii was tested by S.E. Del Blasi, director of civil engineering at the university of Naples, under the direction of the excavator A. Maiuri who reported the tests in 1931. After giving the scientific report (abridged here):

> . . .because of the small number of germs belonging to the same common species; because of the absence of anaerobic *B. colli* and of other bacteria of noted significance; because of the absence of any elements traceable by microscope; and because of the results of chemical analysis, the water is chemically and microbiologically pure. . . but because of its excessive total hardness . . . the water examined cannot be declared potable for continuous or permanent consumption. However. . . it can be drunk temporarily . . .

Maiuri commented that the ancients were quite aware of the unappealing nature of the well water, and found alternatives in natural springs and in rain water collected into cisterns in nearly every house (Maiuri, 1931).

At Athens, through the cooperation of the staff of Mrs. Eva Tolupa, then in charge of the Acropolis, and Alexandra Katsiri, an assistant professor in the Department of Water Resources at Athens Technical University, I was enabled to test the waters from three places on the south side of the Acropolis, in and near the Asklepion. The results were surprising to me. They showed that two springs and a reservoir within 50 feet of each other had three different waters, varying in mineral contents and in pollution. The one that appeared the least appetizing was in fact the safest to drink! (Crouch, 1993, Chapter 19) I suggest that further analysis of water quality could provide fascinating insights into ancient cities.

Thrifty use of water required the allocation of each quality of water to its proper use. Let's consider the qualities of water under their modern classifications of potable, sub-potable, and non-potable. Potable in turn can be separated into preferred and non-preferred. By this I mean that spring water, frequently presented in ornamental fountains, was the preferred drinking water because of its fresh lively taste, a by-product of the calcium carbonate content of the water picked up from the limestone rocks it flowed through. This water could also be accessible in wells.

However, cistern water and well water each were drinkable under certain circumstances. At worst, well water might be off-flavor because of volcanic or salt-water intrusion. At worst, cistern water was rather flat in taste, and perhaps demanded the addition of wine to the water to improve the flavor. Thus began the venerable tradition of watered wine, a custom that also extended the supply.

An extreme example of provision of potable water by pipes was the Ionian city of Priene, known to us in its fourth and third century rebuilding. Because Priene was located on a terrace near the foot of a marble mountain, large amounts of flowing water were available even in mid-summer. Consequently, the city had an elaborate pipe system to supply water. The pipes were fed from a series of reservoirs uphill from the core of the city, from whence the water flowed by gravity to the houses below. The reservoirs in turn were filled by pressure pipes from the acropolis above, where there were additional reservoirs. Excavators Wiegand and Schrader (1904) found lead pipes and stone channels that supplied these acropolis tanks, parts of a pressure system that came from higher up the mountains. As part of our re-study of the hydrogeology of Priene, we have located the springs - shown in Wiegand's maps of a century ago - and the channels and upper reservoirs, but have not found any lead pipes.

This is the only city that I have studied in which the supply of drinking water was so abundant that it could be used for almost all purposes, as modern Americans also do. Only about 20% of the houses, those farthest from the slope of the hill rising behind the city, were equipped with cisterns, This indicates that the pipe system supplied the rest of the houses with enough drinking-quality water for all uses, probably supplemented by small springs belonging to individual houses that backed up to the mountain slope.

Less versitile than potable water was the second category, sub-potable water, which could be drunk in case of necessity but usually wasn't. This water included the still water in both domestic cisterns and municipal or religious reservoirs. Although the cistern was the ultimate security device, the water in it was not merely saved against a rainless day. It was used regularly. At Morgantina, Sicily, for in stance, the cistern water was used for cooking, bathing, cleaning the house, washing clothes, watering house plants and domestic animals, and producing craft items like ceramic figurines (Fig. 7.5).

Judson (1959) has estimated that the rain collected in the cistern of an average house at Morgantina was enough to carry the family for the six months of the dry season, even without special effort to conserve, and not taking into consideration the amount carried from the fountain for drinking purposes.

Figure 7-5. Water uses in a courtyard at Morgantina. At left, the wellhead protecting the cistern under the court. Behind it, a giant pithos filled by the downspout from the roof, used for washing clothes which are drying on lines strung between the posts that hold up the roof over the walkway around the courtyard. At right, a woman washes dishes in a louter, which was also used for washing hands and faces. The family bathtub and toilet are out of sight in a separate room

7-10. RE-USE

In order to assign every quality of water to its appropriate use, it was necessary to use and re-use water. Certain human activites produced lightly used water, which we today call "gray water." For instance, bath water could be reused to wash the floor. Bath or cooking water might be given to animals, or it might moisten house plants, since neither soap nor detergents had yet been invented. Any kind of used water could serve to flush the latrine into the sewer. An example of Greek practice, dating from the fifth century B.C., was the re-use of overflow from Peirene Fountain at Corinth in the latrine and baths to its north. A later example can be found at Hadrian's Bath at Lepcis Magna in North Africa (modern Libya). Here the used water from swimming pools and from plunge baths was channeled through the latrines, to flush them out. This was a common procedure in Roman baths, and like so many of their water management ideas, was derived from Greek practice.

Non-potable water included mainly storm runoff and sewage. Storm water could become irrigation water, especially valuable if it picked up sewage along the way and thus could act also as fertilizer. Houses and public buildings customarily had drains that connected with the sewers under the streets, and these - in the most modern fashion - ran out under the gates and utilized the dry bed of the nearest river as a drainage channel. More commonly, I believe that the combined waste water was first used for irrigation, as suggested above.

In every aspect of water management, the Greeks saw the necessity of cooperation with natural cycles of water in the environment. How do we know this? We are forced to reason backward from the answer to the question, a problematical method at best, but unavoidable in the absence of ancient documents on the subject. Still, the absence of ancient writing about other elements of Greek culture (such as vases) has not prevented their study by modern scholars! So too, we can look at the data on water system elements, observe recurring patterns, and - since water is still behaving as it always has - relate those ancient solutions to the solutions we make today. From there, it is a small step to postulating that the same solutions apply to the same kinds of problems (Fig. 7-6).

7-11. LARGE-SCALE STORAGE

In a Mediterranean climate, there is need to store water for summer use. For example, in Southern California, which has a Mediterranean climate, the Metropolitan Water District has just persuaded their municipal clients that a new reservoir to hold 700,000 acre feet of water must be constructed. So too, the ancients needed .water storage. Some examples are the four long narrow reservoirs behind the Priene Fountain at Corinth, public and private cisterns such as those we mentioned at Pergamon, and the famous Fountainhouse of Megara. In the latter, a rectangular hall, water flowed in all night and was drawn down by users during the day (Fig. 7-5).

This Fountainhouse and the Pergamon cisterns were man-made, while the reservoirs at Corinth were human alterations of naturally occurring seepages and channels along the seam between clay/marl and conglomerate. The Corinthian builders extended the seepage channels and dammed the outer ends, turning the passages into reservoirs. In Hellenistic times, a basin extended all across the front, for easy access to the water. In Roman times, a new outer facade was added in front of the old access basin, plus a paved courtyard with a pool fed by the overflow from the new fountain basin.

The ability to direct and control the flow of surface and of underground waters was well developed. At Agrigento, Sicily (ancient Akragas), for instance, in the early years of the sixth century B.C., the engineer Psiax incorporated karst channels into an extensive

Figure 7-6. Reconstruction of the Fountainhouse at Megara, built probably in the seventh century B.C. by the tyrant Theagenes, but drawn as having columns of a a sixth century type

water system for the city (Crouch, 1993), which we are investigating as part of our restudy of Agrigento. This engineering accomplishment was important enough to be celebrated in an ode of Pindar, the greatest Greek poet of the fifth century B.C. Both surface waters from storms, sewage waters from drainage channels, and untapped karst waters from the hill on which the city was built were directed to form an artificial lake at the south-west foot of the hill. This lake was famous in antiquity for its fish and its swans. Its location is still visible, and recent digging in the area indicates a massive aquifer there, accordint to a private communication from E. and G. Trassati, directors of the local sewage treatment plant.

In the case of Agrigento, modern water management has not yet reached the sophistication of ancient practice. Sewage lines broken in an earthquake about twenty-five years ago but not repaired have been allowed to contaminate the ample underground water of the site. The proposed modern solution is to import desalinated water from

Gela rather than to mend the sewer lines. The ancient Greeks would not have considered this a sustainable solution.

At the domestic scale, in Pompeii, neighbors collaborated to harnass and save water flowing along the surface, for later use in garden irrigation. Jashemski (1979) describes a series of gardens that used gravity flow to direct street drainage into the upper edge of the upper garden. There the water was slowed by steps and narrow channels. At the low point of the first garden, the wall was pierced for the water to flow through into the second garden. Here it filled large jars setdown into the ground, channeled along lines of roof tiles laid into the ground. In both of these gardens the overflow sheeted across the surface of the ground. Again the water pierced the lowest wall and entered the third garden, where a double cistern "typical of the old Samnite houses" (4th century B.C.) received the water. Thus the waters of torrential downpours were conserved, making ingenious use of street gutters and natural topography.

Fresh water also had to be managed for human use, even where it was plentiful. At ancient Delphi, site of the oracle of Apollo and the Delphic games, clever manipulation of aquifers and springs was essential to provide for the numbers of pilgrims, athletes, and spectators who visited the site. The old stadium of the archaic Greek period, for instance, was attached to a fountain that signified a karst resurgence high on the hill of the sanctuary, providing water for contestants and spectators alike. Slightly lower on the hill, but still higher than the temple of Apollo, more fountains and a bath were located. The temple itself was placed on and beside at least two more springs, one being that of the Delphic oracle. Drainage channels and water pipes are frequently visible on the hillside, being features which handled the plentiful output of the perched water tables within the hill and the runoff from rain clouds dumping their moisture on the sanctuary. To the east of the sanctuary, at the level of the road, two more fountains were placed. They occupied the two sides of a cleft between the sanctuary and the lower group of buildings. The early, spring-fed fountain below road level was later replaced by an elaborate nymphaeum carved into the opposite cliff wall. Overflow from the fountains furnished water to the temples, tholos, and bath-gymnasium below. In the absence of excavations outside the sanctuary, we can only imagine further out-croppings of water and additional human structures extending all the way down the hill to the valley and port far below.

7-12. CITIES AND KARST

So abundant is the karstic water supply of the mountains around Delphi that the area is today the major source of water for the city of Athens. Elsewhere in Greece, cities have taken advantage of other karst features. On the island of Rhodes, for example, the water in the central mountains was in the fifth century B.C. and later utilized to supply the new city also called Rhodes, located at the northeast angle of the island. Here the dampened flow of the karst system plus the existing channels that could be reworked as aqueducts combined to make creation of a municipal water system a cost-effective project. Meulenkanp, de Mulder and van de Weerd (1972) report "bioclastic limestones of the Rhodos Formation dipping west [from Paradision on the northeast coast] with intercalation and displaced boulders of bluish clays or marls." From karst formations farther to the south, water was brought northward to the high point of the acropolis of Rhodes City. From the grottos of the acropolis, a series of underground aqueducts ran eastward to supply the city with water. The aqueducts were cut into these bioclastic limestones, with the floor sometimes cut down into the clay/marl units corresponding to the line of seepage. The aqueducts and sewers of the classical city have provided the best clues to the original Greek plan of Rhodes City. It is from the discovery of sectiions of the sewers that the modern excavators have been able to reconstruct the street pattern and thus the size of blocks. Konstantinides' work has shown that the original plan of Hippodamus allowed for smaller blocks near the harbors, and larger ones as the city expanded inland (Fig. 7-7).

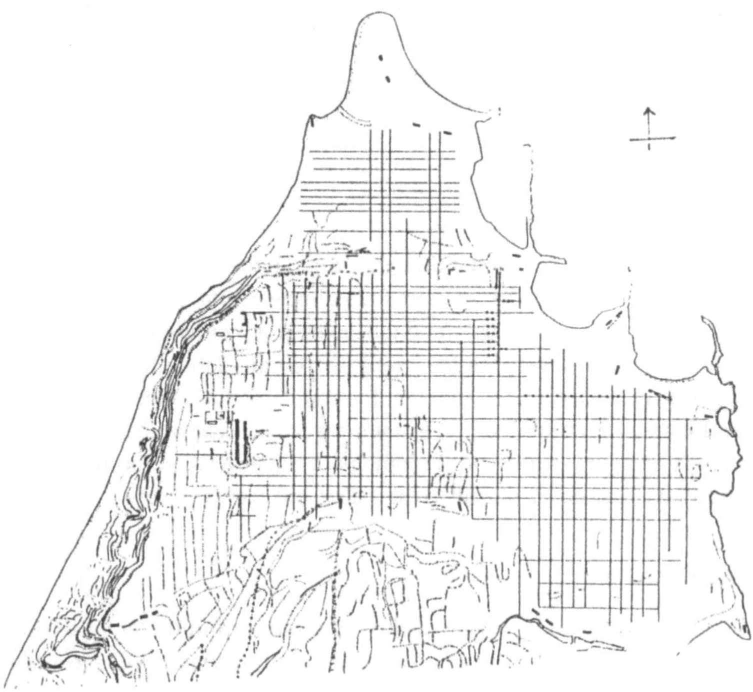

Figure 7-7. Plan of Rhodes city, showing the Hippodamean plan as revealed by the lines of water supply pipes and drains buried under the streets. Aqueducts areshown here as lines of circles. Map based on the work of G. Konstantinedes, former ephor of Rhodes

7-13. DRAINS

In antiquity, drainage was thus as carefully managed as supply. Like modern engineers, the ancient builders were tempted to use existing karst shafts as receptacles for the waste water system of the town. At Priene in Ionia, for example, ancient plans for recharge of the water table are evident in findings from the site. Immediately outside the western gate, the flow from the rain gutter of the main street still today disappears into a pile of stones that remain from the city gate structure, or into a square box-like structure just beyond the gate. Beyond the ruin, there is no sign of erosion, no evidence of flowing water. The explanation, I think, is that an existing karst shaft receives the water and carries it down towards sea level. This is one of the questions we are investigating in our field work.

7-14. CHRONOLOGY

Now that we have sketched out the kind of arrangements for managing water that can be found in specific ancient Greek cities, let us look at the chronology (reprinted, with some revision, from Crouch 1993) of the increasingly sophisticated technological and social arrangements involved.

Chronology of Water Knowledge

Prehistorical period	Springs, rivers.
3-2nd millennium B.C.	Cisterns.
3rd millennium B.C.	Dams.
3rd millennium B.C.	Wells.
[a] Probably very early	Reuse of excrement as fertilizer.
[b] 2nd millennium B.C.	Gravity flow supply pipes or channels and drains.
[b] 2nd millennium B.C.	Minoan use of pressure pipes.
8th-6th *ca.* B.C.	Long-distance water supply lines with tunnels and bridges, as well as intervention in and harnessing of karst water systems.
6th *ca.* B.C. at latest	Public as well as private bathing facilities, with bathtubs or showers, footbaths, washbasins, latrines or toilets, laundry and dishwashing facilities.
6th *ca.* B.C. at latest	Utilization of definitely two and probably three qualities of water, potable, sub-potable, and non-potable including irrigation using storm runoff combined with waste waters.
6th -3rd *ca.* B.C.	Pressure pipes and siphon systems in increasingly widespread use.

[a] indicates an educated guess
[b] indicates an element discovered, probably forgotten, rediscovered later

7-15. COMMENTARY ON CHRONOLOGY

A few comments will be useful to elucidate the significance of the items of this chart. The comments will be confined to evidence from Greek lands. During the prehistorical period, before the mid-third millenium B.C., water was used wherever it was found naturally, mainly at springs and ponds fed by springs, but also at the rare rivers found in these terranes, and at intermittant lakes such as Lake Kopias (Knaus *et al.*, 1980 and Knauss, 1987). Primary to the later development of water management was the amassing of information about the behavior of water, and the beginnings of water tradition. Even today, children who are allowed to play in water similarly amass information about its behavior.

No later than the third to second millennium B.C., people began to save water deliberately in cisterns. Large cisterns are known from the earliest phases of the Minoan palace of Knossos on Crete (late third millenium). The cistern has several advantages that have insured its survival as a water management device until today:

a) It can be built and maintained by a family or an individual.
b) It has no moving parts and requires minimal maintenance (annual cleaning and occasional renewal of the waterproof lining).
c) It utilizes rainwater from the roof, or can be arranged to capture flowing surface water, being cost-free and utilizing water would otherwise go to waste. Storage of water in cisterns also reduces the amount of surface runoff that must be handled by the municipal drainage system, thus reducing the costs and operating problems of such a system.
d) Cisterns attached to public buildings provide water for cleaning, flushing latrines, watering gardens, fire fighting, and in emergencies for drinking.
e) Domestic cisterns provide water for cleaning, cooking, craft operations in the courtyard, watering plants and animals, and emergency drinking - water that does not have to be carried from a distance, freeing the time and energy of the residents for more productive labor.

Cisterns have thus consistently recommended themselves as cost-effective, sustainable means of providing sub-potable and even potable waters.

Also in the third millennium B.C. we find the use of large dams in Egypt and Mesopotamia, and can infer a much older tradition of the use of small dams in all the areas of intermittant torrential rain. Neolithic farmers had already learned the basic rules of water management in a Mediterranean climate: ***slow it down, spread it out, soak it in.*** Numerous small dams of brush and loose stones could be reassembled if broken by the force of the water, and the negative consequences of such a break were much less than what would happen if a giant modern dam was overthrown by its impounded waters. Knauss (1990) has been studying the second millenium Mycenaean dams of Greece, and another group under Gunther Garbrecht (formerly at Braunschweig) have studied dams in Armenia and Egypt (1984).

At least by the third millenium, wells were deliberately being dug in a number of places. Centuries of observation of the behavior of water collecting between certain layers of rock, clay, etc. had supplied the necessary basis for predicting where accumulated water might be found. In addition, certain plants came to be seen as pointers, namely those that love to grow "with their feet in water," such as fig trees and brambles. These plants and others are still useful signals to present-day investigators of water (Meinzer, 1927).

In terrane with calcium carbonate rocks, karst activity produces vertical shafts with water at the bottom (Bogli, 1967). It was only a matter of time before some child, who should have been fetching water for his family, one day was one day idly tossing pebbles into such a shaft, actually heard the spash at the bottom and put that information together with his need for water. Use of natural shafts led inevitably, it seems to me, to the digging of similar wellshafts placed for the convenience of the users. Like ourselves, the ancients would not have had perfect success in digging wells, but by the eighth century if not earlier wells were a standard feature of Greek settlements.

Probably equally early was the reuse of excrement as fertilizer, rising from observations of the amazing growth and fecundity of seeds accidently thrown onto dung heaps. Once when I was a child, a bag of fertilizer got left out in the rain; the paper bag disintegrated and by spring when we planted the tomatoes we had forgotten what this pile of stuff was. But there was no mistaking that the tomatoes planted just there had a decided advantage over the others in the garden. This solution combined two problems, the disposal of excrement and increased yield of food crops. Present-day Chinese agriculture still depends heavily on such fertilizer, while American cities spend enormous amounts ridding themselves of what should be considered an asset. It is intuitively obvious that by the time ten or more families are living in close proximity, especially if they also have animals, careful thought must be taken for waste disposal/ utilization. Surmising it must have been so, and finding out how it was done, however, have been two different activities. My study has shown that waste management was a constituent part of the water management system of ancient Greek cities.

By the second millennium B.C., the Minoans and Mycenaeans who preceeded the Greek "Dark Ages" knew how to use gravity flow as well as pressure supply pipes, together with channels and drains for disposal. The excavations at Knossos, Phaistos, and other Minoan sites on Crete, and at Tiryns, Mycenae and other Mycenaean sites in mainland Greece give us many examples of these water system elements. During the Dark Ages, it is thought, families and individuals retained the knowledge of cistern building and maintenance. The community-based provision for drainage and re-use of storm water in irrigation may well have been lost. Unfortunately, little evidence of Hellenic water management from the tenth and ninth centuries has survived or been rediscovered.

Gradually, beginning in the eighth century B.C., the old centers of culture such as

Athens and Argos revived again. Expanding trade in the eighth and seventh centuries meant larger settlements with greater pressures on existing resources. Of these, water would always have been the crucial and ultimately determining resource. The drought of the eighth century may have spurred not only colonization but more sophisticated management of the water supply at existing population centers. In the same century, Assyrian invaders of Armenia were deeply impressed with the long-distance underground water supply lines they found there. These came to be called *foggara* or *qanat*: The system utilized existing karst channels or built new tunnels to bring water from natural storage in a mountain to the urban area where it was needed. Every thirty-five meters or so, a shaft was pierced to the water channel from the ground surface, to withdraw debris from the tunnel and to bring in air and provide access for the tunnel builders. The debris was used to make an earth berm around the shaft, to prevent contamination by flowing surface waters (Kobari, 1980).

These *qanats* were adopted in Persia in the seventh century and by the sixth century were being built not only in Egypt (temporarily under the control of Persia) but also in the Aegean islands adjacent to Persian Asia Minor. The first physical evidence for the renewed use of pressure pipes and siphon systems comes from the sixth century. The long-distance line at Olynthos had both siphons and pressure pipes, and seems to date from the sixth century with later rebuildings - an extremely early date for a pressure line (Crouch, 1993, chapter 13). The first Greek aqueduct on Lemnos, was about as early as the famous tunnel system on Samos, about 575 B.C. (Kienast, 1986-87). It is likely that these early builders incorporated even earlier experiments with contour lines, bridges, and the intervention in and harnessing of karst water systems.

By the fifth century, long-distance water supply lines are known also at Syracuse (Crouch, 1993, Chapters 8-11). The Galermi line at Syracuse, a pressure system, is noteable for its length (at least 25 miles) and its persistence in use until the present. It would have been impossible to deliver the water from springs high on the mountain to the city without using a pressure line and siphons.

The evidence from vase paintings and archaeological findings shows that by at least the late sixth century B.C. there were public as well as private bathing facilities. These were equiped with bathtubs, showers, footbaths, and both large and small washbasins. Public baths also incorporated latrines or toilets. It was necessary to locate baths near springs or to supply them with piped water; in either case, reservoirs facilitated water storage and later use. No later than the fifth century, bathtubs were either built-in or provided with metal stands. In the bath at Gortys and at other places, heated water was piped to the rows of tubs, being carried in a shoulder-height channel behind each bather.

In houses, both laundry, dishwashing, and bathing activities utilized cistern water. Houses were provided with free-standing wash basins called louters (with a stem and a wide shallow dish, looking like a modern bird bath) and we also find large globular containers that were filled from the roof and seem to have been used as tubs for laundry. Many houses had built-in bathtubs, but free-standing ones on metal legs are known from vase paintings.

7-16. FOUNTAINS

In most cities, fountains became larger, more elaborate, and more numerous as the Greco-Roman period went on. The contrast is evident in two towns of about the same size, Priene of the fourth and third centuries B.C. with its five public fountains and Pompeii of the first century A.D. with its thirty-five neighborhood fountains (Crouch, 1993, chapters 12,13). This is an interesting comparison because Priene is unusual for the Greek period in its reliance on piped water in individual houses for almost all uses, while Pompeii (which had many more wells and cisterns than Priene) seems to have preferred to drink the water from aqueduct-supplied fountains.

Fahlbusch has shown (1982) that increased sophistication of choice about what kind of water line to build was a striking feature of water management during the third century B.C. to third century A.D. For instance, we have seen that already in the sixth century, the Greeks were building pressure lines. These lines were not without their problems in maintenance. As the builders gained experience and codified it into a teachable tradition, they came to evaluate the cost effectiveness of using gravity-flow water channels that followed the contour lines rather than architectural features such as solid walls or piers connected by arches that supported the water channels. Where siphons or inverted siphons were used, the builders supplied relieving tanks where the water could be dumped out into a reservoir and collected again into a channel. These were placed before and after a sharp change in level, such as the river and acropolis at Aspendos. This study by a modern hydraulic engineer has shown that the ancients made the same kinds of choices that a modern engineer would, in selecting from the repertoire of available solutions which would be best for any given section of a water line. Indeed, Fahlbusch told me that when his group was studying the water lines of Pergamon, they might lose sight of where in the countryside a line was running. If they stopped to think, "Where would I put it next?" they would usually find the ancient waterline just in that same place.

Political and technical control of waterlines changed as the Greeks experimented with different forms of government. During all of these centuries, however, a large measure of individual self-sufficiency was maintained through use of the family cistern and private wells. Drainage tended to be a more communal effort, but in some places excrement was collected separately by each household for transport to the fields, while sewers held only used water and rain water.

I have postulated that the water management system of the ancient Greeks was an evolutionary advance, enabling more of the people to survive. In a town, there were multiple sources of water: wells tapping the water table, cisterns collecting rain water from roofs plus garden cisterns (as at Pompeii) collecting surface flow, and spring-based fountains served by long distance water supply lines. It was highly unlikely that all three water sources would fail simultaneously. Hence, the redundency of the supply made dense settlement feasible.

Redundency of supply was extended by re-use of water to the maximum extent possible. Many human activities that use water do not require that it be of drinking quality, as we are now coming to realize. Given these multiplicites of source and of re-use, based on their respect for and knowledge of the climate and the geology, the Greeks experienced only two major droughts that we know of, in the eighth and fourth centuries B.C. This contrasts well with the California pattern of droughts every ten or twenty years. It was not the Greek way to grow rice in the desert nor to plant English lawns as we insist on doing in arid Southern California.

We may conclude, then that the first principle of ancient Greek water management was to recognize that water shortage was an ever-present threat, requiring careful thought about the management of resources. This led to allocation of water to its best use, by quality and by source, as well as to the re-use of water to the maximum extent. Cities were sited for access to water. Water was stored where it would be used, and technology was developed to bring in additional water. Their methods of water management were modest, economical, and sustainable. In water management, as in philosophy and art, we have much to learn from them. That such problems are perenniel is shown by the present predicament of small island states, as discussed in Vol. 6 Number 1 of *Our Planet* (1994). According to Angela Cropper (p.9, 12):

> The very size and virtual isolation of these islands make their ecosystems, both on land and in the sea, particularly fragile. And smallness exacebates this further by creating small feedback loops in ecological processes, accelerating them and magnifying their impact beyond what would be expected in a larger system. External or internal threats - whether from agricultural practices, mining, oil spills, the pattern of coastal settlements or the modification of the climate - cannot readily be contained. Little errors can therefore lead to big consequences both for the environment and for natural wealth. In small islands, with limited landspace and few options, this can be devastating. . . .
>
> No country or set of countries, as yet seems to have any tried and tested approaches to the radical transformations that are required in perception, ethics, planning, organization and management, and patterns of production and consumption. . . . Islands could provide a conducive environment for working out approaches to sustainable living.

Modern islands have many of the same problems that the ancient Greeks faced and solved during the first millineum B.C. Their "tried and tested approaches" could be tried again in our new circumstances, to see whether we really need to re-invent the wheel.

7-17. CONCLUSIONS

The Greek mind did not confine itself to writing drama and inventing philosophy, but also concerned itself with practical matters. Improved methods of waging war and improved disposal of waste waters were as ammenable to thought and experimentation as sculpture and diplomacy. It was probably the rapid sixth century growth in size of settlements that required municipalities to utilize two or three qualities of water, potable, sub-potable, and non-potable (Fig. 7-8).

Larger settlements put increased pressure on natural resources. Water supply arrangements such as the Peisistratid aqueduct at Athens date from this century, like the

Figure 7-8. Great Drain of the west side of the Athenian Agora

qanats mentioned above. Increased and predictable supplies of water, coupled with more intense year-round use of common areas of the city, increased the need to build large drainage channels like the Great Drain along the west side of the Athenian agora. (Fig. 7-8). Once led outside the city, the waste waters combined with storm runoff were a kind of man-made river and could be utilized for irrigation.

Pressure systems were increasingly widespread in the fifth to third centuries B.C. wherever the pressure of population made them necessary and the increase of wealth made possible their costs. Only in a place like Olynthus, distant from the mainstream of technological development and both destroyed and abandoned in the fourth century, has preserved the rough heavy pressure pipes of the early systems. Elsewhere, these were in time replaced by more efficient and elegant versions.

From the fifth century on we see a number of refinements of earlier technologies. Until the fifth century B.C., cisterns show no dominating formal pattern, but by the third century B.C. it was common for flask-shaped cisterns to be set deep into the earth, where the steady temperature of the ground kept the water relatively cool (about 55 degrees F) year round, thus preserving its palatability and reducing contamination (Fig. 7-9).

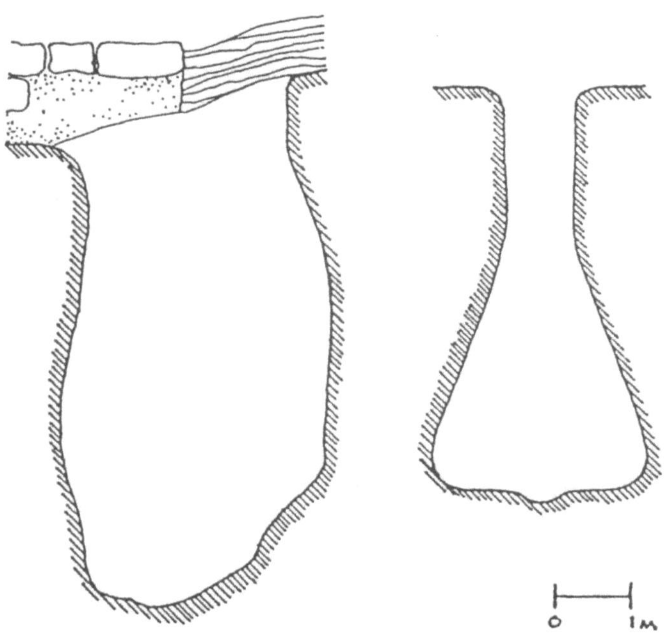

Figure 7-9. Comparison of an irregular fifth century cistern and a regular, flask-shaped cistern of the third century, both from Morgantina, drawn to the same scale, and about six meters deep

A further variation on the form and placement of cisterns is seen at Delos, where drinking water came from wells. This meant that there was little reason to drink cistern water. Therefore the saved rain water was kept in cisterns that were closer to the ground surface and hence the water was warmer. The typical cistern at Delos is rectangular and is roofed by the paving of the courtyard. Bathing, cleaning, laundry was more comfortable in the water that was warmed by the sun-heated stone of the paving. A second reason for the placement of these cisterns so close to the ground surface is that the water table under the island was unusually high owing to a rise in sea level. The kind of deep placement of cisterns that we find at other sites would have meant, at least in the lower areas of the island, increased risk of contamination of the stored water by salt-water intrusion.

REFERENCES

Adams R.McC. (1981). Heartland of Cities: Surveys of Ancient Settlement and Land Use on the Central Floodplains of the Euphrates. Chicago: University of Chicago Press.

Arnold G. (1969). Pneumatik und Hydraulik in der Geschichte der Energeitechnik. Krausskopf-Verlag, Mainz: 5-6.

BogliA. (1967). "Shafts," Actes 3eme Congress Suisse Speleog, pp. 17-18

BrinkerW. (1990). Wasserspeicherung in Zisternen. Ph.D. Thesis, Technical University, Braunschweig, Germany.

Butzer K.W. (1974). "Accelerated Soil Erosion" in Manners, I.R. and Mikesell, M.W., (Eds.), Perspectives on Environment, Assn. of American Geographers, Pub. No. 13: 66-68.

Camp J.M. (1982). "Drought and Famine in the 4th Century," in "Studies in Athenian Architecture, Sculpture, and Topography presented to Homer A. Thompson," Hesperia, Supp. XX, 9-17.

Camp J.M. (1979). "A Drought in Late 8th Century BC," Hesperia, 48: 397-411.

Climates of Change. (1963). Proceedings of the Rome Symposium organized by UNESCO and the World Meteorological Organization. UNESCO, Paris

Cropper A. (1994). "Small is vulnerable," Our Planet (published by United Nations Environmental Program), 6 (1): 9-12.

Crouch D.P. (1969). Palmyra. Ph.D. dissertation, University of California, Los Angeles.

Crouch D.P. (1984). "The Hellenistic Water System of Mogantina, Sicily: Contributions to the History of Urbanization," American Journal of Archaeology, 88 (3): 353-365 and plates 46-67.

Crouch D.P. (1993). Water Management in Ancient Greek Cities. New York: Oxford University Press.

Diakaios P. (1961), Sotira. University Museum, Pennsylvania.

Di Castri F. and Mooney H.A. (Eds.) (1973). Mediterranean Type Ecosystems. New York: Springer Verlag.

Fahlbusch H. (1982). Vergleich antiker griechischer und romisher Wasserversorgungsnlagen. Vol.73. Mitteilungen of the Leichtweiss Inst. für Wasserbau, Technical University, Braunschweig, Germany.

Feldhaus F.M. (1931). Die Technik der antike und des mittelalters. Wildpark-Potsdam: Akademische verlagsgesellschaft Athenaion.

Garbrecht G. (1984). "Reservoirs (Dams) in the Eastern Mediterranean," Mitteilungen of the Leichtweiss Institut für Wasserbau, Technical University, Braunschweig pp. 1-25 and 17 illustrations

Garbrecht G. *et al.* (1984). Historische Talsperren, I and II. Deutscher Verkbund für Wasserwirtschaft und Kulturbau e.V.

Ginouves R. (1959). L'etablisseement thermal de Gortys de Arcadie. Paris: J. Vrin.

Hill B.H. (1964). The Springs. Corinth I, pt. VI. Princeton, N.J.: American School of Classical Studies at Athens.

Jashemski W.E. (1979). The Gardens of Pompeii. New Rochelle, NY: Caratzas Bros.

Judson S. (1959). "Geologic and Geographic Observations at Morgantina,"unpublished manuscript in the Princeton University Morgantina files; forthcoming in a book on the topography of the site by M. Bell.

Kienast H.J. (1986/87). "Der Tunnel des Eupalinos auf Samos," Mannheimer Forum. pp. 179-241; see also his forthcoming monograph on this topic, published at Munich

Knauss J. (1987). Die Melioration des Kopaisbeckens durch die Minyer im 2. Jt. v. Chr. Kopias 2. Bericht Nr. 57 of the Inst. für Wasserbau, Technical University, Munich

Knauss J. (1990). Wasserbau und Gesichichte Minysche Epoche - Bayerische Zeit. Kopias 3. Bericht Nr. 63

Knaus J., Heinrich, B., and Kalcyk, H. (1984). Die Wasserbauten der Minyer in der Kopais - die älteste Flussregulierunf Europa. Kopias 1. Bericht Nr. 50.

Kobari I., (Ed). (1980), Qanawat Romani of Taibe Oasis. Tokyo University Scientific Mission for the Comparative Study of the Foggara Oasis in the Arid Zone of the Old Continent: Report no. 1.

Lambert T. (1971). Article on Assyrian response to Armenian qanats, Los Angeles Times, Sept. 29, 1971, pp. 20. Reported in the New York Times the same week.

Lang M. (1968). Waterworks in the Athenian Agora. Princeton, N.J.: American School of Classical Studies at Athens.

Lindh G. (1987). "Water Resources Management Problems in Urban Agglomerations," *in* Wunderlich, W.O. and Prins, J.E., (Eds.), Water for the Future. Rotterdam: A.A. Balkema: pp. 191-20

Lindh G. and Betrthelot, R.M. (1979). Socio-economic Aspects of Urban Hydrology. Paris: UNESCO Studies and Reports in Hydrology, No. 27.

Maiuri A. (1931). "Pompeii Pozzi e condotture d'acqua . . ." and "Scoperta di grandi conditture in piombo dell'acquedoto urbano," Notizie degli Scavi di Antichicta, 7: 546-76.

Marinos P.G. (1978). "Les conditions hydrogeologiques et la disposition des dechets municipaux dans un haut plateau karstique (Grece -Peloponnese centrale) Implications de l'Hydrogeologie dans les authres Sciences de la Terre (I.H.E.S. Symposium Montpellier, France, Sept. 1978) Memoire hors serie CERGH -USTL Montpellier, pp. 537-551.

Meinzer O.E. (1927). Plants as Indicators of Ground Water. U.S. Geological Survey Water Supply Paper 577.

Morefis A., Paraskevopoulou P., Vrhovsek D., and Zupan M., (1986). "Pollution of ponor and spring waters" in Karst Hydrogeology of the Central and Eastern Peloponnesus (Greece) Riedmüller G. and Zojer H., (Eds.), Inst. für Technische Geologie, Petrographie und Mineralogie, Technical University of Graz, pp. 92-102.

Meulenkamp J.E., de Mulder E.F.J., and van de Weerd A. (1972). "A Sedimentary and Paleogeography of the Late Cenezoic of the Island of Rhodes," Z. Deutschl. Geol. Ges., Hanover, Band 123: 541-53 and 4 maps.

Picard L. (1956-58). "Development of Underground Resources in Greece," five reports prepared for the Government of Greece through U.N. Technical Assistance Program, especially No. 3, 1957.

Picard L. (1964). Final Report: Karst Groundwater Investigations-Greece, Rome: FAO, 1964.

Plutarch, Lives," Solon" XXIII, 6.

Renfrew C. and Wagstaff M., (Eds.) (1981). An Island Policy: The Archaeology of Exploitation in Melos. Cambridge University Press.

Robinson, H.S. (1969). "A Sanctuary and Cemetery in Western Corinth," Hesperia, 38: 1-35.

Roebuck C. (1951). The Aesklepion and Lerna. Corinth, XIV. Princeton, N.J.: American School of Classical Studies at Athens.

Schiøler T. (1973). Roman and Islamic Water-lifting Wheels . Odense University Press.

Vitruvius (late first century BC). De architectura, Bk. X, Ch. 5.

Wiegand T. and Schrader, H. (1904). Die Wasseranlagen. PrieneIV. Berlin: Reimer Verlag.

Wulff H.E. (1968). "The Qanats of Iran" Scientific American, 218 (16): 94-100.

CHAPTER 8

THE STATUS OF WATER RESOURCES IN MINOAN TIMES
A PRELIMINARY STUDY

ANDREAS N. ANGELAKIS[1] AND STYLIANOS V. SPYRIDAKIS[2]

[1]*National Foundation for Agricultural Research, 71110 Iraklio, GREECE, and*
[2]*Dept. of History, Univ. of Calif., Davis, Davis, Ca. 95616, USA.*

8-1. INTRODUCTION

ὥς δ' ὁπότ' 'Ιασίωνι εὐπλόκαμος Δημήτηρ,
ᾧ θυμῷ εἴξασα μίγη φιλότητι και εὐνῇ
νειῷ ἔνι τριπόλῳ· οὐδέ δήν ἦεν ἄπυστος
Ζεύς, ὅς μιν κατέπεφνε βαλών ἀργῆτι κεραυνῷ.
Odyssey, V. 125-128.

Δημήτηρ μέν Πλοῦτον ἐγείνατο δία θεάων,
'Ιασίων ἥρωι μιγεῖσ' ἐρατῇ φιλότητι
νειῷ ἐνί τριπόλῳ Κρήτης ἐν πίονι δήμῳ,
ἐσθλόν, ὅς εἶσ' ἐπί γῆν τε καί εὐρέα νῶτα θαλάσσης
πᾶσαν· τῷ δέ τυχόντι καί οὗκ ' ἐς χεῖρας ἵκηται,
τόν δ' ἀφνειόν ἔθηκε, πολύν δέ οἱ ὤπασεν ὄλβον.
Hesiod *Theogony*, 969-974).

A well-known passage in Homer's *Odyssey*, probably based on an ancient ritual myth, tells the story of Demeter, the Greek corn-goddess and Iasion, the son of Zeus by Electra, daughter of Atlas. The latter was the guardian of the pillars of heaven (*Odyssey*, 1. 53), the Titan who holds the sky up (Hesiod, *Theogony*, 517) and is, thereby, identified with water and rainfall.

The story relates that Iasion lay with the fair tressed Demeter in "a thrice ploughed fallow field", but Zeus, when informed of the act, smote Iasion with his thunderbolt and slew him. Hesiod places the Homeric "thrice ploughed fallow field" on Crete's "fertile island" and adds that the offspring of the union of Demeter with the hero Iasion was Ploutos, "wealth", who travels far and wide bringing riches and happiness to whomever he encounters. It must be noted that in antiquity Ploutos was viewed primarily as a dispenser of abundant crops, a characteristic that explains his close affiliation with

NATO ASI Series, Vol. I 36
Diachronic Climatic Impacts on Water Resources
Edited by Andreas N. Angelakis and Arie S. Issar

Demeter at Eleusis (Deubner, *Attische Feste*, 85 f.). A much later, weak tradition in Servius (on *Aeneid*, 3) relates that Iasion, the father of Ploutos, and his brother Dardanius, were Cretans, thus providing us with evidence that the affiliation of Crete with the myth persisted until late in antiquity.

The basic elements of the myth of Demeter-Iasion-Ploutos with their Cretan associations may well reflect the importance of agriculture in the economy of early Crete and the role of irrigation in the productivity of the island: They are, at least, consistent with Crete's ancient characterization as μακαρόνησος "island of the blessed" (Stephanos of Byzantium, *s.v.* Κρήτη) and Pliny's contention that all products of the Cretan soil are incomparably superior to all others (Pliny, 25.8; cf. Theophrastus, *History of Plants*, 4.1).

Κρήτη τις γαῖ ' ἔστι, μέσω ἐνί οἴνοπι πόντῳ
καλή καί πίειρα, περίρρυτος, ἐν δ' ἄνθρωποι
πολλοί, ἀπειρέσιοι, καί ἐννήκοντα πόλη ες.

Odyssey, XIX. 172-174.

Nearly three thousand years ago Homer sang the beauty and richness of Crete with its ninety cities (elsewhere he mention a hundred), and large population which only a fertile, well forested and watered land could have supported. Crete is by far the largest of the hundreds of islands that constitute the Aegean Archipelago and enjoys a unique geographical position at the crossroad of the three continents that formed the ancient world. In fact, it would not be far from accurate to state that the island is equidistant from Europe, Asia and Africa. Considering its close geographical proximity to the early centers of civilization in Egypt and the Near East, it is not surprising that Crete cradled the first European civilization.

As early as the Neolithic age, the island was inhabited by a scattered population living partly in caves some distance from the coast but also concentrated at Knossos, in one of the largest settlements of that period in the eastern Mediterranean. Although little is known of the origins of these early settlers, the associations suggested by pottery and other artifacts, point to Anatolia, and possibly Egypt, rather than the Greek mainland. Crete's population was reinforced at the beginning of the Bronze Age, *i.e.* shortly after 3000 B.C. with the arrival of new settlers perhaps from Asia Minor. Moreover, linguistic and toponymic affinities suggest that a considerable element in the Cretan population may have been related to the Luwians who, in the Middle and Late Bronze Ages, were established in Asia Minor, southwest of the Hittites (Huxley, *Crete and the Luwians*, 1961).

Cultural advancements can be observed throughout the third and second millennia B.C. but Crete progressed especially in the Middle Bronze Age (*ca.* 2100-1600 B.C.) when the island's population in its central and south regions increased, towns developed, the first palaces were built and Crete achieved a prosperous and uniform culture. By the end of this period, both manufacturing and the arts flourished and the islanders engaged in extensive trade with Egypt, the Aegean Archipelago and the Near East. In the early phases of the Late Bronze Age (*ca.* 1600-1400 B.C.) Crete appears to have prospered even more, as the larger houses and more luxurious palaces of this period indicate. At this time, the flourishing arts, improvements in metal-work along with the construction of better-equipped palaces and an excellent road system, reveal a wealthy, highly cultured, well-organized society and government in Crete, before the island's power collapsed following the destruction of the Minoan palaces *ca.* 1400 B.C.

Sir Arthur Evans, the legendary British archaeologist who was primarily responsible for the recovery and scholarly reconstruction of Minoan civilization, had devised a system of chronology which is largely followed todate, although many scholars strongly disagree with specific aspects of it. Here, in round numbers, adequate for our purposes, is the accepted chronological scheme of Minoan civilization (Evans, 1964; Graham, 1987 and others): Early Minoan (E.M.), *ca.* 2600-2100 B.C., Middle Minoan (M.M.), *ca.* 2100-1600 B.C. and Late Minoan (L.M.), *ca.* 1600-1100 B.C. In addition, through the combined use of relative and absolute dating, the chronological system of the Minoan world works out approximately as the following stages: Prepalatial, *ca.* 2600-1900 B.C., Protopalatial, *ca.* 1900-1700 B.C., Neopalatial, *ca.* 1700-1450 B.C., and Postpalatial, *ca.* 1450-1100 B.C. (*History of the Greek Nation*, 1970).

Before we begin our review of Minoan water resources, a brief survey of the main Minoan palaces, villas and houses frequently encountered in our discussion, might be useful. Major Minoan Palaces are: the Palace of Minos at Knossos, the Palace of Phaistos, the Palace of Mallia, and the Palace of Zakros. Minor Minoan Palaces, Villas, and Houses are: the Palace of Gournia, the Villa of Agia Triadha, the Little Palace at Knossos, the Royal Villa at Knossos, the South House at Knossos, the House of the Chancel Screen at Knossos, the Southeast House at Knossos, the House of the Frescoes of Knossos, the House at Nirou Khani, the Houses at Tylissos: Houses A and C, the Houses at Mallia: Houses Da, Za and Zb, and the "Little Palace," the Villa of the Lilies at Amnisos, the House B at Palaiokastro, the House at Sklavokambos, the Plati in Lasithi Plain, and other minor sites.

Although little is known of the Early Minoan world, extant evidence points to a sustained cultural growth in this period of Crete's past. The veritable "cultural explosion" that occurred at the time encompassed many cultural aspects typical of the modern world, such as architecture, civic planning and development, a prosperous and esthetic way of life, an advanced agriculture, silviculture and navigation, environmental protection concerns, etc. Thus, the Minoan is universally considered to be one of the grandest and most brilliant of all ancient civilizations. Typical considerations for the selection of the principal Minoan centers appear to have been their defensive position, the beauty of their environment, an abundance of food and especially water supplies.

In this preliminary study a review of the status of water resources in Minoan times is attempted. It encompasses various scientific fields of water resources such as wells and groundwater hydrology, water supply, storm water and wastewater sewerage systems, construction and use of fountains, bathrooms and other sanitary and purgatory facilities, drainage and irrigation of agricultural lands and even recreational uses of water. It is evident that in Minoan times extensive systems and elaborate structures for water supply, sewerage systems, irrigation and drainage were planned, designed and built to supply the growing population with water for the cities and for irrigated agriculture. Thus, it is not by chance that the main technical and hydraulic operations of capture, conveyance, raising, storage and measurement have been practiced in varying forms since *ca.* 3000 B.C. (Fig. 8-1).

1. Household - Drinking water	.~~~~~~~~~~~~~~~~~~~~~~~~~>
2. Household - General use	.~~~~~~~~~~~~~~~~~~~~~~~~~>
3. Fishery	.~~~~~~~~~~~~~~~~~~~~~~~~~>
4. Recreation	.~~~~~~~~~~~~~~~~~~~~~~~~~>
5. Navigation	.~~~~~~~~~~~~~~~~~~~~~~~~~>
6. Agriculture - Irrigation	6000~4000 B.C. ~~~~~~~~~~~~~~~~>
7. Agriculture - Animal production	~5000 B.C. ~~~~~~~~~~~~~~>
8. Communities	~3000 B.C. ~~~~~~~~~~>
9. Sewage disposal	~3000 B.C. ~~~~~~~~~~>
10. Trade (production)	~3000 B.C. ~~~~~~~~~~>
11. Industry (production)	~3000 B.C. ~~~~~~~~~~>
12. Water power	~200 B.C. ~~~~~>
13. Industry (cooling water)	~1900 B.C ~~~>

Fig. 8-1. Historical development of water resources utilization (Garbrecht, 1987)

8-2. GEOLOGY, CLIMATE, VEGETATION AND HYDROLOGY

Geology, Hydrogeology and Hydrology

Throughout the geologic history of Crete the following deposits and tectonic units appeared in a more or less successive, overlapping sequence: The Crete-Mani geotectonic unit (semi-metamorphic platy-limestones), the Crystalline carbonate unit of Trypalion, the Phylite-Quartzitic basement and its carbonate alpine cover of Tripolitza-geotectonic zone and the overlapping nappes of Pindus and of the more internal Hellenides geotectonic zones. All pre-alpine formations are intensively folded, thrusted and faulted. The morphologic lower parts and basins of the island are covered by thick clastic sediment of Neogene and Quarternary. Since Minoan times and as far as the geological frame is concerned, the only probable changes could not have gone beyond slightly erosional modifications in the physiography, the growndwater table, the sea level and the recent geological formations.

The main characteristics of the hydrology of Crete are the alluvial deposits and the karstic formations in the valleys and on the hills and the mountains, respectively. Without attempting to go into any details of hydrogeology as perceived in ancient times, it is interesting to point out that the presence of karst terrain in Crete and the surrounding countries had quite an influence on classical thinkers: While the basic idea of hydrology - that all water comes from and returns to the seas - was definitely known to the Greeks as far back as Homer's "Deep-flowing ocean, from whom all rivers flow, and all springs and deep wells" (*Iliad,* Bk. XXI), nevertheless, there were two distinct ways in which the phenomenon was understood and explained. In one case, water is forced from the sea into the rocks, where it is driven upwards and its salts extracted, till it again reaches the surface of the ground; this idea was propounded by Thales who lived *ca.* 624-547 B.C. (Burdon and Papakis, 1963) and espoused by Plato, as well as by Pliny. In the other case, springs are considered to be created from condensation in subterranean caves and caverns, as shown in Aristotle's *Meteorologica,* and in Seneca who also adopts this view. It is clear that this latter idea (which recognized evaporation from the sea surface and subsequent condensation) would only emphasize subterranean condensation because the presence of karst caverns in their regions invited ancient scientists to explore them and conduct scientific investigations there.

Fossil Pollen Studies

Most of the extant information on the paleo-climate of Minoan Crete derives from antiquarian sources, especially fossil pollen studies. There exist 15 sediment cores that preserve significant amounts of fossil pollen in Southern Greece (Moody *et al.,* 1990). Six of these cores come from Crete but only three of them have been published: Aghia Galini (Bottema, 1980), Tarsanas and Limnes (Moody, 1987). It is important to note that all of the published cores from Crete come from marginal or lowland regions and consequently do not appear to reflect accurately mountainous areas which constitute nearly 60% of the island's size. Nevertheless, a study of these three pollen cores suggests that the Neolithic and Bronze Age climate of Crete was more temperate than at present. The occurrence in the cores of numerous species characteristic of central Europe, such as those of the genera *Betula, Fracinus, Cotylus, Carpinus* and *Tilia,* which no longer grow naturally in Crete, provides a reasonable indication of this climatic change. Moreover, the overlap of such characteristic Mediterranean species as those of the genera *Olea,* prickly *Quercus* and *Lens* with these temperate trees, suggests that although the climate of Minoan Crete was wetter than today's, winters could not have been more than 10° cooler (Moody, 1987 and 1990).

Between the end of the Neolithic (*ca.* 3000 B.C.) and the Middle Bronze Age (*ca.* 2100-1600 B.C.), all pollen diagrams studied reveal a marked decline in deciduous oak and in sensitive Central European tree pollen, such as those of the species of the genera *Betula* and *Tilia.* In spite of the fact that the genus *Tilia,* and to a lesser degree the genus *Betula,* are particularly palatable to livestock and are often cut for wood, it is difficult to attribute their simultaneous decline throughout southern Greece entirely to browsing and woodcutting (Turner, 1982). Species of the genus *Tilia* are very resilient trees that coppice easily when cut and must be grubbed out by their roots in order to get rid of them (Rackham, 1980). Although browsing and woodcutting probably contributed to its decline, it is inconceivable that *Tilia* could grow locally in the modern dry climate of Tarsanas or even on the wetter cliffs of Limnes. Today in southern Greece this plant is found only in remote, wet environments, often on cliffs that protect it from browsing livestock (Rackham, 1983). This narrow distribution is in sharp contrast to the broader distributions of species of the genus *Ilex,* deciduous species of the genus *Quercus,* etc; these species are also very palatable to browsing livestock, but grow on both wet and dry, cool and hot cliffs (i.e. in any kind of plant refugia). It is reasonable to infer, therefore, that the disappearance of *Tilia* during the Early Bronze Age must be, in part at least, the result of a change in climate and in particular a change to a drier climate (Moody, 1990).

A similar picture is presented by the species of the genus *Quercus.* Deciduous species dominate the *Quercus* profile throughout the Aghia Galini diagram (Late Mesolithic to Early Bronze Age), but do not occur locally today. At other sites, deciduous species of *Quercus* decline in the Early Bronze Age and reappear briefly during the Dark Ages (*ca.* 970 B.C.), while evergreen species increase throughout the Bronze Age (Moody, 1987). This change from deciduous to evergreen species of Quercus, like the decline in *Tilia* and *Betula,* could be interpreted as a change to a drier climate (Rackham, 1983). Moreover, modern vegetation studies indicate that deciduous species of *Quercus* are very aggressive, and if provided with enough moisture and exposed only to moderate browsing, the species will spread easily (Faure, 1988 and Moody, 1987).

In Minoan times, the climate and water resources of Crete were of vital importance to its people. But, was the island's climate the same as today's? Was it as dry as it is today and more important, was water a scarce commodity? If we were to accept the proposition that in Minoan times Crete was more wet than today, we should take notice of the alluvial soil that has formed little valleys in which rivers under continuous flow existed. Notice also that the little valleys of Crete account for less than 400 km^2 of the island's area, which is more than 20 times the total size of these valleys. However, Moody (1990), proposed that although human interference may have been a factor in the decline of deciduous trees on Crete, such a broad and consistent pattern suggests that a climatic factor was operating as well.

Stability or Limited Change in Climate

There is also the opposite view for no or limited climatic change in Minoan times (Flaccus,1992; Thirgood, 1981; Aiginetes, 1954 and others). The proponents of this view find evidence pointing to the conclusion that the climate of Crete in Minoan times was essentially the same as it is today. This evidence is based primarily on the following factors and considerations: a) Very long and continuous human habitation. Some nine thousand years of human habitation have, no doubt, brought great changes in the plant life of the island. Moreover, in Crete vegetation responds very quickly to changes in human affairs (Rackham, 1990); b) Immigration. The long period of human use of the land in Crete and its inhabitants' contacts with the outside world have resulted in the introduction of a large number of non-native species, such as weedy companions of cultivation, many of which have become naturalized; c) Cultivation, grazing and other human activities. These factors, along with the use of timber for shipbuilding and fuel, must have changed vastly the botanical map of the land, particularly at lower altitudes; d) Geology. As noted earlier, Crete is a mass of limestone rock from which limited soils have developed; e) Fires. As is the case today, in all summer-dry Mediterranean climatic regions the role of man caused fires has had a profound effect on the environment; f) Deforestation. What has probably changed most since Minoan times, with deforestation and various other destructive human activities affecting the coastal plains and the lower slopes of the island's mountains, is the severe reduction of trees, such as various species of the genera *Cupressus, Quercus* (evergreen), *Pinus*, and the tall shrub of *macula* (macchia) vegetation; the most commonly encountered species are: *Arbutus unedo, Ceratonia siliqua, Olea europea* var. *oleaster* and *Quercus coccifera*. A direct result of this was a substantial increase in phrygana growth indigenous to rock slopes.

In conclusion, there are two major interpretations of the status of climatic conditions in Crete from Early to Late Minoan times. Some silvicultural researchers hold that there was no great variation in climate from antiquity to the present. However, there is substantial evidence indicating that climatic conditions vacillated between a temperate-wet and a Mediterranean-dry climate. The major problem here is how to determine the duration of this climatic cycle, a concern that should be addressed by future research.

Similar climatic change scenarios have been reported in Israel (Issar **et al.**, 1989). Thus, a humid stage affected Palestine during the Early Bronze Age. This was proved by environmental data, such as $^{18}0$ and ^{13}C, as well as pollen of olive and oak trees in a core in the Sea of Galilee, records of Dead Sea levels and the level of the Nile and of course the settlement of man in the desert. A humid stage in this region was indicated again during the end of the Late Minoan and the Hellenistic Ages (Issar *et al.*, 1989).

Evidence of Greater Precipitation Supplied by Flora

The occurrence of certain species of plants, now extinct, among the flora depicted in Minoan art also points to the conclusion that a more humid climate prevailed in antiquity. A marked increase in precipitation in Minoan times was probably followed by an afforestation activity on the island. Forest areas securing a more equable distribution of precipitation and probably greater amounts of rainfall, as well as an extended vadoze soil-zone, may well have existed at least until the closing years of the Middle Minoan

Age. It should be noted that in Cyprus, recent afforestation has caused a marked increase in precipitation (Evans, 1964). The exceptional building activities of this period, along with the increase in shipbuilding caused by expanding trade, presuppose a continual use of timber supplies; however, in the course of the succeeding end of the Middle and Late Minoan Ages, the greater distribution of population throughout the island marks the deforestation of Crete (Evans, 1964) and its gradual decline.

Some direct evidence for greater precipitation in the Middle Minoan period than at present, was provided in a 1926 discovery, about 700 m west of the site of Knossos, on the edge of a ravine that forms the upper continuation of the gully through which the Vlychia river flows. Here, recent rainfall had exposed some Minoan blocks shown in an exploratory dig to have been the side-supports of a water-channel supplied by tow sub-channels and extending about 50m beyond this point; directly below it there exists a water source of very good quality, known from the dark pool it forms, as Mavrokolybos (Evans, 1964). Presently, however, the spring is of very low discharge and its water is conveyed for irrigation purposes by a single pipe, small in diameter.

8-3. WATER SUPPLY, DRINKING WATER AND SANITARY FACILITIES

Our knowledge of how Minoan cities were supplied with potable water is almost entirely confined to the Palace of Knossos. A few cisterns, fountains and wells were also found at Zakros, Mallia, Gortys and other Minoan palaces and cities. At Phaistos some cisterns have been discovered as well, but owing to the nature of the ground, no wells or springs, have been found here.

Even at Knossos, the sources of water and the methods used for supplying it, are only partially understood. Several wells have been discovered in the Palace area itself, and a single well slightly to the northwest of the Little Palace. The latter, restored to its original depth of about 12.5 m and 1.0 m diameters, continues to furnish an excellent supply of potable water (Evans, 1964).

The inhabitants of the Knossos Palace, however, did not depend on the water of the wells alone, but were supplied with potable water through a network of terracotta piping located beneath the Palace floors. The pipes were constructed in sections of about 60 to 75cm each. These pipes with their expertly shaped, tightly interlocked sections, date from the earliest days of the building and are quite up to modern standards (Fig. 8-2). They imply a practical knowledge of the hydraulic principle that water seeks its own level; this is also supported by the discovery of the Minoan conduit heading towards the Knossos Palace from the pure limestone spring of Mavrokolybos which suggests a descending and subsequently ascending channel (Evans, 1964 and Hutchinson, 1950). The sections of the clay pipes resemble those used in Greece in classical times, though Evans considered the Minoan to have been designed more efficiently; each section was rather strongly tapped toward one end with the objective of increasing the rate of water

(a)

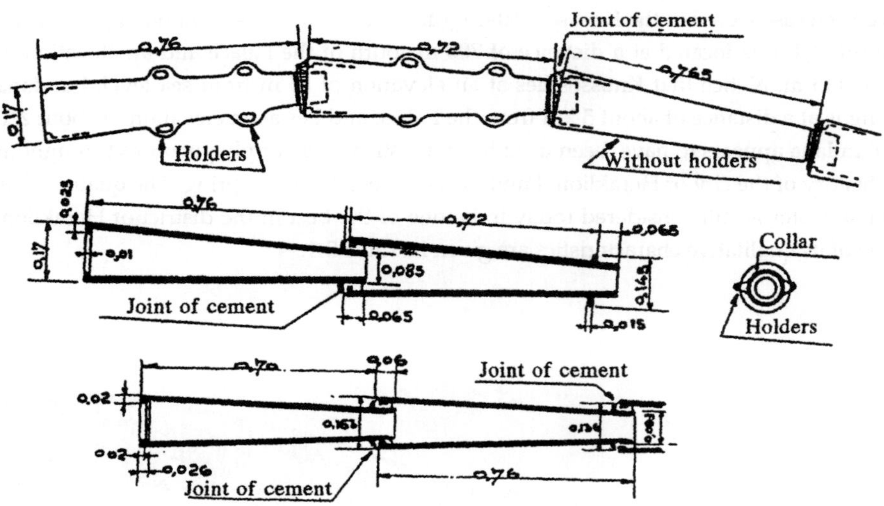

(b)

Figure 8-2. Minoan water supply pipes (Terraccota pipe sections): (a) Overview, and (b) With real dimensions (Buffet and Evrard, 1950)

flow, thus helping to flush any sediment through the pipe (Buffet and Evrard, 1950). Minoan hydraulics apparently were concerned with the solution of some water engineering problems and were able to provide cities and palaces with complete water supply systems. On the basis of their accomplishments it can be assumed that the Minoans were, in a sense, aware of the siphon and Archimedean principles and even of

Pascal's law; however, it appears that they had only a vague understanding of the relationship between flow and friction.

The clay tubes in the Minoan bathrooms must have been filled and supplied by hand (Fig. 8-3). However, in the "Caravanserai", a rest house on the cross-island route just south of the Knossos Palace, a footbath for the weary travellers was supplied by a direct pipe, and the overflow was disposed of by another conduit; a branch of the water channel also served for water supply purposes (Fig. 8-4). Nearby Evans excavated a Spring House consisting of a well-built stone chamber. Within this was a stone-lined basin, 45 cm deep, through whose gravel bottom a spring welled up in antiquity and still does, to some extent, today; the stones of the basin were deeply worn out by those who came there to fill their jars in the days of Minos (Evans, 1964).

There are indications that the water supply system of the Palace of Minos at Knossos was initially dependent on the spring water of Mavrokolybos and later on the Fundana, Karidaki and Paradisi springs. It must be noted that no wells for groundwater supply have been discovered in the Knossos Palace (Marinatos, 1927). As mentioned previously, Mavrokolybos is located at a distance of 700 m south of the Palace and an elevation of about 115 m. Notice that Knossos lies at an elevation of 85 m from sea level; Fundana spring is at a distance of about 5 km from the Palace and lies at an elevation of about 220 m. Fundana appears to have been used for domestic water supply purposes throughout the history of the city of Heraklion. Fundana is a typical karstic spring. The quality of its potable water is still considered today to be one of the best in the district of Heraklion. Some of its qualitative characteristics are given in Table 8-1.

Figure 8-3. Water Facilities - Clay Bathtub

1 Roman basilica and forum; Classical agora probably nearby
2 House of the Frescoes
3 Spring Chamber and Caravanserai
4 Sanctuary of Demeter
5 Middle Minoan tholos tomb

Scale

0 500 1000 m

Figure 8.4. Environs of Palace of Minos

Table 8.1. - Some qualitative characteristics of Fundana spring water

Parameter	Value[a]
pH	7.5
EC (dS/m)	0,68
Alkalinity ($CaCO_3$), total	210
Hardness ($CaCO_3$)	286
Cl^-	53
NO_3^-	13

[a]All concentrations are expressed in mg/L, except electrical conductivity (EC) and pH, which are unitless.

The Minoan's practical knowledge of the fact that water follows the basic hydrostatic law, known today as the principle of communicating vessels, is evident. It is manifested in the water supply of the Knossos Palace through pipes and conduits fed by springs in the neighbouring hills which made the need of digging wells to secure groundwater for drinking less pressing. However, in the Protopalatial stage (*ca.* 1900-1700 B.C.), several wells were used for drawing drinking water. The ready access to cool and fresh water that wells alone could supply, made their construction highly desirable and they were used along with the other sources of water previously described. Their depth did not exceed 20 m and their diameter was no more than 5m (Buffet and Evrard, 1950). At least six such wells have been reported (Evans, 1964). The most important and best known is the one found in the north-west of the Palace in the basement of the House A, which belongs to the first stage of the Middle Minoan period. According to Evans (1964), its upper circuit was mostly a patchwork of rubble masonry, recalling the construction of the numerous Roman wells of this side of the site. However, below its crudely built upper "collar", the well was found to be cased in a series of terra-cotta cylinders of fine clay and of material so hard that it was initially mistaken for some kind of close-grained stone.

These successive "drums", representing in principle an advanced technique of resisting the pressure of the surrounding soil, had been sunk into the natural "kouskouras" clay below the Minoan level to a depth of about 10.40 m; originally there were at least 17 cylinders, each 60 cm high (Fig. 8-5). Every cylinder was composed of three nicely compacted sections and in each section there was a triangular hole which provided a place of support for the feet and hands of the individuals who had to descend the well for cleaning or other purposes (Fig. 8-6). A remarkable feature of the arrangement was the incision of linear signs on the expanding upper rims of each "drum", at the joints of their three sections, apparently designed to identify the cylinder to which they were to be fitted. Two of these signs, a sloping form of N and a kind of W

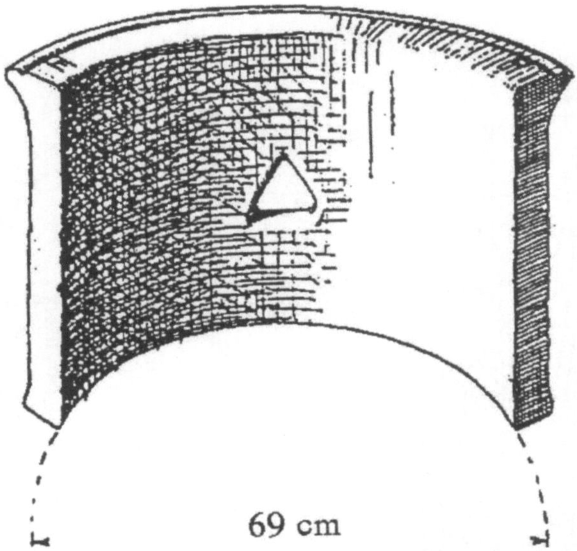

Figure 8-5. Section of terracotta "drum" of Minoan well (Evans, 1964)

and a double X, were simple linear forms of no particular significance. Another was a more complex form . The , however, repeated in Figure 8-5 represents one of the most characteristic signs of the two advanced Cretan linear scripts (Evans, 1964).

In the Zakros Palace the water supply system depended on groundwater. Here the potable water came from the main spring. In the southwest corner of the Cistern Hall an opening leads into a small chamber where the water was collected and channelled into a square underground fountain built on the south; this was thought to correspond with the celebrated man made fountain of the *Odyssey* known as "Τυκτή" fountain (Platon, 1974). The fountain with its walls was built of regular limestone, and there is a descending staircase with fourteen steps (Fig. 8-7b). The room may also have served as a shrine. The water of the fountain is brackish today, of about 13.00 dS/m EC, due to its position and the intrusion of sea-water. However, this may well be an indication that some reduction in the distance of the Palace from the coast has occurred.

In Zakros, there is a comparable chamber with a well-spring located near the southeast corner of the Central Court; here again steps, now damaged by visitors' feet, lead down into a chamber whose function was similar to that of the Spring House in the "Caravanserai" in Knossos.

The wood of the windlass was found in the water, along with an offering cup containing olives; this is a unique, remarkable find, since the olives were perfectly preserved, as though they had just been picked from the trees; unfortunately they

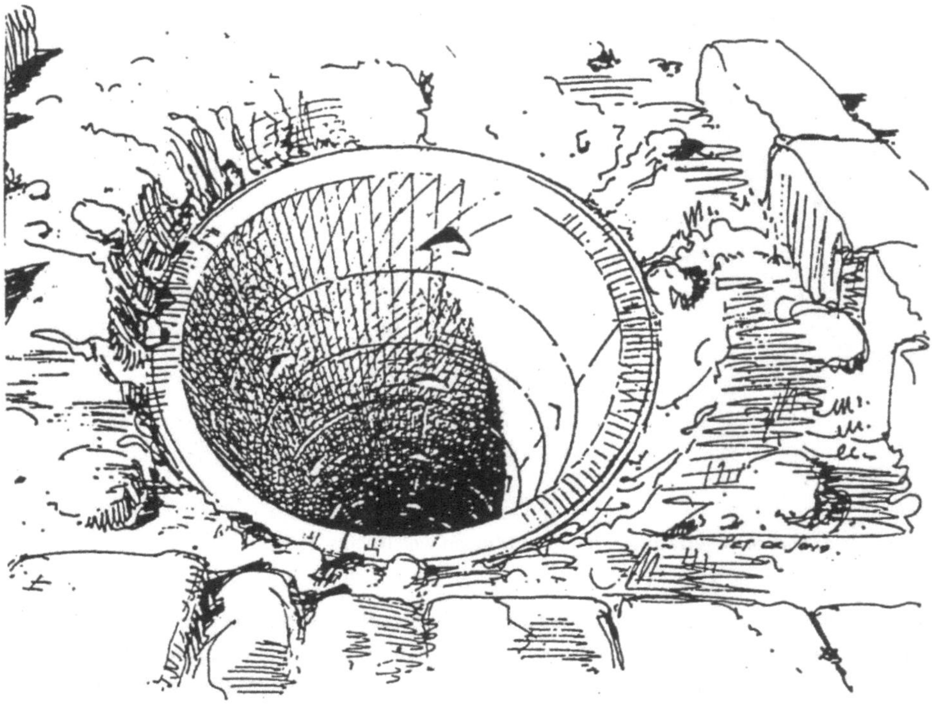

Figure 8-6. Perspective view of well below House A, NW of Knossos Palace (Evans, 1964).

maintained their relative freshness for only a few minutes after they were taken out of the water (Platon, 1974). Its water quality today should be considered as less brackish (of about 6.40 dS/m EC) than that of the main spring, due to its relative position to the latter. A view of this well-spring is given in Figure 8-8.

Although the supply of water at the Knossos Palace was mainly dependent on the Mavrokolybos and perhaps on the Karidaki and Fundana springs, in Zakros it was entirely groundwater. On the other hand, in Phaistos the water supply system was dependent directly on precipitation: here, rainfall-water was collected in special cisterns from the roofs and yards of buildings. Special care was given to securing clean surfaces in order to maintain the purity of water. Also, coarse sandy filters were used to treat the rainfall water before it flowed into the cisterns which were also used for washing clothes and other cleaning tasks.

In the Minoan palaces, bathroom sewers were not necessary, merely convenient. Actually, most palaces did not have them. Obviously, the installation and maintainance

(a)

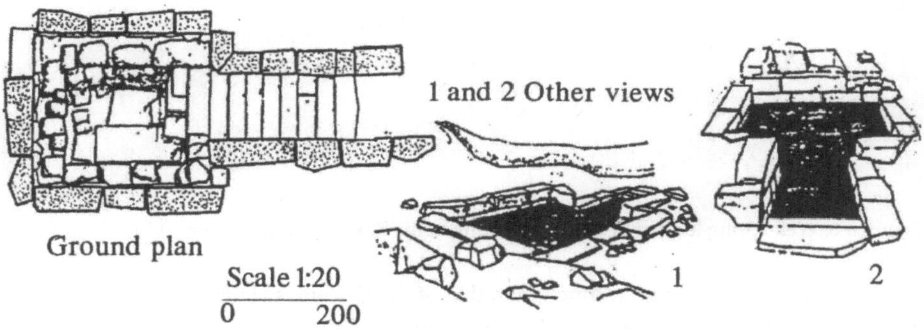

(b)

Figure 8-7. Views of the "Τυκτή" fountain: (a) Overall view and (b) Scheme (Platon, 1974)

of functional sewers in the low-floored structured Minoan rooms would have been extremely difficult. On the other hand, their tubs had handles or ledges for easy carrying and could have been easily taken in and out of rooms. Evans, identified three rooms at the Knossos Palace as bathrooms. The main type which resembles the bathrooms discovered at Phaistos and Mallia, is that found near the Queen's Hall (Fig. 8-9). One point of difference, however, is that the floor was not at a lower level. Remains of a clay tube were found just outside the door of the room. Water was poured through a hole in the floor immediately outside the lavatory door; an under-floor channel linked the hole with the vertical clay pipe under the toilet seat (Castleden, 1993). The toilet could thus be flushed even during a rainless summer, either by an attendant outside the lavatory or by the user (Fig. 8-9).

Figure 8-8. Well-spring located in the eastern wing of the Zakros Palace (Platon, 1974).

The toilet also is similar in function to that of the Queens's Hall; this is the case in the Knossos, Phaistos and Mallia palaces and in some of the houses as well. Fortunately, one of the houses near the Palace at Mallia, known as Da, contains a toilet seat practically in perfect condition, since it was made of stone, not wood, like the seat of the Palace of Minos (Fig. 8-10). This stone seat is about 68.60 cm by 45.70 cm wide front to back and its surface is some 35-38 cm above the floor. It is built directly against an outside wall through which a capacious sewer passes. Like the Knossos find, it was evidently intended to be used as a seat rather than a stand; thus, it resembles the Egyptian toilet more closely than the so-called "Turkish" type found in the Palaces at Mari and Alalakh (Graham, 1987). A similar toilet has been discovered in the west side of the so called "Queen's Apartment" at Phaistos. It was connected to a closed sewer, part of which still exists. Another toilet sewer was discovered in House C at Tylissos.

At certain times of the year the sewers in the Palace of Minos may have been adequately flushed out by the rain that fell into the light-wells; for the most part, however, it appears that water was poured into the toilets to flush them. In fact, Evans acutely noted the existence of sufficient space for placing a large pitcher at one end of the seat at Knossos and concluded with evident delight: "As an anticipation of scientific

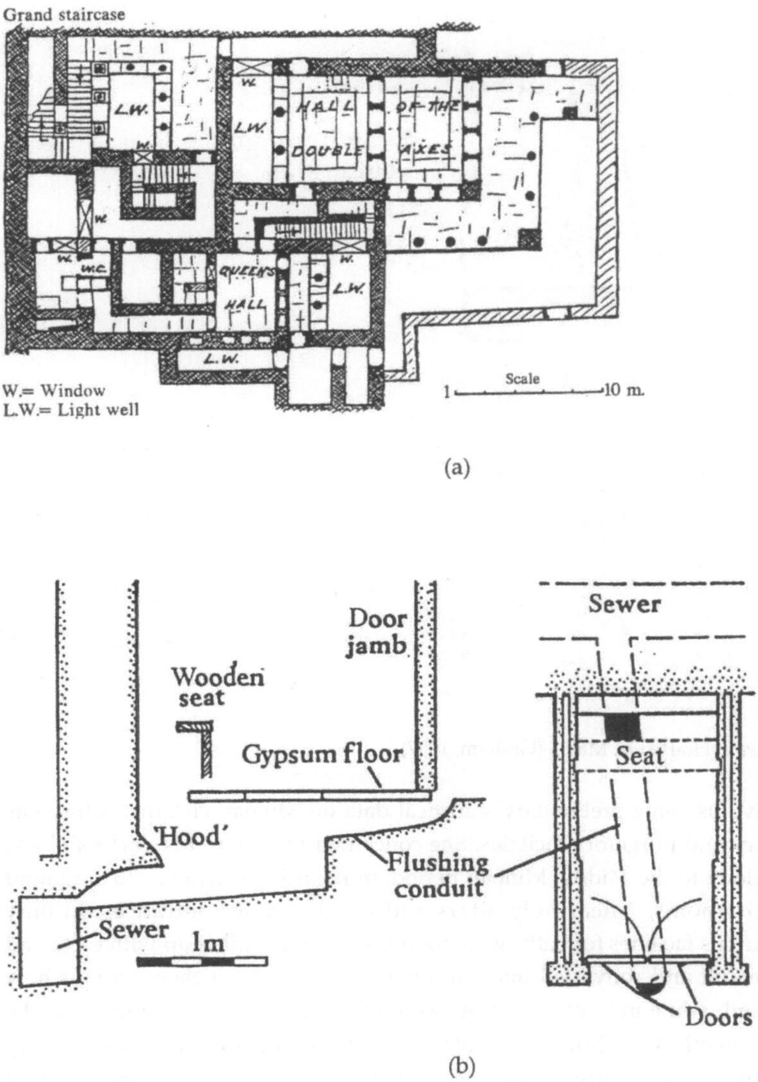

Figure 8-9. Residential Quarter, Palace of Minos, North at top: (a) General plan, (b) Detailed (section and plan of ground-floor toilet)

methods of sanitation, the system of which we have here the record, has been attained by few nations even at the present day" (Evans, 1964).

Similar bathrooms have been reported in other Minoan sites. One of them, newly discovered in the Palace of Zakros by Platon (1974), appears in Figure 8-11. Recently, M.

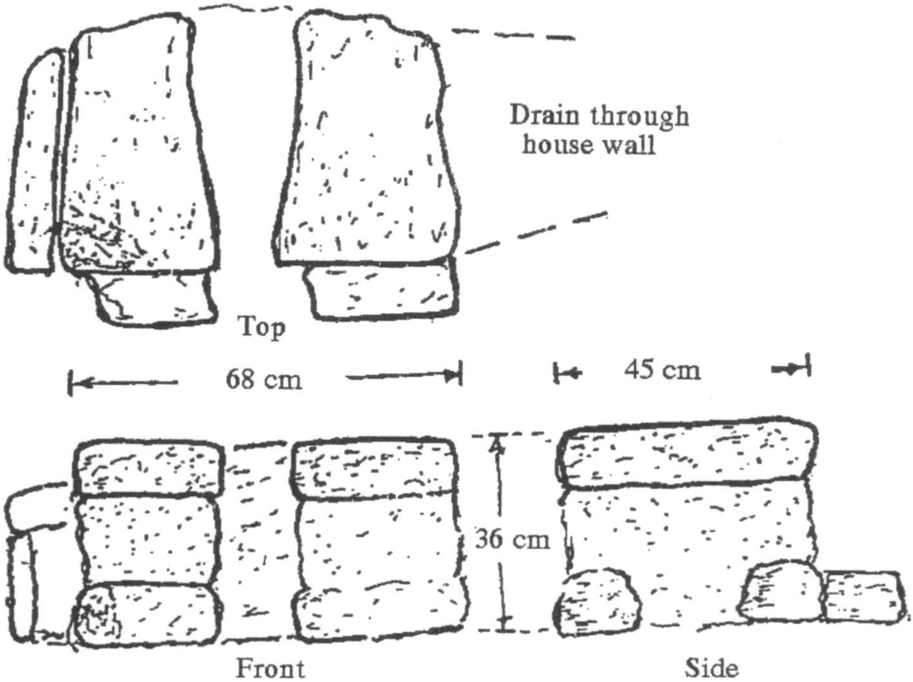

Figure 8-10. Toilet in House Da, Mallia (Graham, 1987)

Platon (1990) gave us some preliminary statistical data on Minoan cisterns, bathrooms and other sanitary and purgatory facilities. She concluded that, in terms of chronology, most of them belong to the Middle Minoan period; in regards to location, 16 are found next to domestic rooms, 7 near holy altars and two in palace entrances. In only two instances various facilities for baths were found, seven were filled up with earth and two had been rebuilt and converted into bathrooms. Also, in 14 of them various holy objects were found, while in 9, cement coats were indicated (M. Platon, 1990). Finally, frescoes related mainly to various holy subjects, were found only in two purgatory cisterns located in the south house of the Knossos Palace and in one cistern-bathroom at the northeastern part of the Zakros Palace. It must be noted that, in spite of the view of some scholars (Alexiou, 1964), the absence of bath facilities in some purgatory cisterns and their existence in other places, should not be considered coincidental. Graham (1987) and Platon (1974), have reported that purgatory cisterns were used for the cleansing of both body and soul. Note also that most Minoan baths were connected to independent septic systems in the outside, a practice indicative of the advanced water resources management and environmental techniques of that period.

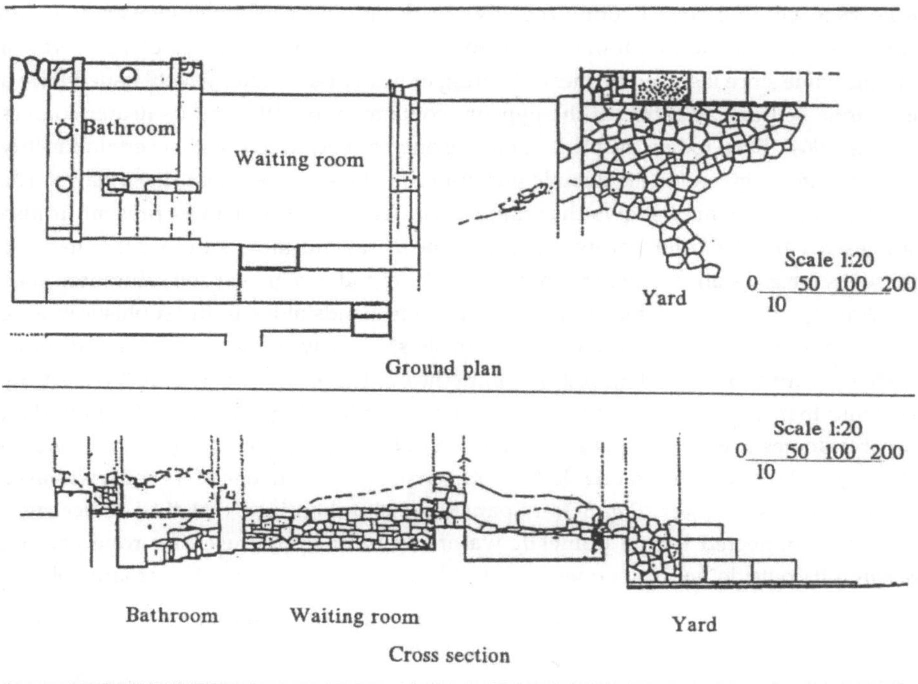

Figure 8-11. Bathroom in the Palace of Zakros (Platon, 1974)

In Minoan Crete, matters relating to the flow of water were practically addressed since early times but it is difficult for us today to fully understand hydrostatic problems empirically solved by the Minoans. Already, in the elaborate construction of the water supply systems of the earliest age of the Knossos Palace, advantage had been taken of the natural law by which water finds its own level, and water pipes had been devised that anticipate the devices of modern water resources engineering technology. At the beginning of the New Era, during which the "House of the Frescoes" was constructed, such a refinement of the water resources science as the use of parabolic paths for water-channels, is repeatedly illustrated among the remains of the Palace. It has been noted with regard to the foot-washing basin of the "Caravanserai" that its builder positively revelled in the use of pipes for water supply and his provisions for their subsidiary distribution of water for overflow, as well as for evacuation and sewage (Evans, 1964). However, it should be remembered that Minoan engineering was a creative art rather than a profession based on science.

A plan of the small section of the Minoan water-course explosed by de Jong is given in Figure 8-12 (Evans, 1964). It seems that the conduit, which was roofed over by slabs, follows a winding course along the edge of the gully. It heads, however, towards Knossos, and was no doubt one of the sources of water-supply. The conduit was of unique value since it served the nearest spring of nearly pure water which issued from a limestone bed and was free of the gypsum contamination affecting its nearer sources (Evans, 1964). Even today, the purity of the spring has endowed it with a certain sanctity and its water is given for therapeutic purposes to patients in neighbouring villages. The Minoan water-channel shows that there was a greater flow than at present; it also suggests that the earlier spring must have been several meters higher up than the present source, a sign of a higher groundwater level and of a greater soil saturation zone.

Among all ancient civilizations Minoan Crete stands alone in the sophistication of its water resources technology which contrasts sharply with the crude practices of its contemporary and succeeding cultures, until the modern era (Marinatos, 1927). It is truly amazing that the most common water quality modification technique for providing suitable domestic water supplies was known to Minoan engineers. Thus, according to Dafner (1921), a strange, oblong device with small openings in one of its ends, was used to treat domestic water (Fig. 8-13). Spanakis (1982) theorized that this device was probably connected with a domestic water reservoir by the use of a rope passing through its outside holds. Its operation was based on coarse screening of large objects

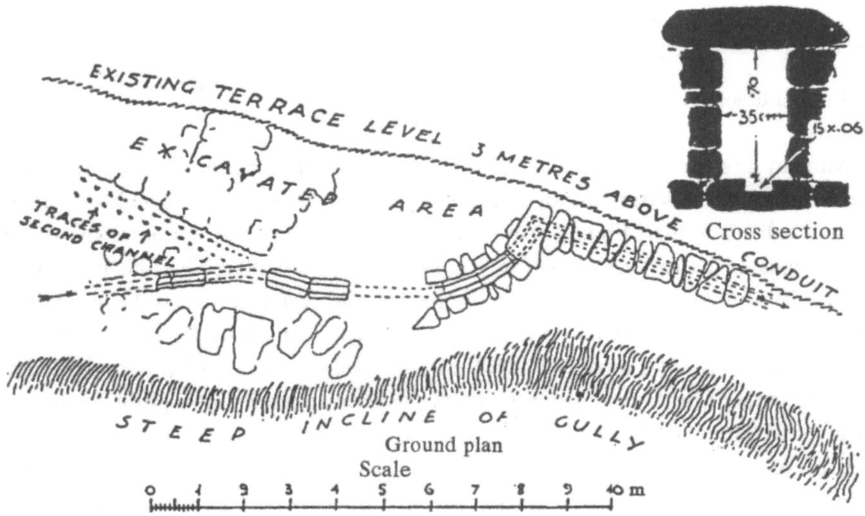

Figure 8-12. Plan of section of a Minoan conduit near the Mavrokolybos spring, west of Knossos (Evans, 1964)

Figure 8-13. - Water treatment device (Defner, 1921)

and grit. For cleaning purposes it was possible to release it from the pipe-end by loosening the rope in the holds. This device was constructed in a similar manner and with the same material as the terracota water pipes.

8-4. MUNICIPAL SEWERAGE SYSTEMS

One of the salient characteristics of Minoan civilization was the architectural and hydraulic function of its storm water and wastewater sewerage systems in palaces and cities. In the entire structure of the Minoan palace nothing is more remarkable than the elaborate sewerage system that runs throughout its domestic quarter and adjoining halls (Evans, 1964). Evans (1964) and MacDonald and Driessen (1988), described the course of these sewers and drew plans of what they considered to have been their original form. Such a plan of the entire sewerage system of Knossos is illustrated in Figure 8-14. This plan should provide the visitor with a basic orientation of the site and help him view the entire network in an integrated manner. A stone- by stone description of the sewerage system with reference to the architecture above has been given as well (MacDonald and Driessen, 1988).

Today, the sewers can be reached by a series of modern entrance holes, marked A - E on the plans. Entrance Hole A is actually the east exit of the system. In some manner, the sewers probably discharged their contents into the Kairatos River below. However, the east side of the Kephala hill is so badly eroded today that the precise manner of discharge can no longer be determined. Just inside the East Light Well of the Hall of the Double Axes, the main sewer is met by the North-South Tributary, visible from today's ground level through Entrance Hole B. The main sewer itself continues west, then west-

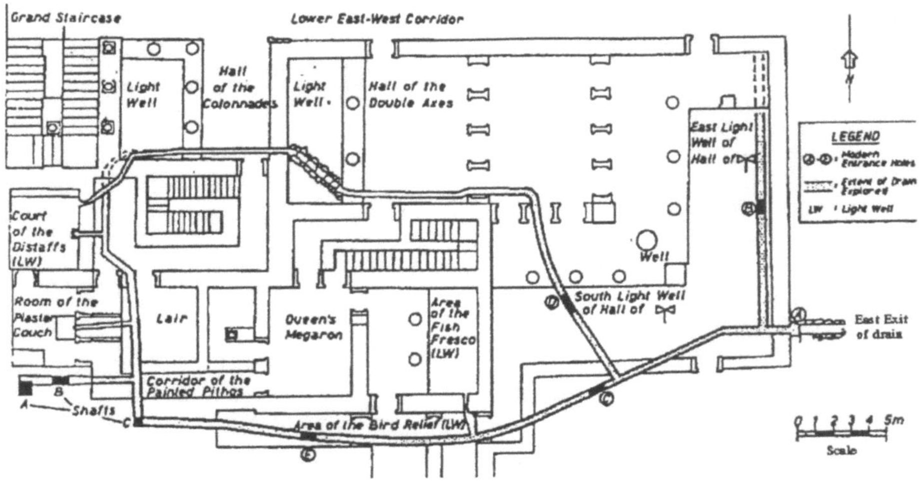

Figure 8-14. The Sewerage System beneath the Domestic Quarter (MacDonald and Driessen, 1988)

south-west to be joined by what has been termed the return of the main sewer, since the whole system from this point on forms a somewhat angular loop as it travels beneath the areas it served (MacDonald and Driessen, 1988). The main sewer curves until the outlet from the East Light Well of the Queen's Megaron enters it at the point where it is once more aligned on an approximate east-west axis. Farther to the west, the cover slabs of the sewer are visible form above, just east of Entrance Hole E. The conduit then disappears beneath the great west wall of the South Light Well of the Queen's Megaron, as part of the so-called Great Cutting. It soon meets Shaft C where it turns abruptly north, past the side sewer from Shafts A and B, and the outlet from the Toilette in the Room of the Plaster Couch. After a zigzag, the sewer once more becomes visible from above in the room east of the Court of the Distaffs. The outlet from the Court joins the main sewer just south of the point where it travels beneath the stone U-shaped channel. In the Light Well of the Hall of the Colonnades, only the U-shaped channel is visible, although the main sewer lies directly beneath. By the time the West Light Well of the Hall of the Double Axes is reached, the U-shaped channel disappears and the main stone-built sewer is once again visible, its cover slabs having been removed. Here, its orientation is north-west, south-east, but as it enters the Hall of the Double Axes, it changes to east-west. In the middle of the south side of the *polythyron* of the Hall, it turns south-south-east, passes Entrance Hole D and returns to the main sewer, thus completing the angular loop previously mentioned (MacDonald and Driessen, 1988).

The total extent of the sewerage system, including outlets and tributaries, exceeds one hundred and fifty meters. The extent to which each section has been explored by us is indicated on the general plan (Fig. 8-14). The diminutive size of the channels and other obstacles prevented a more thorough examination.

In several Minoan cities and palaces there existed well established sewerage systems, which are in good functional condition even today. Stormwater from the flat roofs of the Palace at Knossos was carried off by vertical pipes; one of these, located in the eastern wing, emptied on to a stone sewerhead from which a stone channel carried the flow of storm water (Evans, 1964).

Surface water from a part of the Central Court of the Palace of Minos was handled by a very capacious underground channel built of stone and lined with cement; it ran beneath the passage leading from the North Entrance and received several flows from various quarters. The most fully explored part of the Palace sewerage system is the portion which ran beneath the floors of the Residential Quarter (Fig. 8-9). This formed a great loop with its high point located under the Light-Well, next to the Grand Staircase, and emptied via a combined channel down the slope to the east of the Palace (Evans, 1964). In the area of the Hall of the Double-Axes and the Queen's Hall with its associated chambers, it received the wastewater of no less than five light-wells; it also served a toilet on the lowest floor, and was connected with three vertical shafts. The latter, evidently, received storm water from the roof and were probably connected with toilets on the upper floors (Fig. 8-9). The sewer was built of stone blocks lined with cement and measured about 79 by 38 cm per section. The sewers, then, were large enough to permit men to enter them for cleaning or maintainance; in fact, manholes were provided for that purpose. Airshafts at intervals also helped to ventilate sewers (Graham, 1987).

The sewerage system of Zakros was quite dense and of high water-engineering standards (Platon, 1974). As at Knossos and other Minoan cities, here we have well-preserved remains of sophisticated networks in which descending shafts and well-constructed stone sewers, large enough to permit the passage of a man, play a part. Yet, there is evidence that the entire system was inadequate in times of intense or extended storms. However, due to the privileged location of the site in a natural slope, the final disposal of wastewaters and storm waters at sea was easily attained.

Platon (1974) finds three basic types of conduits in the sewerage system of Zakros: a clay-conduit in reversed Π -shape, another built up with stones and a third, narrow type, constructed with stones but open at the top. A small section of the Zakros sewerage system as it looks today is shown in Figure 8-15.

The most advanced of all Minoan wastewater and storm water sewerage systems was discovered in Agia Triadha. It has been reported that, at the beginning of this century, when the Italian novelist Angelo Mosso visited the villa of Agia Triadha during a heavy storm, he noticed that all the sewers of the villa functioned perfectly and was amazed to see storm water come out of sewers through which a man could pass (Fig. 8-

Figure 8-15. Section of sewerage system at Zakros Palace

16). Mosso, who was also a reputable scientist, concluded that in the entire history of humankind there exists no other example of a sewerage system still functional today, 4,000 years after its original construction (Spanakis, 1982).

Among the advanced water resources techniques practiced in Minoan Crete is the storm water sewerage system found in the outer stairway of the north-east wing of the Palace at Knossos near the Kairatos river (Evans, 1964). On one side of the stairway there is a small open sewer that follows a parabolic flow instead of the line of the stairs. In this manner the storm water not only flows down smoothly without causing any erosion, but does not become a nuisance to people using the stairs (Fig. 8-17). These runnels with their parabolic curves following the turns of outer stairways, and the further arrangement of their channels, secure the precipitation of sediment in intermediary little tanks, placed at proper intervals, to allow the sediment to settle to the bottom (Michailidou, 1992). This elaborate method of collecting the maximum amount of rain-water, freed from impurities, in a downstream tank for washing or other purposes, is only one aspect of the highly skilled hydraulic knowledge attained by the Minoans. The special suitability of rain-water for washing linen, encourages the assumption that the tank was used for this purpose and that Minoan maidens may have made their way here from the Palace halls above.

8-5. DRAINAGE AND IRRIGATION OF AGRICULTURAL LAND

It is well known that agricultural developments in the Bronze Age were responsible

Figure 8-16. A section of the main sewer in Agia Triadha

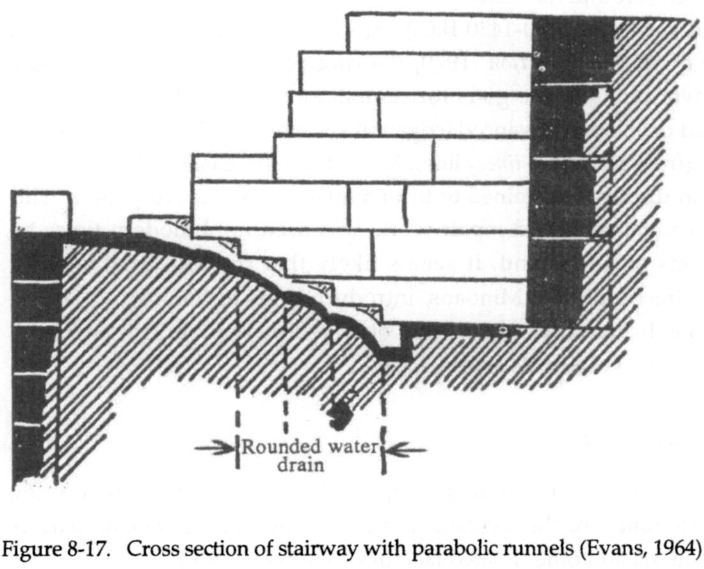

Figure 8-17. Cross section of stairway with parabolic runnels (Evans, 1964)

for the growth of large populations. Moreover, there is evidence that during the first millennium B.C. agricultural productivity in Europe for the first time reached levels capable of sustaining highly stratified societies (Sallares, 1991). Increases in agricultural productivity are the essential prerequisite for economic surpluses on which civilization is built. Thus, the beginnings of the agricultural developments that led to the creation of classical civilization should be traced to the Mycenaean and Minoan states (Lebesi, 1985).

The techniques of the practice of drainage and irrigation are probably as old as those of agriculture although the first recorded examples of drainage and irrigation do not precede the Roman Imperial epoch (Luthin, 1957). There are some indications, however, that drainage and irrigation practices were known long before the Romans. Drains and/or furrows and trenches, conduits, water-courses named ἀμάραι are mentioned in Homer (*Iliad*, 259). Thus, the names of the Cretan villages Αμάρι (Amari) and Νέο Αμάρι (Neo Amari) in Rhethymnon probably originated at a time when drainage and irrigation projects were constructed by the Minoans in that region (Papadimou, 1975). However, no traces of such projects have been found todate, although Platon (1974) believed that the Minoans had practiced irrigation. He assumed that the main region of agricultural production of the time was the Messara valley where irrigation and reclamation projects would have developed. In addition, according to Marinatos (1927) many of today's agricultural crops such as vegetables, cereals, olives, grapes and aromatic species were grown in Minoan Crete. One of the main aromatic species grown in Minoan times was *Origanum dictamus.* Vegetables were considered irrigated crops and were grown in the gardens around the residential areas.

In the Neopalatial stage (*ca.* 1700-1450 B.C.), irrigation practices were of special significance (*History of the Greek Nation,* 1970). During periods of drought people performed various invocations to the gods for rainfall and there is little doubt that farmers at that time had developed advanced irrigation systems. Best known of these are the so-called "Linies" (from the Latin *linea*=line), located in the Lasithi plateau. Here, numerous channels and ditches were joined to form a huge checker board pattern. This system, with proper maintainance and repair work, was used until modern times by succeeding civilizations on the island. It seems likely that the Minyans, another prehistoric people of Greece like the Minoans, introduced this technology to Orchomenos in central Greece. In fact, traces of it were observed there by British engineers during the reclamation project of Kopais lake in 1886.

8-6. RECREATIONAL USES

Minoan engineers had known the basic principles of water resources engineering and had developed techniques for the recreational use of water. In the Minoan palaces water jets, fountains, aquaria and other related facilities have been found.

A Minoan Fountain

Perhaps the most remarkable of all the fragments from the "House of Frescoes" are those restored to what appears to have been their original form. Although the actual summit of the fresco was not found, the object depicted in the upper part of the field is clearly some type of fountain or "jet d'eau" (Evans, 1964). The fragment below, with the same forked base and falling drops seems, moreover, to be the base of another column of water drawn in similar conventional manner, but with a small section of the contour of the ground projecting an undulating bank beneath it (Fig. 8-18).

The background here is white, the central column of water and the falling drops on either side a deep blue, while the drops falling in front of the main jet are painted white, thus becoming quite distinct and visible. As noted previously, the spout of the water itself is made to rise from a forked base (Evans, 1964). Remnants of it are exhibited in the Archaeological Museum of Heraklion.

Since no geysers or sulphurous ebullitions, like those of Palici in Sicily, exist in Crete, it is highly unlikely that these designs were copied from any natural fountains found on the island. Here we are clearly dealing with artificial "jets d'eau", of a type that could have been seen only in some Pompeian basin, since it was unknown in classical antiquity before the Hellenistic Age. Note also that such fountains were entirely foreign to ancient Egypt and Mesopotamia.

The Central Cistern of Zakros

In the Zakros Palace on the side of the Central Court, across the so-called "King's Apartment", there was a colonnade with a central pillar on a high base, while on the east a large wooden pier-and -door partition led into the spacious rectangular Hall of the Cistern. In the center of the Hall there is a circular underground cistern 7 m in diameter, with steps and waterproof cement walls, to collect water from a spring (Fig. 8-19). A screen or parapet projected from the floor, supporting a row of at least five columns set in a circle. The area above the cistern was uncovered. This installation is unique in Minoan architecture (Platon, 1974).

There are many theories regarding the function of the cistern: was it a swimming pool or an aquarium, or even a ceremonial room in which, as in Egypt for example, the sacred boat was kept? Moreover, Platon (1974) considers it very likely that the throne-room was also here.

In antiquity, it seems that the cistern served multiple purposes, including re-creational ones. Today its water is brackish (of about 8.00 dS/m EC) due to sea water intrusion resulting from the unregulated pumping and excessive use of groundwater in the area south-east of the Palace. However, there is evidence that in Minoan times groundwater in the areas adjacent to the Palace was plentiful and of very high quality.

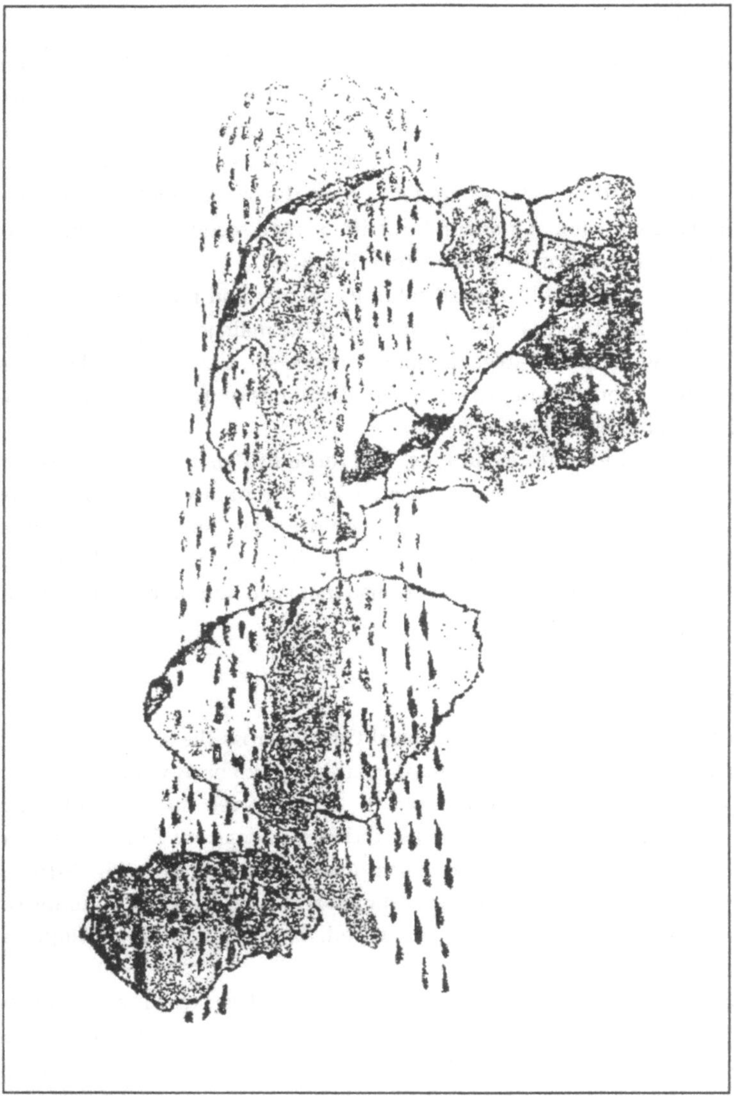

Figure 8-18. Some type of fountain (jet d'eau); fragment from the fresco heap (Evans, 1964)

8-7. CONCLUSIONS

Although little is known of the Early Minoan world, extant evidence points to a sustained cultural growth in this period of Crete's past. Paleo-vegetation information

Figure 8-19. East side with view of the cistern and the main spring of the Zakros Palace (Platon, 1964).

from pollen cores analysis shows that the climate of Crete in the Neolithic and Bronze Ages was more temperate than now. The possible disappearance of certain plant species during the Early Bronze Age must be, in part, the result of a change to a drier climate. In addition, archaeological evidence suggests that there was greater precipitation in the Middle Minoan period than now. This is manifested on the edge of a ravine about 700 meters west of Knossos which forms the upper continuation of the gully through which the Vlychia flows. Moreover, the Minoan water-channel in Knossos shows not only that there was a greater flow than at present, but also that the earliest spring was several meters higher than the present source, a sign of a greater soil saturation zone.

It is premature as yet to conclude whether the non-shortage of available water in Minoan times was due to either higher precipitation rates and better distribution, as various archaeological reports indicate, or, less likely, to a limited water demand, since the population of central and south Crete increased in the Middle Bronze Age. However, substantial evidence indicates that some oscillations of climatic conditions between the Early, Middle and Late Minoan periods occurred although their chronological duration has not been defined as yet.

Independently of these questions, during the Middle Bronze Age (*ca.* 2100-1600 B.C.) a "cultural explosion", unparalleled in the history of other ancient civilizations, occurred on the island. A striking indication of this is manifested, *inter alia,* in the advanced water management techniques practiced in Crete at that time. These included various scientific fields of water resources such as wells and ground-water hydrology, water distribution, domestic water supply, wastewater and storm water sewerage systems, construction and use of fountains, bathrooms and other sanitary and purgatory facilities, drainage and irrigation of agricultural lands and even recreational uses of water. It must be concluded, therefore, that Minoan hydrologists and engineers were aware, to some degree, of the principles of siphon and communicating vessels, Pascal's law, Archimedes' principle and other fundamental laws and basic concepts of what we call today water sciences. In the history of water resources development, pragmatic hydraulic engineering mandated by the basic needs of life and sometimes by the struggle for bare survival, is thus traceable to Minoan times.

REFERENCES

Aiginetes B. (1954). "Το κλίμα της Κρήτης και η Σταθερότης του Κλίματος της Ελλάδος". Πραγματείαι Ακαδημίας Αθηνών, 18(3): 1-46.

Alexiou S. (1964). "Μινωϊκός Πολιτισμός". Υιοί Σπ. Αλεξίου, Ηράκλειο, Ελλάς.

Bottema S. (1980). Palynological investigations on Crete. Review of Paleobotany and Palynology, 31: 20-40.

Buffe B. and Evrard R. (1950). L' Eau Potable a Travers les Ages. Editions Soledi, Rue de la Province, 37, Liege, Belgium.

Burdon D.J. and Papakis N. (1963). Handbook of Karst Hydrogeology With Special Reference to the Carbonate Aquifers of the Mediterranean Region. FAO, United Nations Special Fund, Inst. for Geology and Subsurface Res., 1 Hippokratous str., Athens, Greece.

Gastleden R. (1993). Minoans: Life in Bronze Age Crete. Routledge, 11 New Fetter Lane, London, UK.

Defner M. (1921). "Διυλιστήριο Υστερομινωϊκής Εποχής", Αρχαιολογική Εφημερίδα, 78, Αθήνα, Ελλάς.

Evans S.A. (1964). The Palace of Minos at Knossos: A comparative account of the successive stages of the early Cretan civilization as illustrated by the discoveries. Biblo and Tannen, 63 Fourth Ave., New York.

Faure P. (1988). "Η καθημερινή ζωή στην Κρήτη τη Μινωϊκή Εποχή". Μεταφρ. Ε.Ι. Αγγέλου, Νέα Έκδοση, Εκδόσεις Δ.Ν. Παπαδήμα, Αθήνα, Ελλάς.

Flaccus E. (1992). The climate and vegetation of Crete. *In*: The Aerial Atlas of Ancient Crete, I. W. Myers, E.E. Myers and G. Cadogan, (Eds.). Univ. of California Press, Berkeley and Los Angeles, California.

Garbrecht G. (1987). Hydraulics and Hydraulic Research: A Historical Review. Intern. Assoc. for Hydr. Res. 1935-1985. A.A. Balkema, Rotterdam, Boston.

Graham J.W. (1987). The Palaces of Crete. Revised Ed. Princeton University Press, Princeton, New Jersey.

Hutchinson R.W. (1950). Prehistoric Town Planning in Crete. Town Planning Review, 21: 199-220.

Huxley G. (1961). "Crete and the Luwians". Oxford, U.K.

Issar A., Tsoar H., and Levin D. (1989). Climatic changes in Israel during historical times and their impact on hydrological, pedological and socioeconomic systems. *In*: Paleoclimatology and

Paleometeorology: Modern and Past Patterns of Global Atmospheric Transport, M. Leinen and M. Sarnthein, (Eds.) Kluwer Academic Publishers, The Netherlands.

History of the Greek Nation. (1970). Εκδοτική Αθηνών, Αθήνα, Ελλάς.

Lembesi A. (1985). "Το ιερό του Ερμή και της Αφροδίτης στη Σύμη Βιάννου. Ι. Χάλκινα Κρητικά Τορεύματα". Εν Αθήναις, Αρχαιολογική Εταιρεία, Αρ. 102, Αθήνα, Ελλάς.

Luthin J.N. (1975). Drainage of Agricultural Lands. Amer. Soc. of Agron., Madison, Wisconsin.

MacDonald C.F. and Driessen J. M. (1988). The Drainage System of the domestic quarter in the Palace at Knossos. British School of Athens, 83: 235-358.

Marinatos S. (1927). "Ο Αρχαίος Κρητικός Πολιτισμός". Ξεν. Ε. Σεργιάδου, Αθήνα, Ελλάς.

Mellaart J. (1958). Myceneans and Minoans. Amer. J. Archaeol., 62: 9-33.

Michailidou A. (1992). Knossos. A Complete Guide to the Palace of Minos. Ekdotike Athenon S.A., Athens, Greece.

Moody J. (1987). "The Environmental and Cultural Prehistory of the Khania Region of West Crete: Neolithic Through Late Bronze Age. Univ. of Minnesota, Ph.D. Dissertation, Univ. Microfilm, Ann Arbor, Michigan.

Moody J., Rackham O., and Rapp G. (1990). Paleoenvironmental Studies in West Crete. Πεπραγμένα ΣΤ' Διεθνούς Κρητολ. Συνεδρίου. Φιλολ. Συλ. Χρυσόστομος, Χανιά, Ελλάς, A2: 9-27.

Papadimou D.L. (1975). "Τα Υδραυλικά Έργα Παρά τοις Αρχαίοις". Τόμος Γ', ΤΕΕ, Αθήνα, Ελλάς.

Platon M. (1990). "Νέες Ενδείξεις για τα Προβλήματα των Καθαρτηρίων Δεξαμενών και των Λουτρών στο Μινωϊκό Κόσμο". Πεπραγμένα του ΣΤ' Διεθνούς Κρητολογικού Συνεδρίου, Φιλολ. Σύλλογος "Ο Χρυσόστομος", Χανιά, Ελλάς, A2: 141-155.

Platon N. (1974). "Ζάκρος". Το Νέον Μινωϊκόν Ανάκτορον". Η Εν Αθήναις Αρχαιολογική Εταιρεία, Αθήνα, Ελλάς.

Rackham O. (1980). Ancient Woodland: its History, Vegetation and Uses in England. Edward Arnold, London.

Rackham O. (1983). Observation on the Historical Ecology of Boeotia. Annual of the British School at Athens, 78: 291-351.

Rackham O. (1990). The Greening of Myrtos. *In*: Man's Role in the Shaping of the Eastern Mediterranean Landscape, Bottena,Entjes-Nieborg and van Zeist, (Eds.). Balkema, Rotterdam, The Netherlands.

Sallares R. (1991). The Ecology of the Ancient Greek World. Cornell University Press, Ithaca, New York.

Σπανάκη Σ.Γ. (1981). "Η Ύδρευση του Ηρακλείου 828-1939". ΤΕΕ/ΤΑΚ, Ηράκλειο, Ελλάς.

Thirgood J.V. (1981). Man and the Mediterranean Forest: A History of Resource Depletion, 19-26. Academic Press, New York and London.

Turner J. (1982). The *Tilia* Decline: an anthropogenic interpretation. New Phytologist, 61: 328-341.

CHAPTER 9

RESPONSE OF SOME HYDROGEOLOGICAL SYSTEMS IN CENTRAL ITALY TO CLIMATIC VARIATIONS

WALTER DRAGONI

Dipartimento di Scienze della Terra, Università di Perugia ,
Piazza Università 1, 06100 Perugia, ITALIA

9-1. INTRODUCTION

Aims of the Work

One of the aims of this article is to present a series of information on the overall variations of the water surplus in Central Italy over the last 3000 years. Water surplus is understood here to be the water in the hydrological balance which is available as surface runoff and aquifer recharge. The other aims of this work are to describe the present climatic trends and, on the basis of these and that which has taken place in the past, to attempt to get an idea of what is likely to occur in the near future. A schematic map of central Italy and the location of the places mentioned in following is given in Figure 9-1. The data and conclusions given in this article refer only to the area west of the Apennine chain, roughly at a latitude between 41° and 43° 30′ N: at this time the results given here cannot be considered as applicable outside this area, especially north of the Apennine water divide.

Physical Setup of Central Italy

A schematic lithological map of central Italy is shown in Figure 9-2. In central Italy there are physical environments which differ considerably from each other, as indicated in Figures 9-1 and 9-2. From a geological standpoint, there are volcanic, terrigenous and calcareous rocks. The elevations go from sea level to the almost 3000 meters above sea level of the highest Apennine mountains. Average temperatures and precipitations differ considerably from one place to another: this is shown in Figure 9-3 which gives the average monthly rainfall and temperature for two representative stations located close to the sea and high in the Apennines. The hydrological systems in central Italy therefore have widely differing climatic conditions and differing rain-yield transformation mechanisms: thus the effects of the climatic variations should be expected to be considerably different from one system to the next.

NATO ASI Series, Vol. I 36
Diachronic Climatic Impacts on Water Resources
Edited by Andreas N. Angelakis and Arie S. Issar

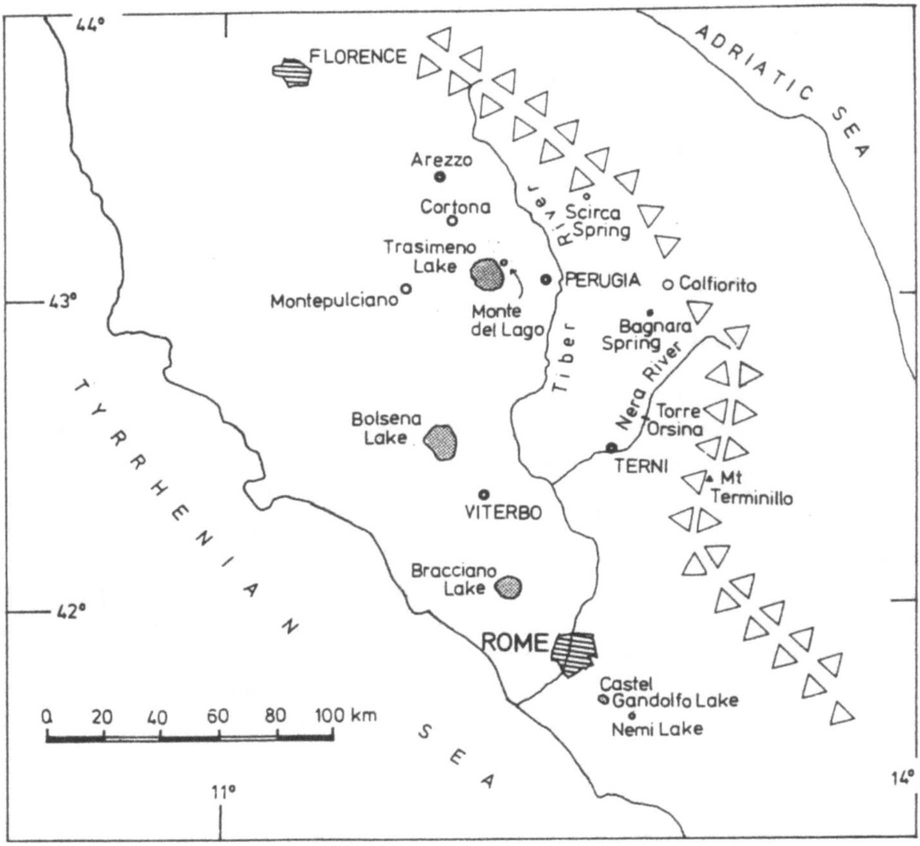

Figure 9-1. Sketch of central Italy. All the localities mentioned in the text are shown

9-2. THE PAST

General Considerations

A large amount of information regarding the climatic events of the past and their effects on the hydrological systems,can be had by researching diaries, reports and historical accounts regarding the weather or factors directly connected with it, such as the dates of crop or grape harvests (cf. for instance Le Roy Ladurie, 1972).

In this regard the information given in the following is only an infinitesimal part of that which could be of use in reconstructing the climatic variations and their effects on

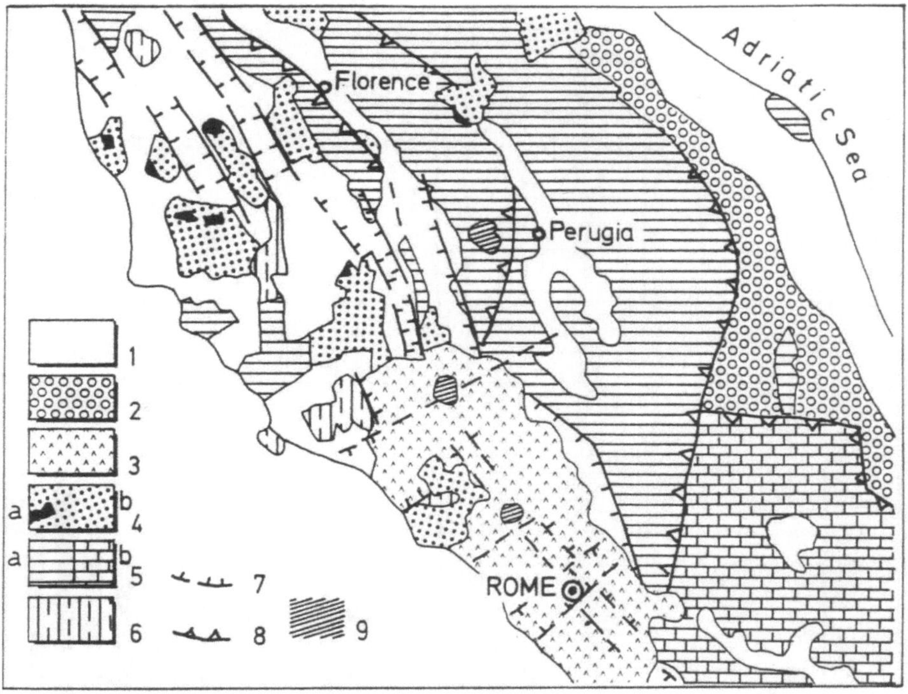

Figure. 9-2. Geological Sketch of Central Italy. 1) Post-orogenic succession; mainly unconsolidated sediments. 2) Late orogenic succession; mainly sandstone and flysch. 3) Quaternary volcanic rocks. 4) Mesozoic pelagic succession, with alpine ophiolites (a). 5) Continental margin deformed by alpine orogenesis; 5a = Mesozoic basins and pelagic shallow (and related Cenozoic flysches); 5b = Mesozoic carbonate platforms (and related Cenozoic flysches). 6) Pretriassic basement, metamorphic. 7) Regional normal faults. 8) Overthrusts. 9) Lakes (Redrawn after Parotto & Praturlon, 1979; Praturlon, 1980)

the environment in Italy: besides the possibility to collect new data, there is a huge amount of dispersed palynological, sedimentological and archaeological data, only part of which is in print. If these were put together and studied from the viewpoint of climatic variations and water resources, most probably they would provide results relevant to the understanding of that which took place in the past and that which could take place in the near future. Sometimes monuments can also provide information on climate: a nice example of this is the following. In the main square of Perugia there is the "Fontana Maggiore", a fountain built between A.D. 1277 and 1278 (Nicco Fasola, 1951). The fountain is decorated with a series of bas-reliefs depicting saints, arts and

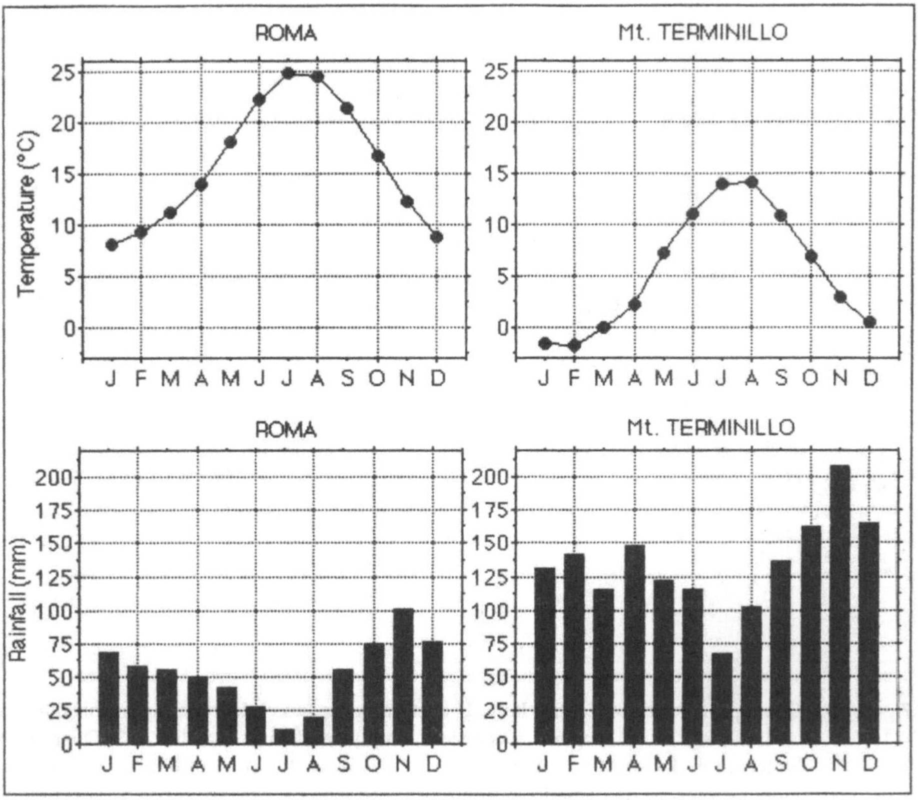

Figure. 9-3. Average monthly temperature and rainfall in Rome (51 m a.s.l.) and Mt.Terminillo (1750 m a.s.l.)

sciences,and the 12 months of the year with the main activities of each month. The panels for September indicate that one of the main activities for September is the grape harvest (Fig.9-4). Grape harvest time depends mostly on the average temperature and the insolation of the late winter - early autumn months; (Le Roy Ladurie, 1972), at the present time, the grape harvest usually takes place in October, which is considered to be the "month of the grape". The Fontana Maggiore of Perugia therefore suggests that at the time of its construction the average weather between late winter and early autumn was warmer than it is today. In any case, this suggestion should be submitted to further research, taking into account the changing of grape varieties, and that up until 1582 the

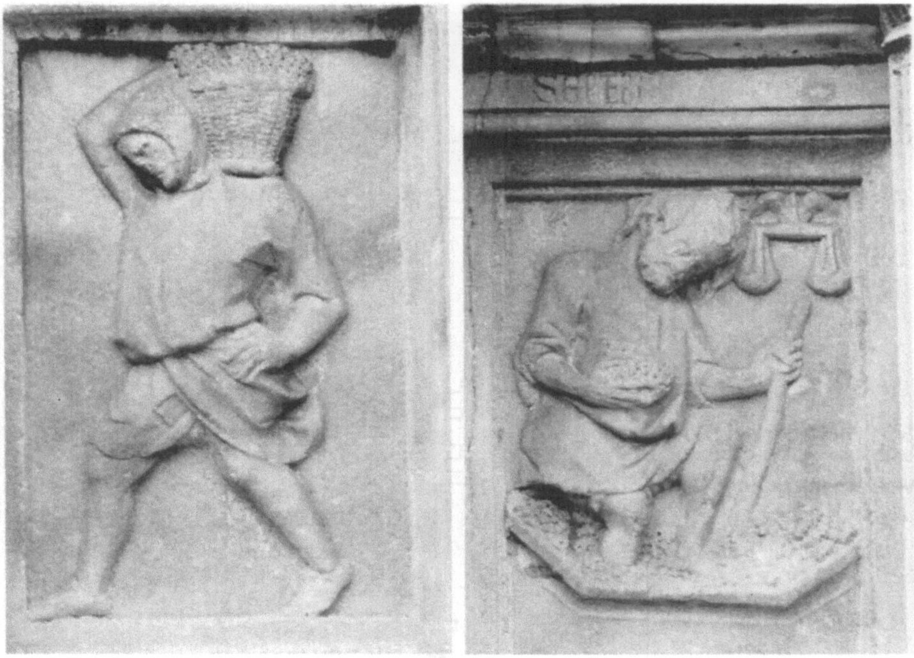

Figure 9-4. Bas-reliefs of the "Fontana Maggiore" in Perugia (built around A.D. 1277). The panels show that at that time the grape harvest took place in September. Presently the grape harvest takes place in October; this shows that at the time the fountain was built climate was warmer than today

Julian calendar was used, which differs by seven days from the calendar in use today: in the present calendar, September 1, 1280 would today correspond roughly to September 7 (Moyer, 1982).

Another important source of information for the reconstructing of climatic variations and their effects on the hydrological systems is the study of lacustrine systems. In central Italy, the craters and calderas in volcanic areas and karst depressions in carbonate zones have allowed for the presence of numerous lakes, many of which have no natural surface outlets. In a given period of time, the average level of a lake (especially if there is no outlet) is strictly tied to the average water surplus, i.e. to the average rainfall and temperature conditions. Information is provided in following regarding the past levels of various lakes in central Italy. This information, put into terms of "wet" or "dry" in relation to present conditions, is compared in Figure 9-5 with a time sequence obtained by assembling data provided by Dansgaard *et al.*(1969), Issar (1992) and Cremaschi (1992).

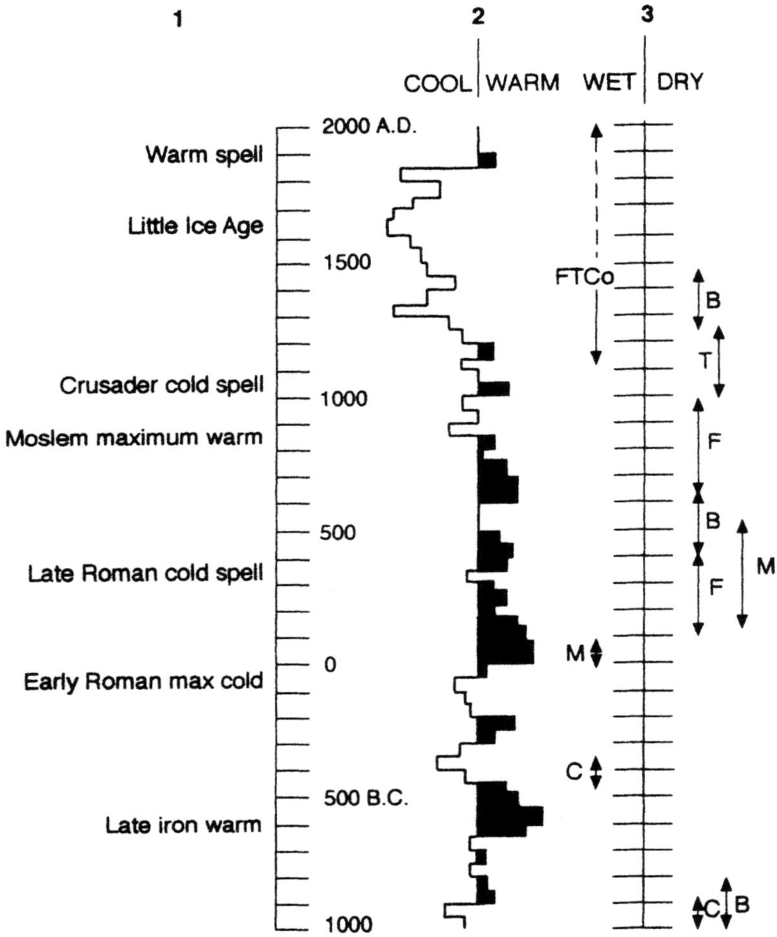

Figure 9-5. Climatic changes in the last 3000 years: 1) Climatic changes according to the "Base section for climate change in the Levant" (redrawn after Issar, 1992). 2) Climatic changes according to the Camp Century ice core (redrawn after Dansgaard *et al.*, 1969; Cremaschi, 1992). 3) Response of the lakes in central Italy according to the available data: C = Castel Gandolfo L.; M = Martignano L.; B = Bolsena L.; F = Fucino L.; T = Trasimeno L.; FTC_O = Fucino L., Trasimeno L. and the Colfiorito area

Castel Gandolfo Lake

Rome is located in a volcanic area whose center is the volcanic complex today known as the "Colli Albani" (Alban Hills), as shown in Figure 9-6. The main volcanic cone is surrounded by several craters, which were naturally destined to become lakes. These crater-lakes have no natural outlets and have a very small basin. In order to control the levels of the water or to drain out the smallest lakes, the people living in the area dug outlet tunnels, which can be dated between B.C. 500 and 300. The lake of Castel Gandolfo (or Lake Albano) is the largest lake of the area, and has an underground outlet still in use today, built around B.C. 400 (Castellani and Dragoni, 1992). According to Guidi (1986), at a depth of 8-10 m below the present water level, there are traces of a late bronze settlement (B.C. 1150-900). The bottom of the outlet tunnel is about 3 to 4 m above the present level of the lake: given that the tunnels were designed to carry water with a free surface, and that no pumps were available, it must be supposed that the elevation of the bottom of the tunnel is some meters lower than what the lake's average level was at the time of construction. This means that between B.C. 900 and 400 the level of the lake increased by many meters. A shifting of the climate towards wetter conditions some time after the late Bronze Age is documented independently by other archaeological evidence (Potter, 1979).

Martignano Lake

The small Martignano Lake is located N of Rome, very close to the much larger Bracciano Lake (Fig.9-7). In the year B.C. 2 an aqueduct was built to carry water to Rome (Aqua Alsietina). The aqueduct begins with a tunnel, the bottom of which is about 12 m above the present level of the lake. As in the case of Castel Gandolfo Lake, it must be supposed that the elevation of the bottom of the tunnel is some meters below the average level of the lake at the time of the digging of the tunnel. At a depth between 8 and 17 m below the present level of the water there are remnants of trees on the shore of an old coastline. Carbon dating tells us that the trees were alive between A.D. 200 and A.D. 500 (Liberati Silverio, 1986; Mocchegiani Carpano, 1976). This means that between B.C. 2, and about A.D. 200 the level of the lake dropped approximately 30, with a shift towards a "dry" condition in the climate. This situation persisted at least until A.D. 500. In order to reclaim some land, at the beginning of the last century some hydraulic works were carried out in the area and thus the relationship between the average level of the lake in B.C. 2 and the present average level cannot be attributed to climatic variations.

Bolsena Lake

Lake Bolsena is located N of Rome, in a volcanic caldera (Fig. 9-8). The morphometric analysis of the lake points out the existence of an ancient shore, now submerged under several meters of water. Near Gran Carro, at a depth of about 5 meters below the present level of the lake there is a protohistoric settlement, discovered by A.

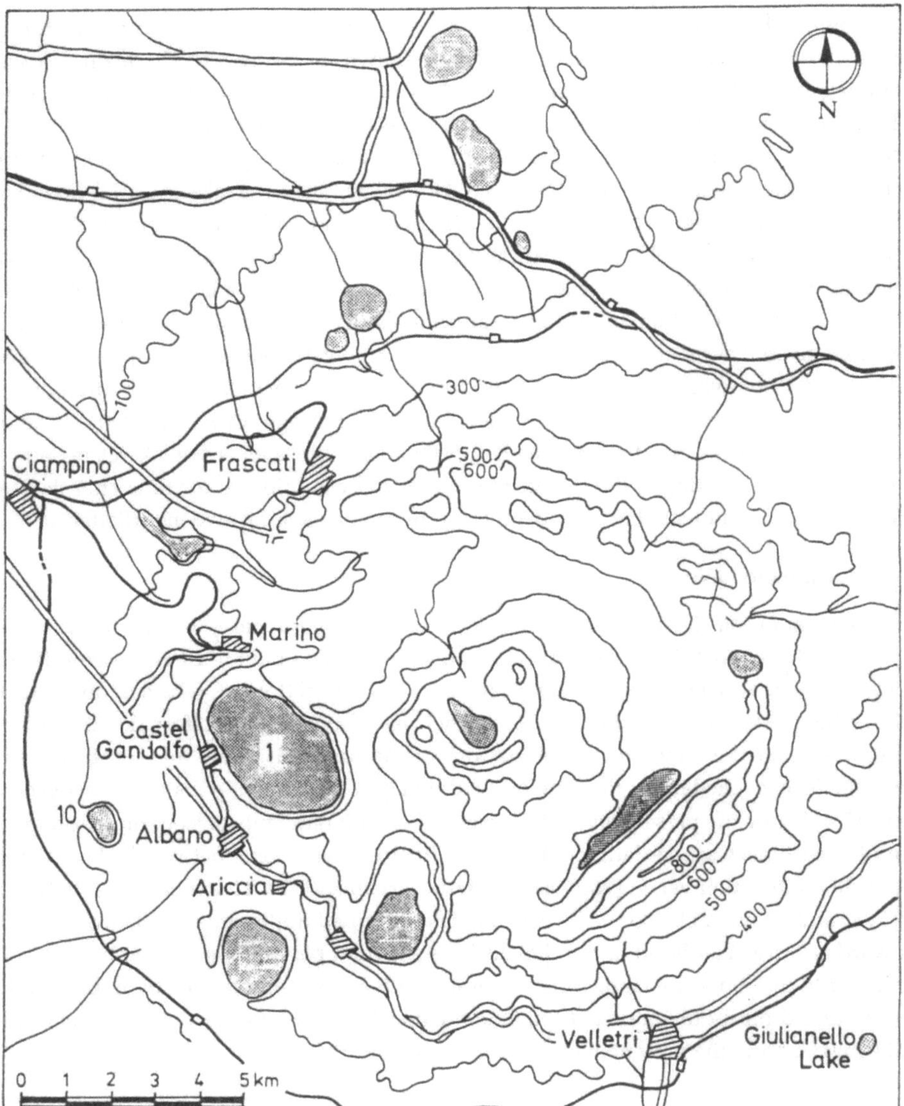

Figure 9-6. Alban Hills area. All the craters shown in the map have an artificial outlet. 1) Castel Gandolfo Lake. In the majority of cases the digging of the underground or surface canals should have occurred between B.C. 500 and 300 (Castellani and Dragoni, 1991)

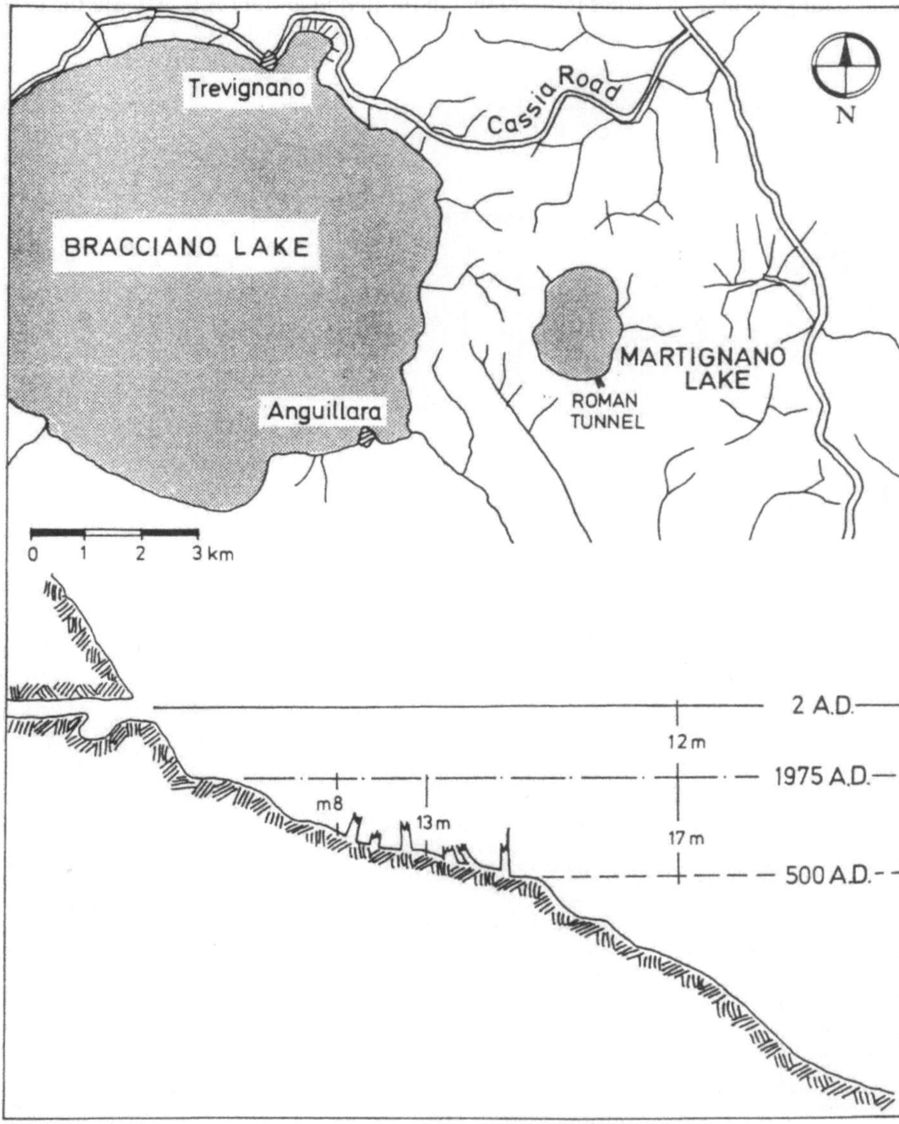

Figure 9-7. Martignano Lake (redrawn after Mocchegiani Carpano,1976)

Fioravanti in 1959 (Tamburini, 1986). Numerous carbon datings carried out by the University of Rome (Belluomini *et al.*,1992) date the material over a wide range, from B.C. 1670 to 780, with a particular abundance of remains between B.C. 1000 and 900. Other datings in other localities indicate that in the period around A.D. 415-615 the level

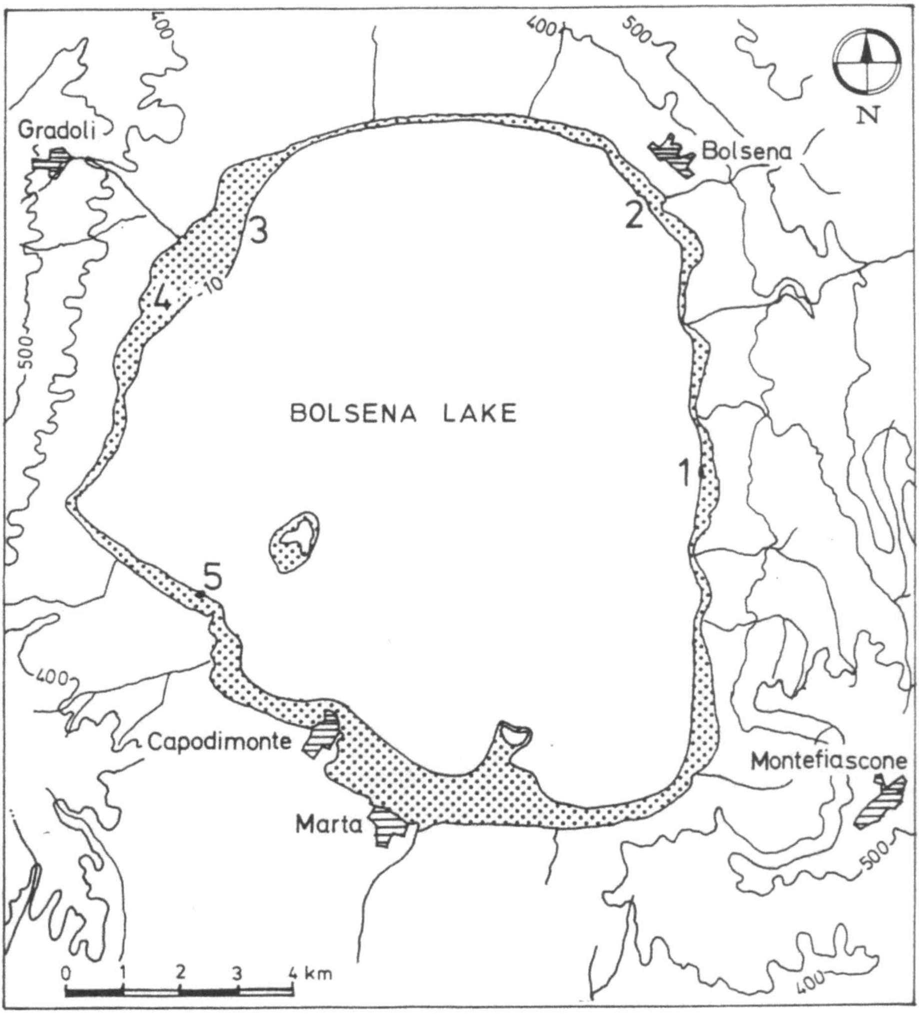

Figure 9-8. Bolsena Lake. Dotted area shows an old shore.Archaeological sites: a) Gran Carro; b) Civita; c) Ragnatoro; d) Mt.Senano; e) Bisenzio sub (Tamburini, 1986)

of the lake was 6 m lower than today. Another set of samples indicates, on the average, that between A.D. 1250 and A.D. 1450 the level was 7.5 - 9.6 m lower than today (Belluomini *et al.*,1992).

Fucino Lake, Trasimeno Lake and Engineering Works after A.D.1400

Starting from about A.D. 1300, there is a fair amount of documentation on works (or work plans) for draining water from lakes and swamps in central Italy. What is most striking is the fact that for a few centuries before this time, colder than the preceding times, there are no reports of attempts of this kind. This could be interpreted as a "casual" lack of evidence; however,it is this author's opinion that in some circumstances the total lack of evidence is evidence in itself.

Lake Fucino and Lake Trasimeno are the two largest lakes that the Romans tried to regulate by means of artificial outlets. Lake Fucino, which was drained in 1876, had a basin of 840 km^2 and (in the last century) had an area of about 150 km^2. Lake Trasimeno has a basin of 309 km^2 and an average area of approximately 122 km^2. The technical capabilities in ancient times made practically impossible the digging of tunnels with sections and flow capacities sufficient for controlling the levels of lakes with basins and areas of this size. This implies that the fluctuations in the levels of Lake Fucino and Lake Trasimeno were due essentially to natural causes rather than to the effects of known ancient outlets (Castellani and Dragoni, 1990 and 1992).

The Roman tunnel from Lake Fucino was completed in A.D. 52. Roman historians tell that the tunnel did not work well, and, with some reason, they attribute this to poor execution of the work. Later on (around the end of the 1st century A.D. and the first half of the 2nd century A.D.) it is reported that the tunnel was restored by the Emperors Traianus and Adrianus; the latter lowered the level of the intake gate, and new farming land was reclaimed. As mentioned before, the lowering could not have been caused by water drained out by the tunnel, so it must be concluded that it was caused by a "dry" period. For several centuries there is no direct information regarding the level of the lake. According to Brisse and De Rotroù (1876) various ancient reports indicate that in 7th, 8th and 9th centuries A.D. the levels of the lake must have been low. According to Muratori(in D'Amato, 1980) in A.D. 1177 there was an exceptional increase in the water level. All this suggests that during the warm Middle Ages period the level of the lake was low, and that around A.D. 1177 the climate was beginning to shift again towards wetter conditions. In 1239-1240 Emperor Frederick II ordered that an attempt be made to put the Roman outlet back into operation so as to lower the lake level. From that point on the level of the lake remained almost constantly above the level it had had during the warm Middle Ages period, and numerous studies were made for lowering the level. Brief periods with low levels were reported around the middle of the 17th century and in 1752. In 1816 the lake reached the highest level recorded in the 19th century: 16 m above the low level of 1752. This matches the definition of 1816 as the "year with no summer" (cf. Bourke, 1984).

Regarding Lake Trasimeno, there is no definite information on the Roman outlet, other than a quotation of Strabo, (*ca.* A.D. 7-20) and the discovery of traces of ancient works at the end of the 19th century, when a new outlet (which now controls the level of the lake) was built (Dragoni,1982). It appears that before the 12th century A.D., the area of the lake was smaller than it was in A.D. 1300-1400 and than it is today. According to Cialini (1991), this is indicated by old documents reporting the names of villages which later disappeared (presumably under the water) and by the amount of wheat produced annually by farming land now flooded. All this is supported by the presence of early medieval ruins, today under water, but still visible in exceptionally dry periods and described in the last century. Between 1300 and 1400, frequent flooding (a rising, in fact, of the average lake level) had devastating effects on the area's economy, and around 1420 a new outlet was dug in a fruitless attempt to control the level. Given that the flooding continued, around 1490 some streams that emptied into the lake were diverted towards the nearby Chiusi lake. This attempt also failed and, in alternating phases, the average lake level remained higher than it was in the early Middle Ages and is today. There were particularly high levels in 1602, between 1762 and 1773, and between 1810 and 1820, the latter two in concomitance with particularly cold periods recorded throughout Europe (Lamb, 1977). There are various reports of other works for controlling the levels of lakes and closed depressions which took place in Central Italy starting around the 13th century A.D.. In many cases the documentation is uncertain, but it all enters within the overall picture of the "little ice age" and of a consequent "wetter" period which occurred at the end of the Middle Ages. As a last example, there are the works carried out in the area of Colfiorito and Camerino, in the middle of the Appennines (Marchegiani, 1976; Metelli, 1983 and Sensi, 1984). Here, between A.D. 1458 and 1471 a tunnel was dug in order to drain out the Plestina plain (Fig. 9-9), and other minor works were carried out in nearby localities (reclamation of the Montelago plain). Around A.D. 1570 a channel was dug in order to lower the water level of Colfiorito swamp. It has to be noted that most of these works were successful because the drained areas measured at most a few square kilometers.

Some Conclusions

The pieces of information given above have little significance individually: however taken as a whole they give indications which are fairly consistent and which to a large degree can be correlated with the known climatic variations.

The available data clearly indicates that, due to unknown natural causes, in the last 3000 years considerable variations took place in the hydrological cycle in central Italy. In addition, Figure 9-5 suggests that, in the area in question, the warm phases have corresponded to a decrease in water surplus, whereas the cold phases have corresponded to an increase. This indication is not entirely trivial, given that in other areas in

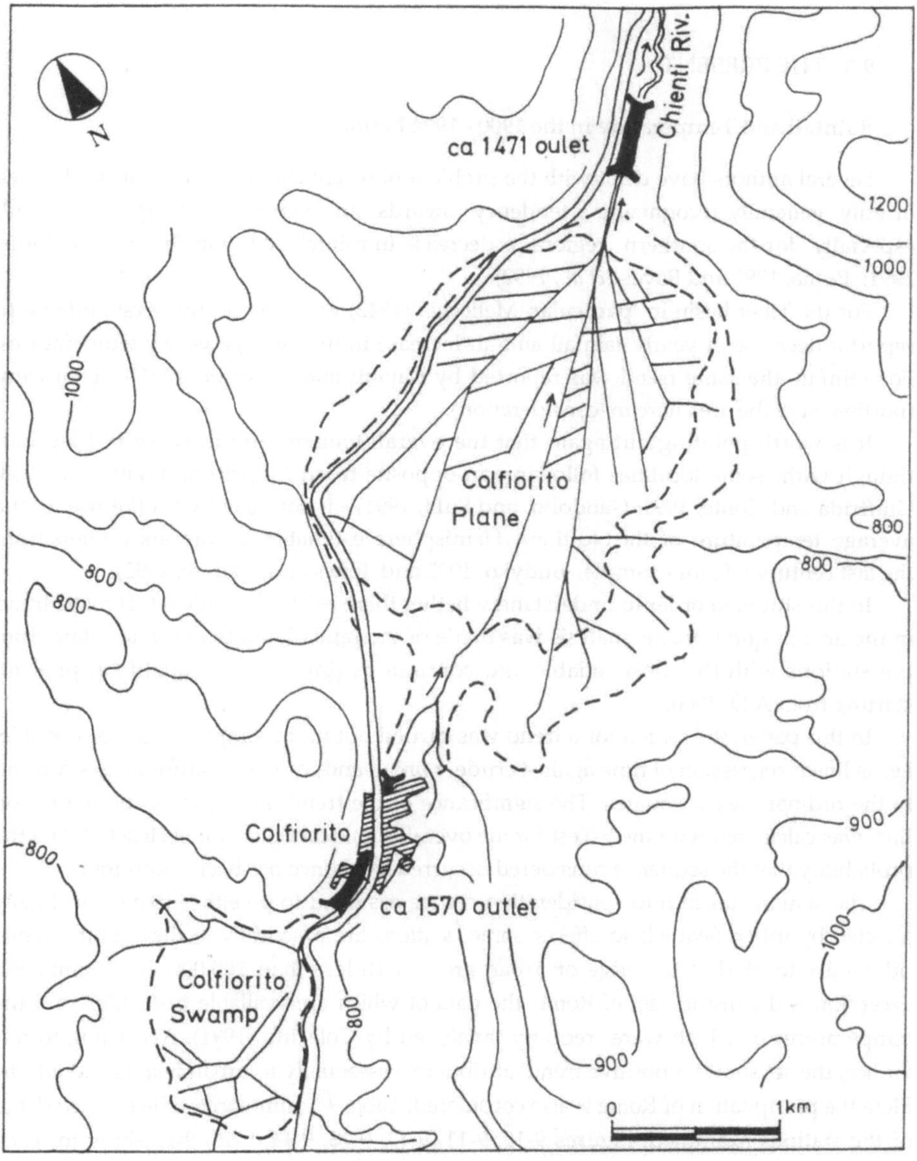

Figure 9-9. The Colfiorito area and the outlets dug in A.D. 1458-1471 and A.D. 1570

the Mediterranean, phases warmer than the present have corresponded to wetter periods (Lamb, 1977).

9-3. THE PRESENT

Rainfall and Temperature in the 1900 - 1992 Period

Several authors have dealt with the problem of recent climatic variations in the area of Italy, generally recognizing a tendency towards an increase in temperature and, especially for the southern regions, a decrease in rainfall (cf. Giuffrida and Conte, 1991; Pinna, 1991 and Pavese *et al.*, 1992).

For the Tiber basin in particular, Melicchia (1942) was one of the first authors to report a decrease in yearly rainfall and an increase in the average yearly temperatures. For rainfall, the same trend was reported by Rapetti and Vittorini (1991) for Tuscany (northwest of the area here in consideration).

It is worth pointing out again that the overall temperature increase in Italy - although with some localities following an opposite trend (Rapetti and Vittorini, 1988; Giuffrida and Conte, 1991; Gandolfo and Sulli, 1991) - is consistent with the rise in the average temperature of the Northern Hemisphere estimated by various authors from the last century (cf., for example, Budyko, 1982 and Jones and Bradley, 1992).

In this study, in order to understand whether there exists a significant climatic trend in the area in question, an analysis was made of the rainfall and temperature data from the stations with the most reliable and continuous time series available at present, starting from A.D. 1900.

In this paper, the search for a trend was carried out in the simplest manner possible, i.e. as linear regression of time against crude rainfall and/or temperature data according to the ordinary least squares. The significance of the trend indicated by the regression line was calculated with the F-Test for the overall significance, i.e. for each set of data the probability that the sequence considered occurred by chance has been computed.

The stations taken into consideration can be assumed to be either free or negligibly affected by urban heat island effects: some stations are practically in the country, while others are located at the edge of urban areas with less than 100,000 inhabitants. An exception is the historic set of Rome, the data of which are available from 1782, and the temperatures of which were recently analyzed by Colacino (1991). According to this author, the set shows a positive trend, attributable essentially to urban heat island effects. Here the precipitation of Rome is also considered. Table 9-1 summarizes the characteristics of the stations examined. Figures 9-10, 9-11, 9-12, 9-13, 9-14, 9-15, 9-16 show for each station the yearly rainfall and the average yearly temperatures, their mobile average of order 6 and, to visualize the trend, the relative regression line. Table 9-2 give the average of the first and the second halves of the available data, the regression line

Table 9-1. Characteristics of the examined meteorological stations

Station	Lat.N	Long.E	Elevation	Time spell	T	R
Arezzo	43° 28′	11° 53′	277 m	1900-92	14.0	812
Perugia	43° 27′	12° 21′	493 m	1900-92	13.2	878
Cortona	43° 16′	12° 04′	393 m	1925-92	14.9	816
Monte del Lago	43° 09′	12° 10′	295 m	1924-89	13.7	749
Montepulciano	43° 05′	11° 47′	605 m	1928-92	13.8	-
Terni	42° 34′	12° 39′	170 m	1951-92	15.4	943
Roma	41° 54′	12° 29′	51 m	1900-92	16.0	698

T= average yearly temperature (°C); R= average yearly rainfall (mm)

gradients and the overall significance of the regression line for each station. Surprisingly, in spite of the urban effect, Rome does not show any steeper temperature gradient. Figures 9-17, 9-18 and 9-19, show a plot of average yearly temperatures against yearly rainfall for all stations where both these data are available: for 5 stations over a total of 6 stations the regression is significant. The exception is the Monte del Lago station, located on the shore of Trasimeno Lake, which does not show any significant trend. As a whole, the stations analyzed indicate a regional tendency towards a rise in temperature and a decrease in rainfall.This result is in agreement with that found in the literature, even when research on the trend was carried out on fewer stations and using techniques different than those used here, such as the Spearman and Mann-Kendall tests (Giuffrida and Conte, 1991).

Table 9-2. Trend of temperature and of rainfall

	A		B		gradients		l.s.	
Station	T	R	T	R	T	R	T	R
Arezzo	13.9	809	14.1	815	0.005	0.197	0.016	0.738
Perugia	12.9	912	13.4	842	0.009	-1.74	0.001	0.009
Cortona	14.8	848	15.0	785	0.005	-3.61	0.357	0.003
Monte d.Lago	13.5	721	13.7	773	0.002	0.787	0.775	0.496
Montepulciano	13.6	-	14.0	-	0.008	-	0.075	-
Terni	15.2	989	15.7	898	0.017	-5.56	0.009	0.016
Roma	15.9	786	16.0	612	0.005	-3.82	0.012	0.016

T = average yearly temperature (°C); R = average yearly rainfall (mm); A = first half of the data; B = second half of the data; l.s. = level of significance. The regression line gradients are given in °C/year and mm/year.

Water Yield Trend in Some Hydrogeological Systems

The existence of a tendency towards an increase in temperature and a decrease in rainfall implies a decrease in the water surplus.To understand the importance of this

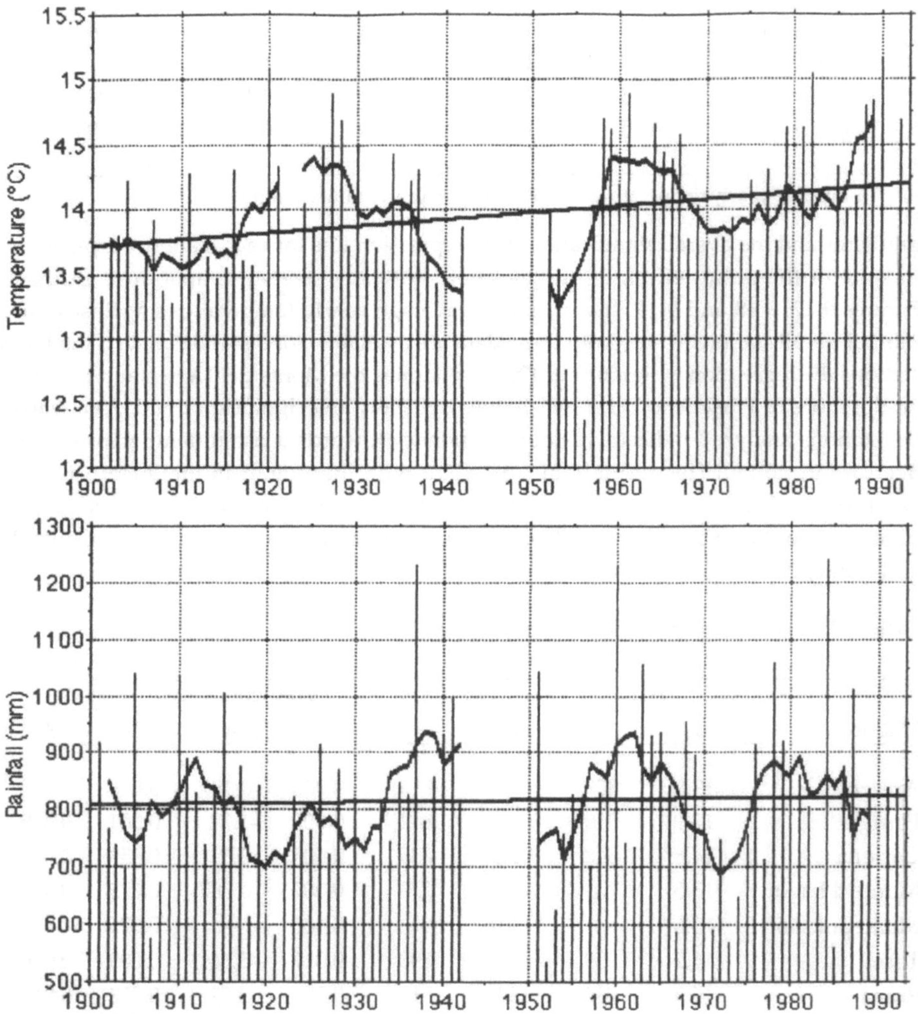

Figure 9-10. Yearly average temperature and yearly rainfall of the Arezzo station

from a hydrological viewpoint, an analysis was made of several decades of data on the water yield of three hydrogeological systems (two springs and a river) where the regimen of the flows is essentially natural (Angelini, 1992; Cencetti *et al.*, 1989). The three systems are the following:

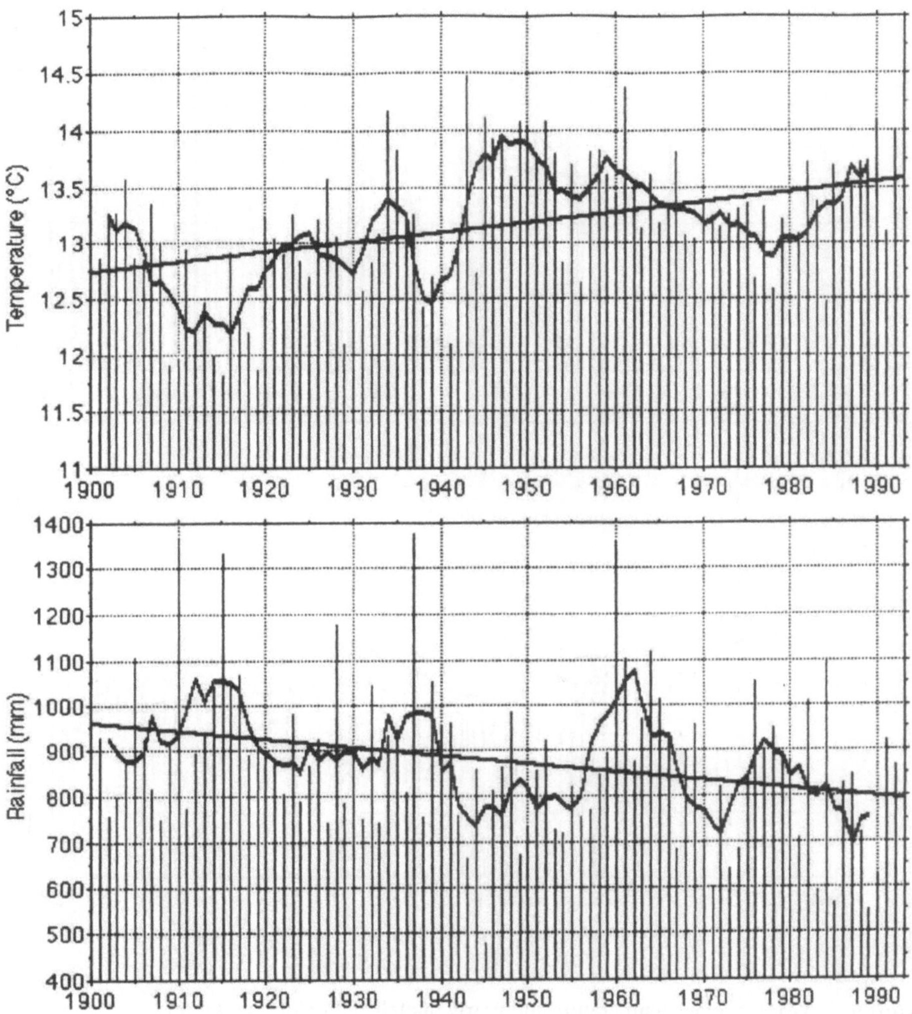

Figure 9-11. Yearly average temperature and yearly rainfall of the Perugia station

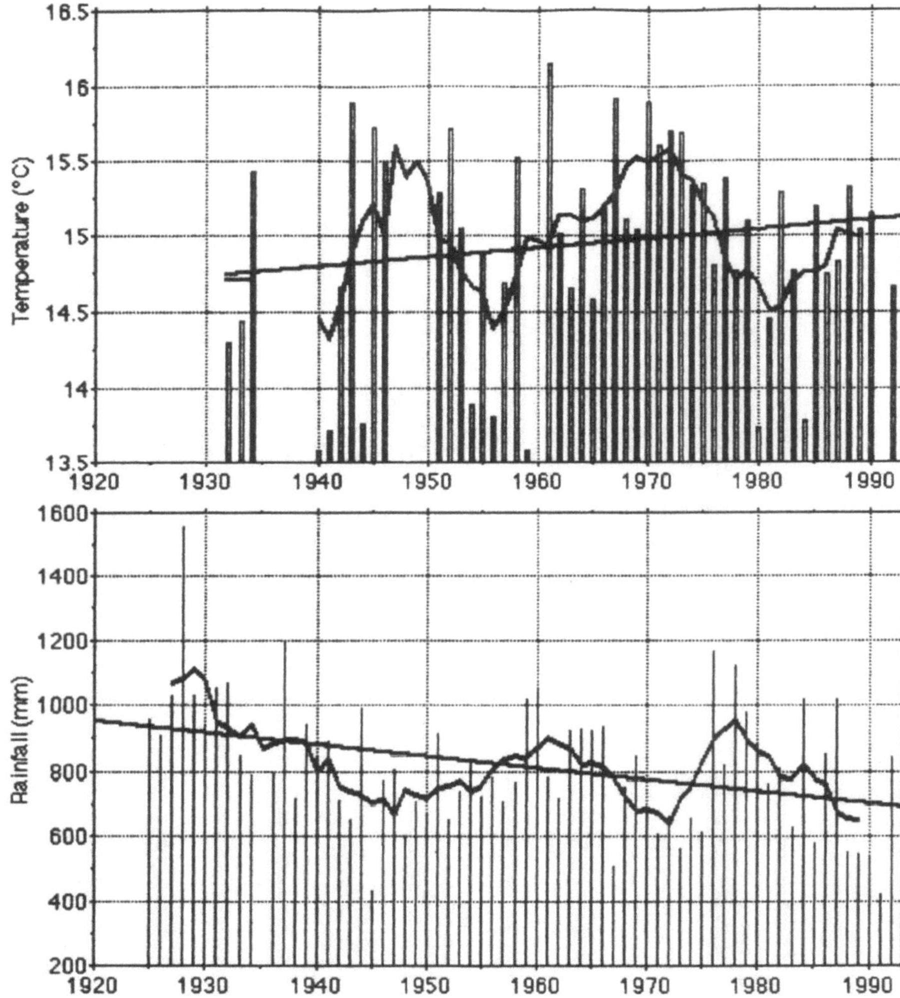

Figure 9-12. Yearly average temperature and yearly rainfall of the Cortona station

a) Scirca spring: located at 43° 22′ N, 12° 44′ E, at 582 m a.s.l.. Maximum elevation of the recharge area is 1566 m a.s.l. Recharge area not known. Aquifer made up of karstified limestone. Depletion constant = 0.012 $days^{-1}$.

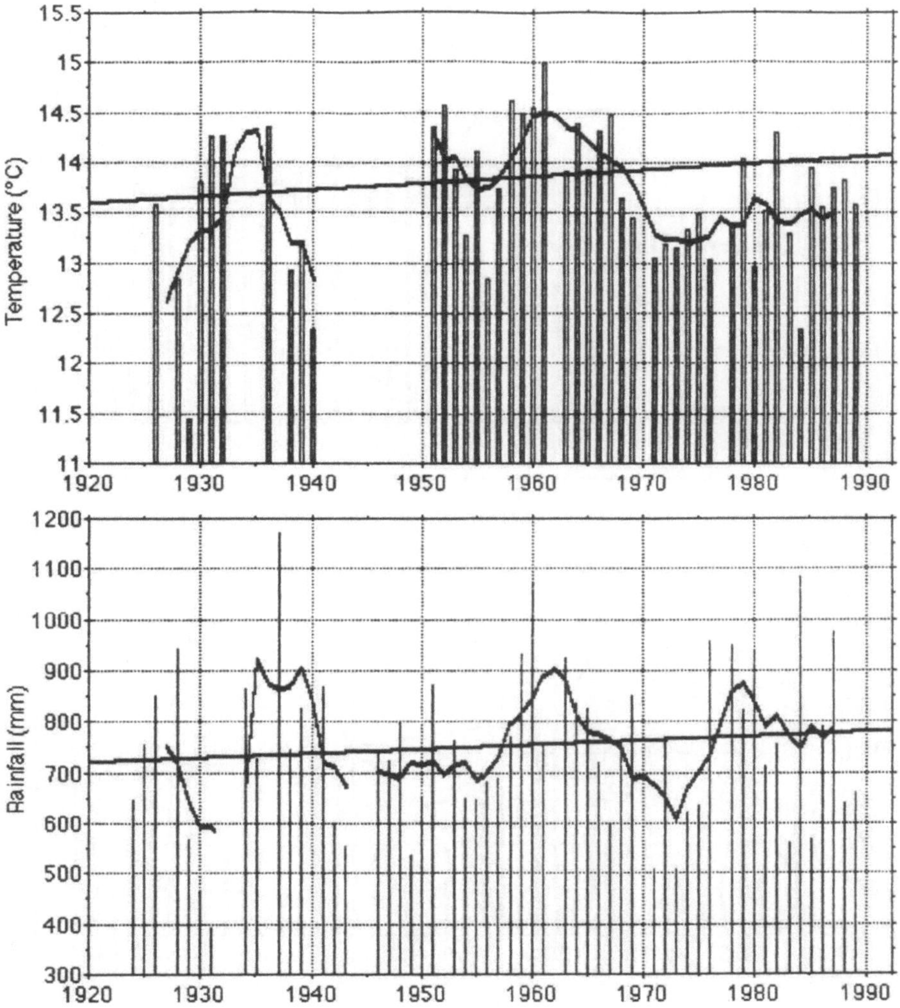

Figure 9-13. Yearly average temperature and yearly rainfall at Monte del Lago station

b) Bagnara spring: Located at 43° 7′ N, 12° 51′ E at 630 m a.s.l. Maximum elevation of the recharge area is 1571 m a.s.l. Recharge area of about 12 km^2. Aquifer made up of limestone. Depletion constant = 0.011 days^{-1}.

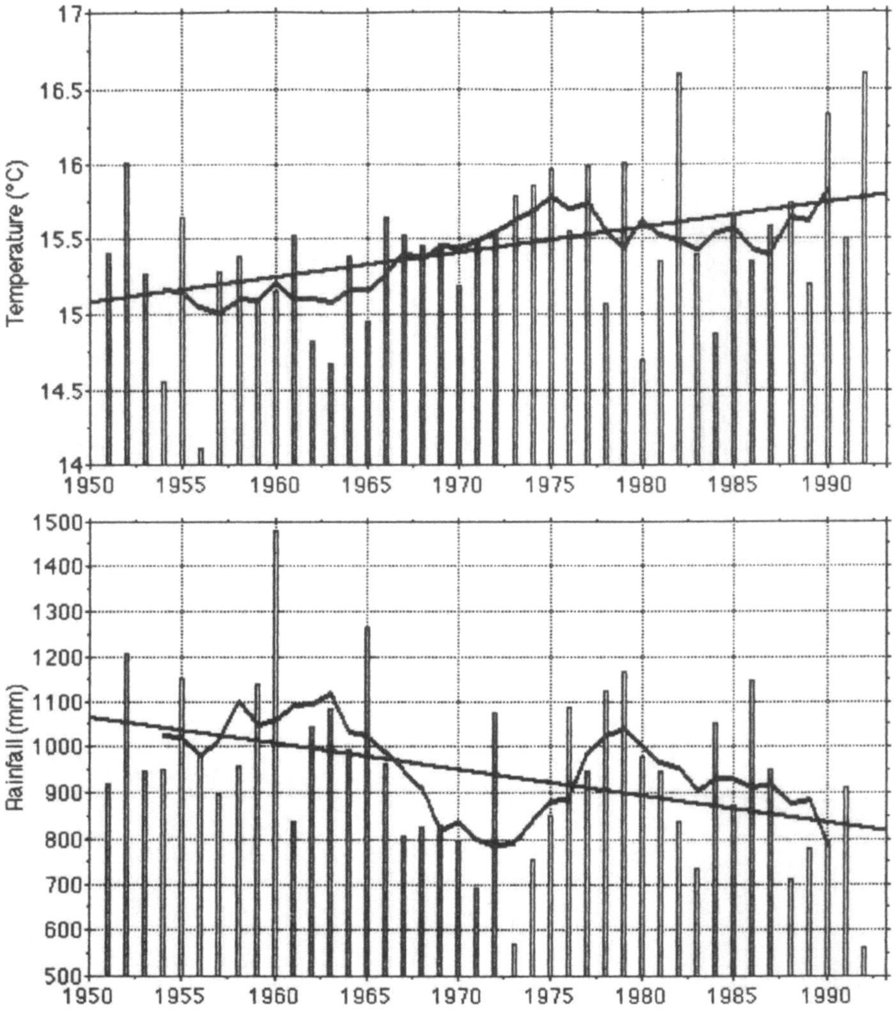

Figure 9-14. Yearly average temperature and yearly rainfall of the Terni station

c) Nera river at Torre Orsina, located at 42° 34'N, 12°33' E; 210 m a.s.l. Catchment area of 1445 km^2. Maximum elevation of the catchment is 2476 m a.s.l. Rocks in the area consist mainly of limestone, plus some marly formations. Depletion constant 0.006 - 0.0008 $days^{-1}$.

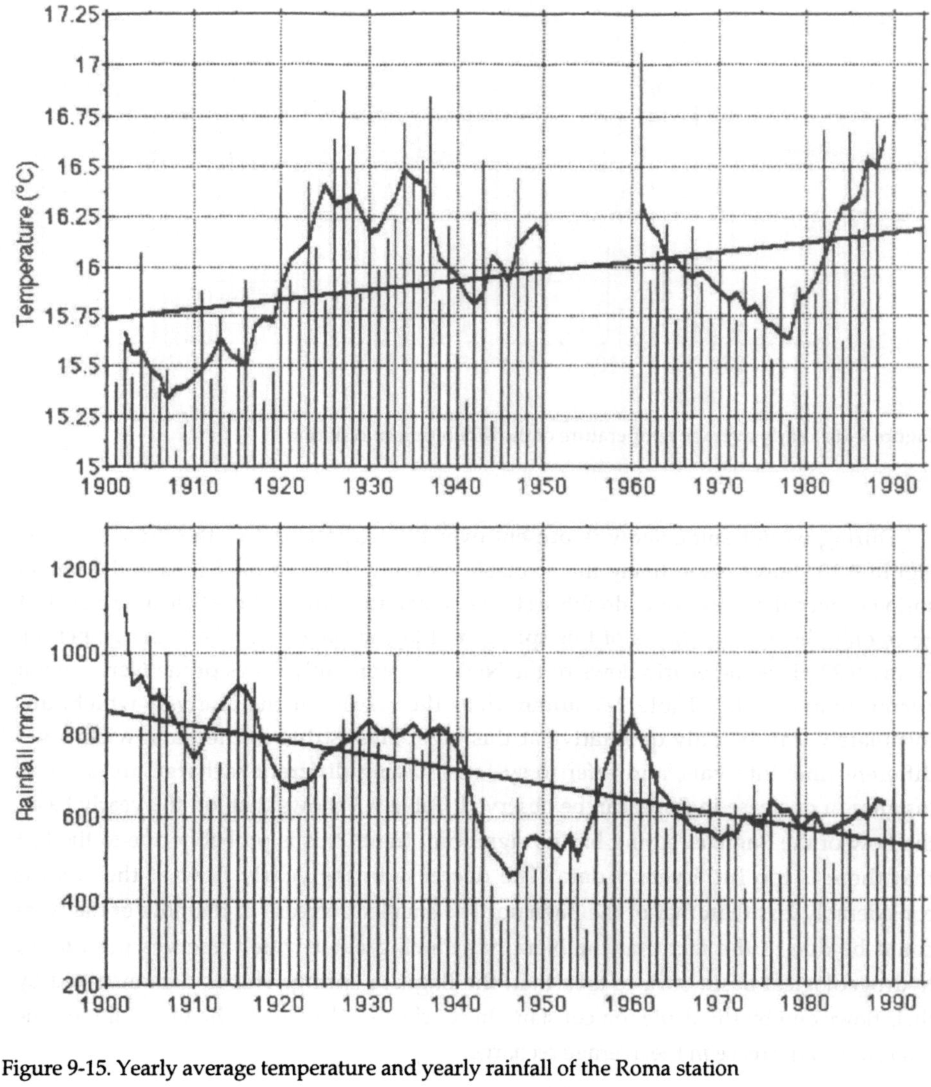

Figure 9-15. Yearly average temperature and yearly rainfall of the Roma station

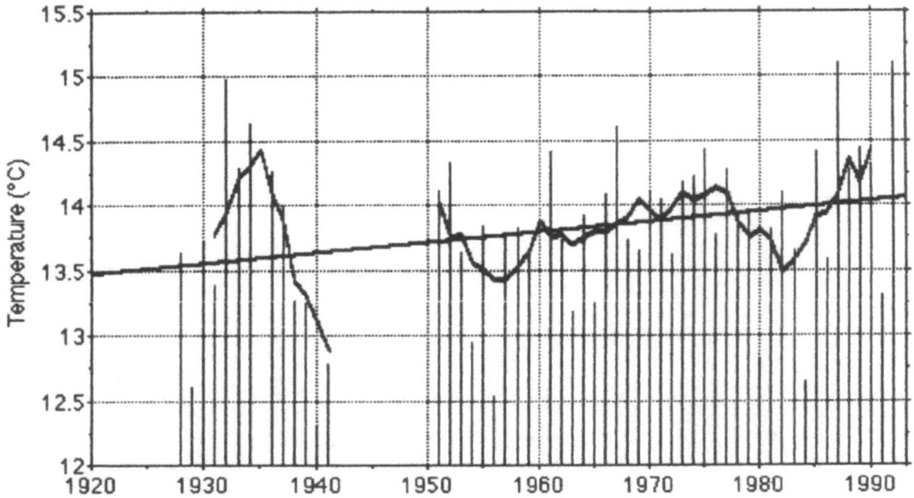

Figure 9-16. Yearly average temperature of the Montepulciano station

During winter some snow is present over the highest part of the three systems. Figure 9-20 shows respectively the available monthly flows of the Scirca and Bagnara springs. Since there are some doubts as to the maximum flows of the Scirca, Figure 9-21 gives only the average flows of this spring in May, at the end of the recharge period. Figure 9-22 gives the yearly flows of the Nera river (monthly flows of the Nera are not currently available). Table 9-3 summarizes the results of the analysis which, unfortunately, can be only qualitative at this tage. This is due to the data, wich cover different time intervals, and refer to systems having different recharge areas. In all instances a decrease in flows can be observed. Table 9-3 shows that, on the yearly basis, the flow of the Bagnara spring has no significant trend: this is probably due to the fact that there is too little yearly data. The decrease in the yearly flow of the Nera is considerably less than that of the Bagnara. The smaller decrease in the flow of the Nera could be due to the fact that the Nera has much greater water reserves per unit of hydrogeological basin surface area than the Bagnara spring. This is demonstrated by their flows and by the depletion constant. In conclusion, despite all the uncertainties, the data show a decrease in the average recharge.

Table 9-3. Trend of flow of Scirca spring, Bagnara spring and Nera River

System	interval	gratient	l.s.	gradient	l.s.
Scirca	1942-91	-0.170 1/s/month	0.0001	-1.108 1/s/year	0.0001
Bagnara	1974-92	-0.058 mm/month	0.0026	-5.266 mm/year	0.478
Nera	1926-91	-	-	-1.942 mm/year	0.0216

(l.s. = level of significance)

9-4 SOME CONSIDERATIONS REGARDING THE FUTURE

Climatic Trends Observed and Future Evolution of Climate

The question remains open on how much the trends observed are due to natural climatic fluctuations only and how much they are due to the combination of these with the anthropic greenhouse effect. The former (i.e. that the increase in greenhouse gases in the atmosphere due to anthropic causes has not produced warming) cannot be excluded, because considerable oscillations in climate have already taken place in the past, without artificial production of greenhouse gases. Besides this, from a statistical viewpoint, there is no proof that the recent climatic oscillations (including warming) are significantly greater or more rapid than those which took place since the beginning of instrumental observations (cf. for instance Hasselman, 1990; Palmieri, 1991; Wigley *et al.*, 1989; Wigley, 1992). The lack of any greenhouse effect, despite the increase in greenhouse gases, could be due to a set of feedback phenomena (very difficult to estimate) or to oceanic thermal inertia. If, in the area considered here as well as on a global scale, the climatic trends observed were casual oscillations or were produced by unknown causes, any forecasting or defining of future scenarios would be entirely meaningless. However, if the problem is looked at from other perspectives, there are very strong signs of the "inevitability" of warming due to the greenhouse effect. For example, a recent analysis of the data provided by the Vostok ice core indicates that greenhouse gases contributed to approximately 50% of the climatic variations taking place in the last 160,000 years (Lorius *et al.*, 1991). It should be pointed out that right now the concentration of greenhouse gases is the highest that it has been in the last 160,000 years and, unless something unforeseeable takes place, they will continue to increase over the next few decades. Moreover, even if the concentration of greenhouse gases were to remain at the present level, we should in any case expect a warming due to oceanic thermal effects (Wigley *et al.*, 1989). It should be kept in mind that in the past the possible effects of negative feedback did not prevent temperature increases of a few degrees Celsius, coupled to considerable variations in the hydrological cycle.

According to the General Circulations Models (GCMs) there is little doubt that global temperature and global rainfall increases should occur. However, there are strong

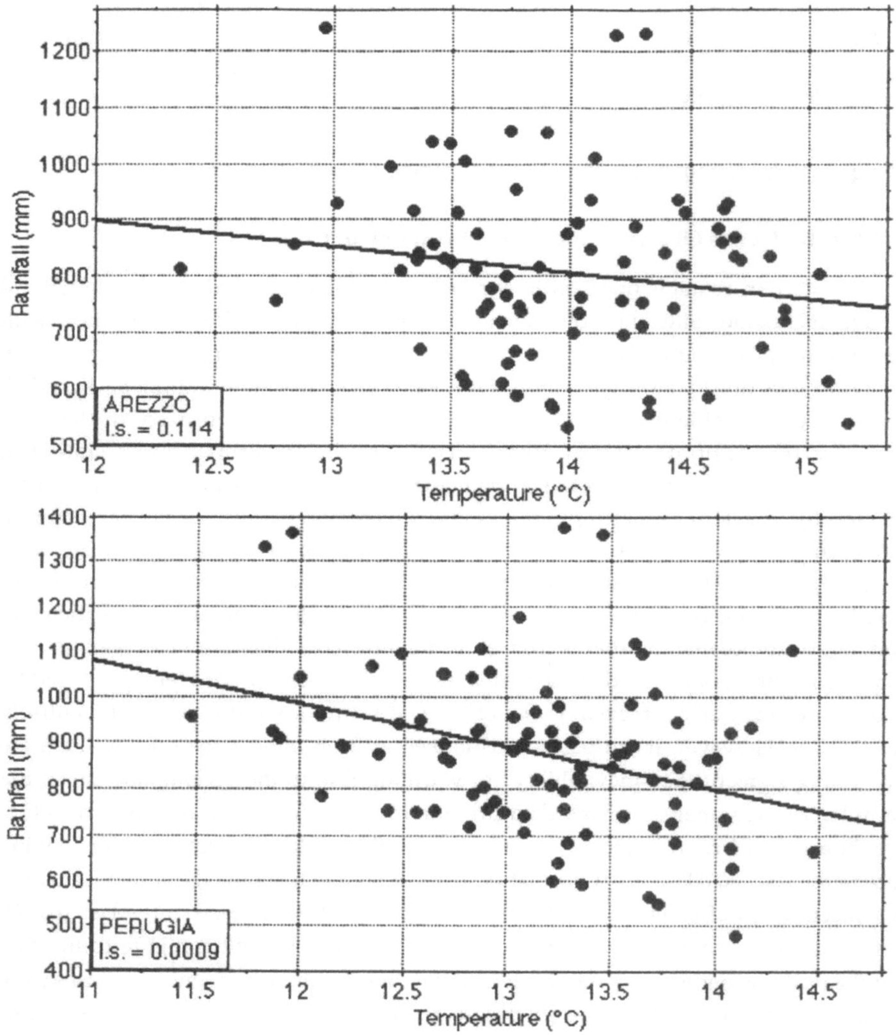

Figure 9-17. Relationship between average yearly temperature and yearly rainfall of the Arezzo and Perugia stations (l.s.= level of significance)

discrepancies in the variations forecasted, especially on the regional scale and concerning the rainfall (Conte, 1991; Duplessy, 1984; Hasselman, 1990, and Zimmermeyer, 1990).

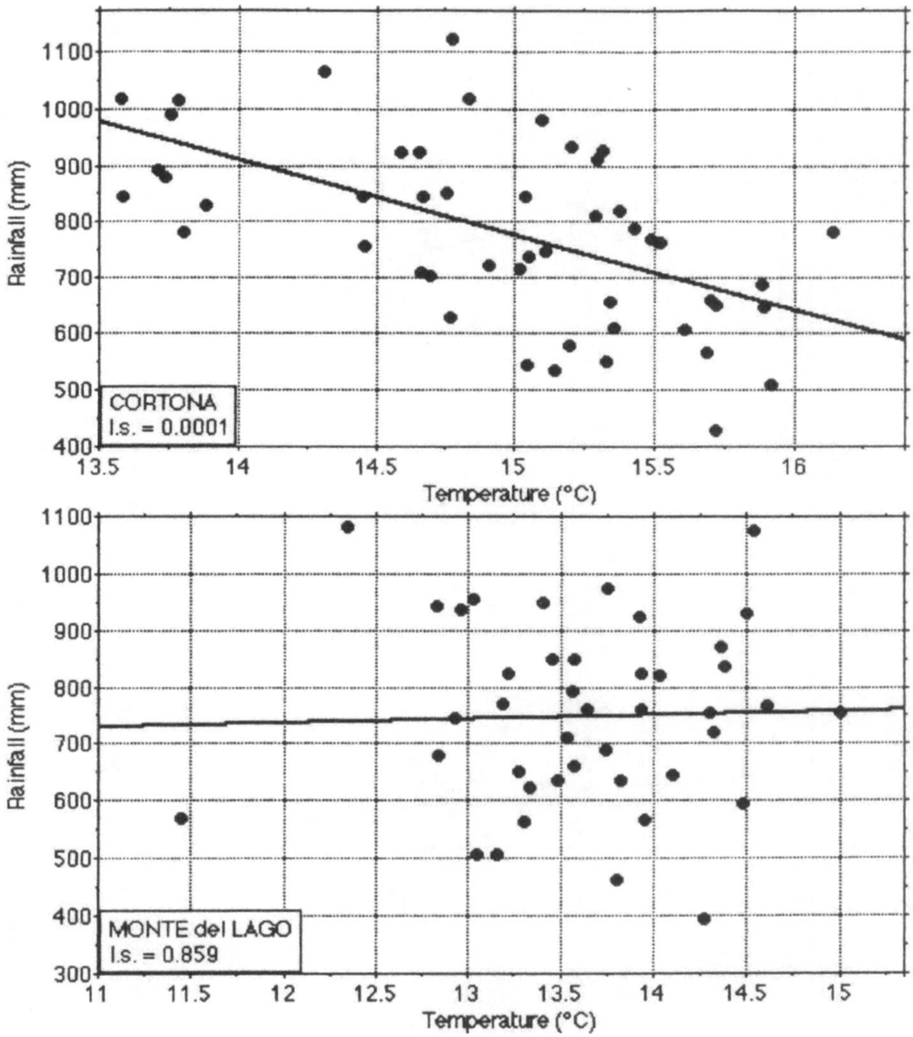

Figure 9-18. Relationship between average yearly temperature and yearly rainfall of the Cortona and Monte del Lago stations (l.s.= level of significance)

It is generally assumed that by 2050 the greenhouse gases will have doubled. According to a recent paper by Wigley (1992) for the Mediterranean area the corresponding temperature increase computed by GCMs is in the range of 1.2-3.5° C.

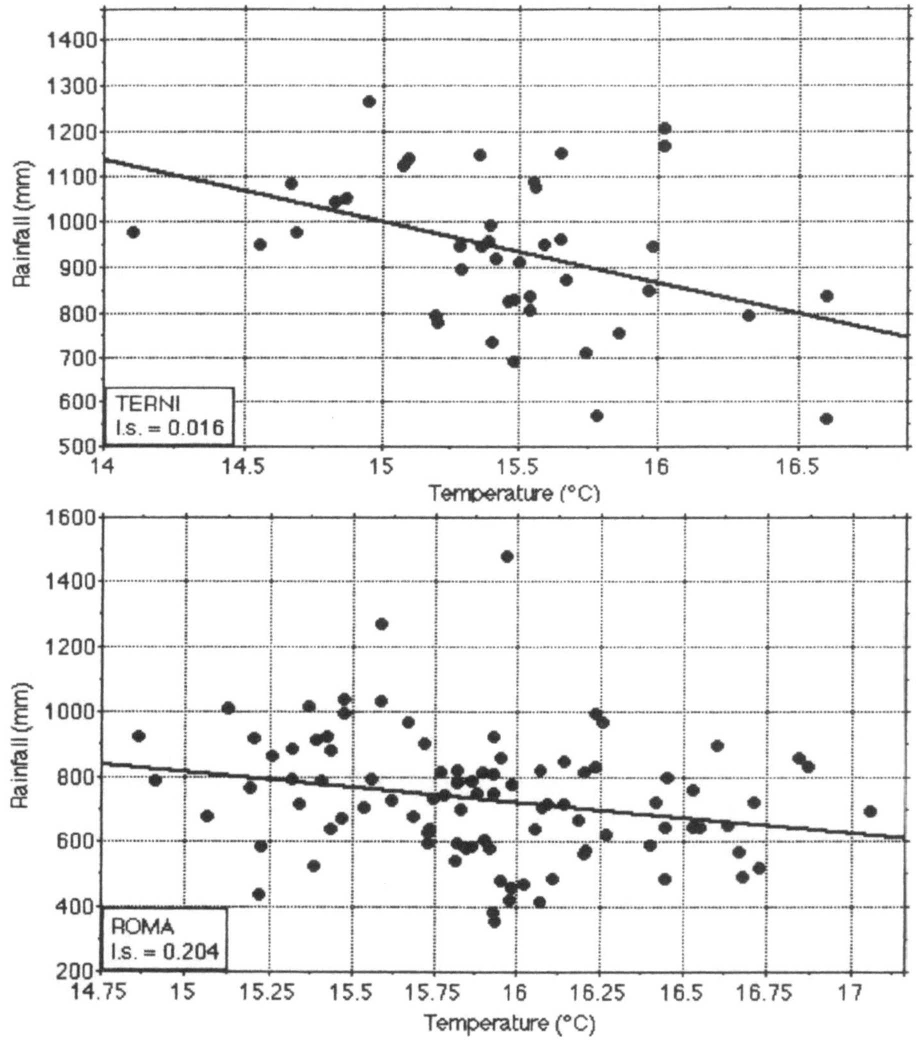

Figure 9-19. Relationship between average yearly temperature and yearly rainfall of the Terni and Roma stations (l.s.= level of significance)

The doubts regarding rainfall variations are much worse: the GCMs "forecast" a variation range between plus or minus 30% of the present rainfall (Wigley, 1992). In

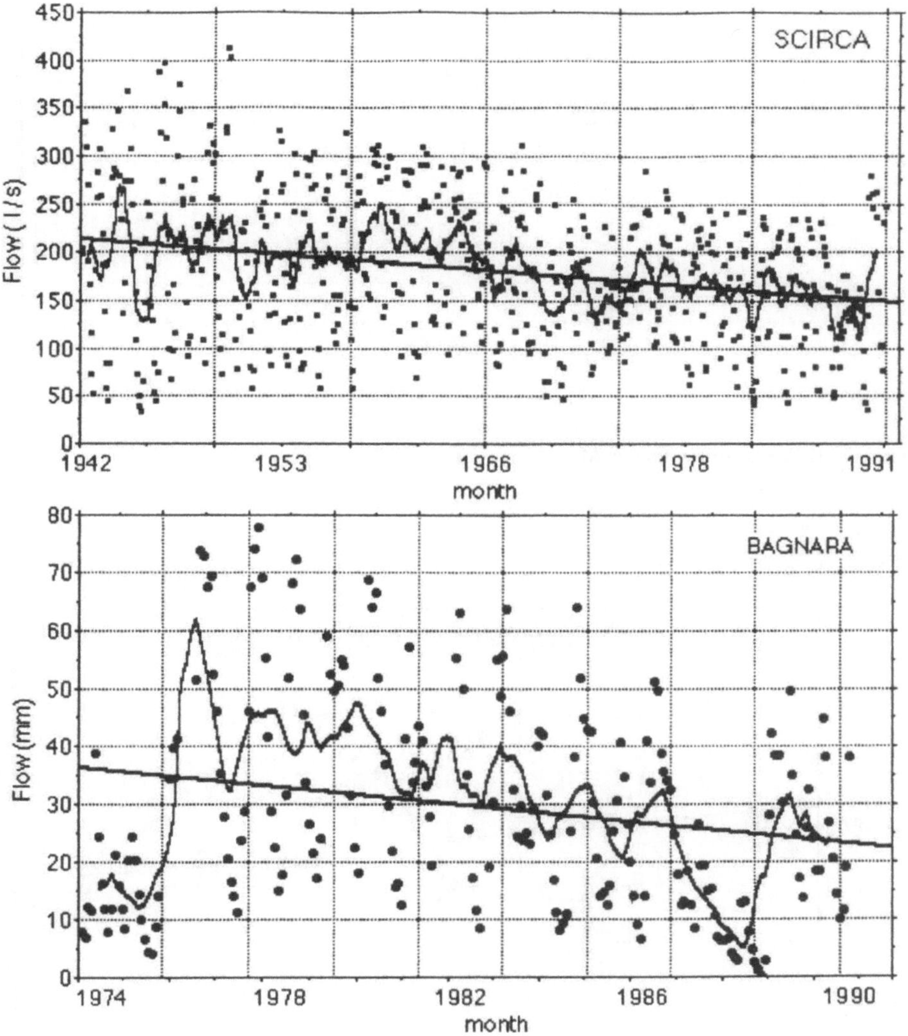

Figure 9-20. Monthly flow of the Scirca and Bagnara springs

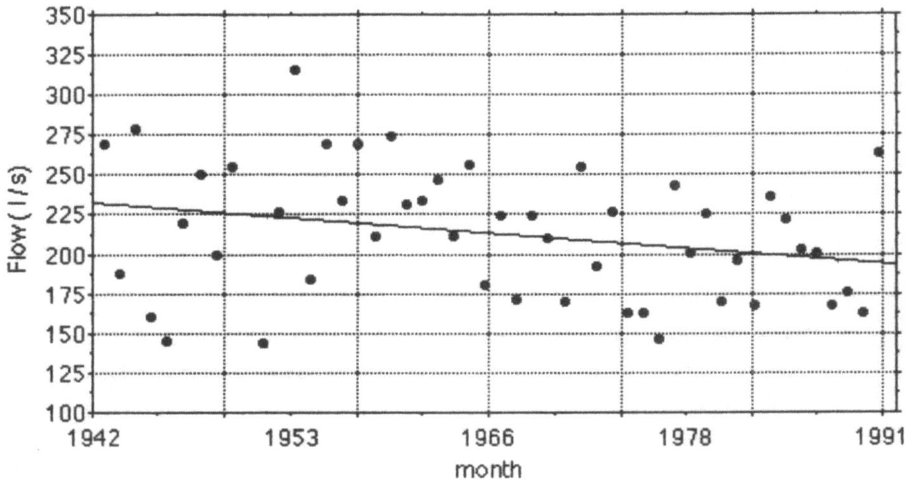

Figure 9-21. Scirca spring: flow in May

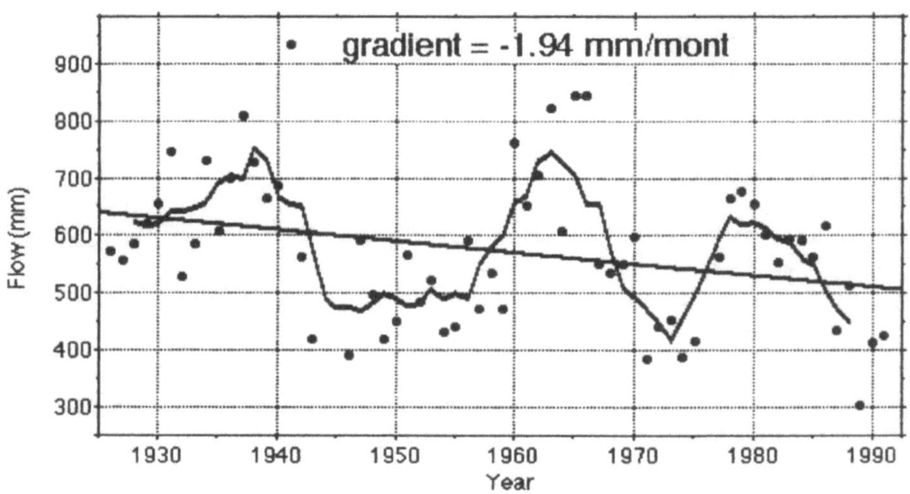

Figure 9-22. Yearly flow of the Nera River

particular, according to the majority of GCMs an increase in rainfall should be expected in northern Italy and a decrease in the south. The boundary between the two regions is different from model to model: according to the previous analysis, in the area here considered, this boundary could be located somewhere north of Cortona, Arezzo and Montepulciano.

In the following, in order to have an idea about the future, the observed climatic trends have been extrapolated to the time intervals 2020 - 2030 and 2040-2050. In order to be conservative, only the data of Perugia, having a yearly trend with a significance greater than 99% for both rainfall and temperature, have been considered (Table 9-2).

The Perugia data imply that if the observed climatic trend should continue, by 2050 the average temperature would increase by 0.52 °C, and rainfall would decrease by about 100 mm (i.e. about 12% of the 1980-1990 time interval). It is interesting to note that this extrapolation gives variations much less intense than those forecasted by some GCMs. For instance, in the case of doubled CO_2, the British Meteorological Office (BMO) model gives for central Italy an increase in temperature of 3 °C, and for rainfall a decrease of 220-290 mm (Conte, 1991). It is necessary also to stress that according to the calibration, carried out on data from between 1901 and 1980, the BMO model is one of the best models for the Italian area. In other words it seems that variations similar to these here extrapolated are reasonable and not too pessimistic.

Water Yields Extrapolations

In the hypothesis that the observed trend in Perugia will continue up to 2050, the extrapolation of monthly and yearly data of rain and temperature allows us to have an idea of the future average water surplus. The computation has been carried out by means of the Thornthwaite-Mather method and the Turc formula, for the intervals 2020-2030 and 2040-2050.

A conventional maximum soil storage capacity of 100 mm has been assumed, while no feedback effects have been considered. The water yield has been computed as the difference between rainfall and Actual Evapotranspiration (AE). The results are reported in Table 9-4: they show that water resources can be heavily affected by the climatic variations here considered. Unfortunately the Turc and Thornthwaite-Mather methods indicate that for the "forecasted" range of climatic variation, the future water yield depends much more on rainfall variations than temperature variations. For instance, and referring to the second column of Table 9-4, if the rainfall were to remain constant, it would be necessary for the temperature to have an unrealistic increase of about 5.7 °C to get a water surplus of 190 mm: thus, from the point of view of water resources, the real problem is to improve the knowledge of what the future rain trend will be.

Table 9-4. Average actual evapotranspiration (AE), average water surplus (S), and average potential evapotranspiration (PE) according to Turc and Thornthwaite in Perugia area. Units are mm/year

	1980-90	2020-30	2040-50
AE Turc	568	550	541
S Turc	242	190 (-21 %)	165 (-32 %)
PE Thornthwaite	760	771 (+1 %)	776 (+2 %)
AE Thornthwaite	595	587	583
S Thornthwaite	215	153 (-29 %)	122 (-43 %)

In order to have an idea also of the possible regimen variation of the water courses, the temperature and rain trends of Perugia have been extrapolated to an experimental watershed at about 30 km from Perugia, Rio Acquina, whose characteristics are given in table 9-5.

Table 9-5. Characteristics of the Acquina River

<u>Gauging station:</u> Carestello. Location: Lat. 43° 17′ N, Long. 12°31′ E.
<u>Area</u>: 11 km^2. <u>Average altitude</u> : 652 m a.s. l.
<u>Hydrometeorological Data</u>: February 1985 - January 1989
<u>Geology</u>: flysch, clays, argillites and fine sandstone. The main rock formation is the Marnoso-arenacea, of the Lower-Middle Miocene.

On the basis of the actual monthly flow for the years 1985-1989, and monthly temperature and rainfalls of nearby meteorological stations, the average monthly flow for the periods 1980-1990, 2020 - 2030 and 2040-2050 were simulated. The simulation was carried out using a variation of a very simple model (De Felice and Dragoni,1992).

The model, made up by two tanks, computes the mean monthly flow of streams, known rainfalls and mean monthly temperatures in an average year. The first tank represents the soil and governs the rainfall- evapotranspiration relationship in a manner very similar to the Thornthwaite-Mather method; the second reservoir, fed by the first, takes on the "invariant" characteristics of the basin, i.e. essentially lithology, morphometry and area. This tank is defined by a single parameter, β, which appears to be more closely linked to the overall permeabilities of the rock formations outcropping in the area than to the other "invariant" characteristics of the watershed. Considering that - especially for small basins in areas with highly permeable rocks - underground exchanges with adjacent systems can be considerable and that the computing of real evapotranspiration (based on temperature readings alone) is always approximate, a coefficient C was introduced in the model. The coefficient C (called the "exchange coefficient") is constant for all months, and it always makes the computed total annual volumes produced by the basin coincide with the volumes measured.

Thus we have:

$$Q_{(i)} = \beta\,(W_{(i-1)} + C \cdot S_{(i)}) \qquad (9\text{-}1)$$

$$W_{(i)} = (1 - \beta)\,(W_{(i-1)} + C \cdot S_{(i)}) \qquad (9\text{-}2)$$

$$\sum_{i=1}^{i-12} Q_{(i)} = \sum_{i=1}^{i-12} F_{(i)} \qquad (9\text{-}3)$$

where

$Q_{(i)}$, $F_{(i)}$ = computed and measured flow during month (i), (mm),
$W_{(i)}$ = water stored at the end of month (i), (mm),
$S_{(i)}$= computed water surplus during month (i), (mm),
β, C = coefficients (dimensionless).

The model is calibrated by simply giving to coefficients C and β the values which

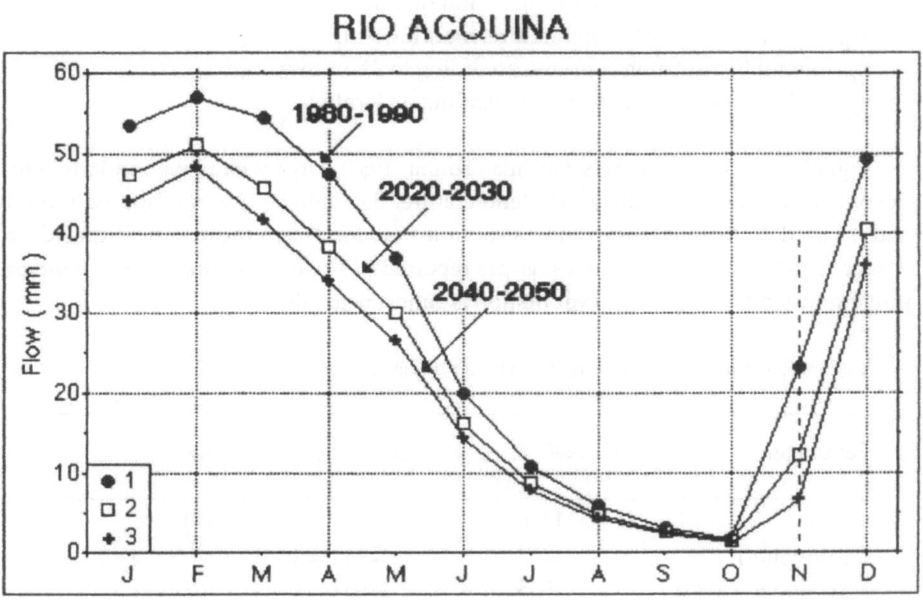

Figure 9-23. Estimates of average flow of Acquina River according to the present climate trend of the Perugia station. a) 1980-1990; b) 2020-2030; c) 2040-2050

minimize the mean square deviation between measured flows and computed flows. The assumed climatic variations would considerably alter the average yearly runoff, with an extending by about one month of the low flow season, as shown in Figure 9-23.

Evaporation Variations

The measured and extrapolated monthly temperature at the Perugia station have been used to have an indication on the evaporation increase that would occur should the temperature trend continue. For this purpose the following two empirical formulas were used:

$$E_a = C_1 * T_a + C_2 \tag{9-4}$$

$$E_m = b * (i_m)^{a1} * (T_m)^{a2} \tag{9-5}$$

where

E_a = average yearly evaporation (mm),
E_m = monthly evaporation during month (m), (mm),
T_a, T_m = average yearly temperature, monthly temperature (^oC),
i_m = monthly insolation constant, according to Thornthwaite,
a_1, a_2, b, C_1 and C_2 = constants, depending on location.

Equation (9-1) is Visentini's formula (Tonini, 1959), largely employed in Italy when the only available data is the yearly temperature. Equation (9-2) was proposed recently and is based on the analysis of three evaporimeters located in the area here considered (Dragoni and Valigi, 1992). The results are reported in Table 9-6: in the range of temperature forecasted the increase in evaporation is almost negligible.

Table 9-6. Estimated evaporation from water bodies in Perugia area.

Parameter	1980-90	2020-30	2040-50
T	13.5	13.8	14.0
E (Visentini)	1213	1244 (+2.6 %)	1259 (+3.8 %)
E (Dragoni-Valigi)	1055	1065 (+1.0 %)	1070 (+1.5 %)

T= average yearly temperature (^oC); E = average yearly evaporation (mm).

9-5. CONCLUSIONS

Results

The available data and information suggest the following:

a) In central Italy, in the last 3000 years, cool periods corresponded generally to an increase of water surplus. More research is necessary to quantify this relationship and to increase the resolution of our observations.

b) The future water yields seem to depend more on rainfall variations than on temperature/evaporation variations. Given that the GCMs "forecasted" rain has a range between plus or minus 30% of the present rainfall, any substantial improvement regarding the future of water resources must pass through an improved knowledge of the future rain trend.

c) The meteorological data of this century show that, in central Italy, when a data set has a trend, this corresponds to an increase in temperature and to a decrease in rainfall. The linear trends are within or below the ranges furnished by GCMs.

d) The analysis of the water yield of some hydrogeological systems shows that the water yields have a decreasing trend.

Final Remarks

Whatever its main causes are, it is inevitable that the warming process, if it continues, will influence the water cycle and water resources. These effects will be particularly important for an area like the Mediterranean basin which, to a large extent, is already subject to strong soil erosion and water shortages, while water demand is increasing.

Any variations of runoff, aquifer recharge and evapotranspiration, are controlled both by climatic conditions and factors such as geo-lithological, pedological and vegetation characteristics: in spite of this, there have not yet been many persistent attempts to understand how certain typical systems (for example a karstic spring, a watershed located on low permeability rocks, etc.) will respond to future climate conditions. For instance it is not clear how the distribution between surface runoff and groundwater recharge could vary. It must be stressed that to answer these questions is not only an academic issue, but the only way to take reasonable steps to try to minimize the negative effects of the probable future climatic changes: the past tells that these effects are something more than a remote possibility.

The monitoring and modeling of some representative hydrogeological systems would help in defining climatic trends and their impacts on water resources. The choice of which systems to monitor appears to be a delicate task: the continuous increase in water consumption makes it difficult, or impossible, for many hydrogeologic systems which already have a long set of data, to distinguish between a reduction of water resources through human activity and a decrease caused by climatic changes.

It appears that, as there are so many uncertainties about future rainfalls, temperature and feedbacks, there is no point in using complicated models for the rain-water yield transformation: in any case it will be possible to have only a good idea of what will happen. In spite of all the uncertainties, the problem is so important that it cannot be neglected on the grounds that it is not possible to be sure of how good and useful the results of the researches will be.

Acknowledgments

Financial support was provided by CNR (Environment Committee, contract. 92.02055.CT13/115.23232) and by CIPLA (Perugia University and L.U.I.S.S - Rome). Information about the Martignano Lake and the Colfiorito area were given by G. Cappa, A. Felici, L. Cianetti and M. Mariani. Access to unpublished data was given by A. Ammannati, R. Giola, P. Pampana of the Hydrographic Service of Rome and Pisa, and M. Pitzalis (Director of the Institute of Agricultural Ecology of Perugia University). A.M. De Felice helped the author in collecting, discarding and elaborating an endless amount of raw data. Sincere thanks are given to all these Institutions and friends.

REFERENCES

Angelini P. (1993). La sorgente di Bagnara: portate giornaliere. Perugia.

Belluomini G., Manfra L., and Proposito A. (1992). Datazione con il carbonio-14 di reperti provenienti dai fondali del Lago di Bolsena (Viterbo). Atti del 1° Seminario di Geoarcheologia, Bolsena 30 agosto 1992 (in print).

Bourke A. (1984). Impact of climatic fluctuation on european agriculture. *In:* The Climate of Europe: Past, Present and Future, Flohn H. and Fantechi R. (Eds). Reidel Publishing Company.

Brisse A. and De Rotroù L. (1876): Dessechément du Lac Fucino. Roma.

Budyko M.I. (1982). The Earth's Climate: Past and Future. Academic Press, New York.

Budyko M.I. and Sedunov Yu S. (1990): Anthropogenic Climatic Changes. *In:* Climate and Development, Karpe H.-J., Otten D., Trinidade S. C. (Eds). Springer-Verlag, Berlin.

Castellani V., Dragoni W. (1990): Contribution to the History of Underground Structures: ancient Roman tunnels in Central Italy. Proc. International Symposium on Unique Underground Structures, June 12-15, 1990, Denver, Colorado.

Castellani V. and Dragoni W. (1992). Opere Arcaiche per il controllo del territorio: gli emissari sotterranei artificiali dei laghi albani. Atti del Convegno " Gli Etruschi Maestri di Idraulica", Perugia, Febbraio 1991, ELECTA.

Cencetti C., Dragoni W., and Nejad Massoum M. (1989). Contributo alle conoscenze delle caratteristiche idrogeologiche del Fiume Nera (Appennino Centro-Settentrionale). Geologia Appl. e Idrogeologia, Bari, 1989, Vol. XXIV.

Cialini G. (1991). Memoria storico-artistica della Chiesa di Santa Maria di Ancaélle. Collegio Pio della Sapienza, Università degli Studi di Perugia.

Coarelli F. and La Regina A. (1984). Abruzzo-Molise. Guide Archeologiche Laterza, Laterza, Bari.

Colacino M. (1991). La temperatura dell'aria a Roma negli ultimi duecento anni: analisi della serie storica del Collegio Romano. *In:* Le Variazioni Recenti del Clima (1800-1990) e le Prospettive per il XXI Secolo. Proceedings of the Italian Geographical Society Conference, Rome, 5-6 April 1990. Mem. Soc. Geog. Ital. Vol. XLVI.

Conte M. (1991). Previsioni climatiche con modelli dinamici. *In:* Le variazioni recenti del clima (1800-1990) e le Prospettive per il XXI Secolo. Proceedings of the Italian Geographical Society Conference, Rome, 5-6 April 1990. Mem. Soc. Geog. Ital. Vol. XLVI.

Cremaschi M. (1992). Mutamenti del clima nel quaternario: le linee generali. *In*: Italia Preistorica, Laterza, Bari.

D'Amato S. (1980). Il primo prosciugamento del Fucino. Avezzano.

Dansgaard W., Johnsen S.J., Moller J., and Langway C.C. (1969): One thousand centuries of climatic record from Camp Century on the Greenland ice sheet. Science, 166.

De Felice A.M. and Dragoni W. (1991). Studio idrogeologico di un bacino sulla Marnoso-arenacea. Atti del 1 Convegno Nazionale dei Giovani Ricercatori in Geologia Applicata, Gargnano (BS - Italia), 22 e 23 Ottobre 1991.

De Felice A.M., Dragoni W., and Giglio G. (1992). Comparison of basin hydrological characteristics using only one lumped parameter: preliminary note. Conference on Methods of Hydrologic Basin Comparison, Oxford, UK, 28 Sept.- 2 Oct. 1992. Robinson M. (Ed), Institute of Hydrology, Wallingford, UK.

Dragoni W. (1982). Idrogeologia del lago Trasimeno: sintesi, problemi, aggiornamenti. Geogr. Fis. Dinam. Quat., 5.

Dragoni W. and Valigi D. (1992). Contributo alla stima dell' evaporazione dalle superfici liquide nell'Italia Centrale. 2 Convegno Nazionale dei Giovani Ricercatori in Geologia Applicata, Viterbo, 28-31 Ottobre 1992 (in print).

Duplessy J.C. (1984). Carbon cicle, greenhouse effect and other anthropogenic impacts on climate. *In* The Climate of Europe: Past, Present and Future, Flohn H., Fantechi R. (Eds.), Reidel Publishing Company.

Gandolfo C. and Sulli M. (1991). Vallombrosa 1872-1989. una serie ultracentenaria di misure di precipitazioni e di temperatura in ambiente forestale. *In*: Le variazioni recenti del clima (1800-1990) e le prospettive per il XXI secolo. Proceedings of the Italian Geographical Society Conference, Rome, 5-6 April 1990. Mem. Soc. Geog. Ital. Vol. XLVI.

Giuffrida A. and Conte M. (1991). Tevoluzione a lungo termine del clima italiano. *In*: Le Variazioni Recenti del Clima (1800-1990) e le Prospettive per il XXI Secolo. Proceedings of the Italian Geographical Society Conference, Rome, 5-6 April 1990. Mem. Soc. Geog. Ital. Vol. XLVI.

Guidi A. (1986). Gli insediamenti perilacustri di riva d'etˆ protostorica nel Lazio centro-meridionale. *In*: Atti dell'Incontro di Acquasparta 1985 "Gli insediamenti perilacustri dell'eta del bronzo e della prima eta del ferro: il caso dell'antico Lacus Velinus", Perugia, 15-17 Novembre 1985, Carancini GL (Ed). Quaderni di Protostoria,Istituto di Archeologia, Università di Perugia.

Hasselmann K. (1990). Climate and development: scientific efforts and assessment - The state of the art. *In*: Climate and Development, Karpe H.-J., Otten D., Trinidade S. C. (Eds). Springer-Verlag, Berlin.

Issar A. (1992). The impact of climate variations on water management systems. Report of UNESCO/IHp, Project IIc

Jones P.D. and Bradley R.S., (1992). Climatic variations in the longest climatic records. *In*: Climate since A.D. 1500, Bradley R. S. and Jones P. D. (Eds). Routledge.

Lamb H.H. (1977). Climate: Present, Past and Future. Vol. 2: Climatic History and the Future. Methuen & Co Ltd, London.

Le Roy Ladurie (1972). Times of Feast, Times of Famine: A History of Climate Since the Year 1000.

George Allen & Unwid Ltd, London.
Liberati Silverio A. (1986). Aqua Alsietina. *In*: Il Trionfo dell'Acqua - Acque ed Acquedotti a Roma, Paleani, Roma.
Lorius C., Jouzel J., Raynaud D., Korotkevich Y. S., and Kotlyakov V. M. (1991). Greenhouse warming, climate sensitivity and Vostok data. *In*: Glaciers-Ocean-Atmosphere Interactions. Kotlyakov V. M., Ushakov A., Glazovsky A. (Eds). IAHS Publication No. 208.
Marchegiani L (1976). Dall'origine della signoria dei Varano ad oggi. *In*: Camerino, Ambiente, Storia, Arte", Marchegiani L (Ed). Misici-Falzi, Camerino.
Melicchia A. (1942). Variazioni climatiche nell' Italia centrale e loro rapporto col regime del Tevere. *In*: Ricerche sulle Variazioni Storiche del Clima Italiano, 8. C.N.R.- Com. Naz. Geog.. Bologna XX.
Metelli G. (1983). Il "lago" di Colfiorito nelle vicende della famiglia Iacobilli. Bollettino Storico della Citt^ di Foligno, Accademia Fulginia di Lettere, Scienze ed Arti.
Min. LL. PP. (1926-1980). Annali Idrografici, parte 1a e 2a. Sezione Idrografica di Roma. Istituto Poligrafico dello Stato, Roma.
Moccheggiani Carpano C. (1976). Osservazioni sui mutamenti di livello delle acque del Lago di Martignano. Prospezioni.
Moyer G. (1982). Il calendario gregoriano. Le Scienze (Italian edition of Scientific American), 167, XV, vol.XXIX.
Nicco Fasola G. (1951). La fontana di Perugia. Libreria dello Stato, Roma.
Palmieri S. (1991). Aspetti evolutivi del clima negli ultimi decenni. *In*: Le Variazioni Recenti del Clima (1800-1990) e le Prospettive per il XXI Secolo. Proceedings of the Italian Geographical Society Conference, Rome, 5-6 April 1990. Mem. Soc. Geog. Ital. Vol. XLVI.
Parotto M. and Praturlon A. (1979). Structural sketch of orogenic belts in central and in eastern Mediterranean. Rome.
Pavese M. P., Banzon V., Colacino M., Gregori G.P., and Pasqua M. (1992). Three historical data series on floods and anomalous climatic events in Italy. *In*: Climate since A.D. 1500, Bradley R. S. and Jones P. D. (Eds). Routledge.
Pinna M. (1991). Le variazioni recenti del clima (1800-1990) e le prospettive per il XXI secolo. *In*: Le variazioni recenti del clima (1800-1990) e le prospettive per il XXI secolo°. Proceedings of the Italian Geographical Society Conference, Rome, 5-6 April 1990. Mem. Soc. Geog. Ital. Vol. XLVI.
Potter T.W. (1979). The Changin Landscape of South Etruria. Paul Elek Limited, London.
Praturlon A. (1980). Cadre Geologique General de l'Italie - Sa Place dans le Cadre Europeen et Meditrraneen.In " Introduction a la Géologie Générale d'Italie et Guide a l'Excursion 122 A", Fagnani G., Zuffardi P. (Eds.). Società Italiana di Mineralogia e Petrologia, Museo di Storia Naturale, Milano.
Rapetti F. and Vittorini S. (1991). Studio di serie pluviometriche secolari in Toscana. *In*: Le variazioni recenti del clima (1800-1990) e le prospettive per il XXI secolo. Mem. Soc. Geog. Ital. Vol. XLVI.
Sensi M. (1984). Vita di pieta e vita civile di un altopiano tra Umbria e Marche. Edizioni di Storia e Letteratura, Roma.
Tamburini P. (1986). Il villaggio del Gran Carro: conoscenze attuali e proposte di ricerca. *In*: Atti dell'Incontro di Acquasparta 1985 "Gli insediamenti perilacustri dell'et^ del bronzo e della prima eta del ferro: il caso dell'antico Lacus Velinus" Perugia, 15-17 Novembre 1985, Carancini G L (Ed). Quaderni di Protostoria, Istituto di Archeologia, Universita di Perugia.
Thornthwaite C.W. and Mather J.R. (1957). Introduction and Tables for computing potential

evapotranspiration and the water balance. Drexel Institute of Technolgy - Laboratory of Climatology, Publications in Climatology, Vol. X, 3, Centerton, New Jersey.

Tonini D. (1959). Elementi di Idrografia ed Idrologia. Vol. I, Libr. Univ., Venezia (reprint 1983).

Wigley T.M.L., Jones P.D., and Kelly P.M. (1989). Warm world scenarios and the detection of climate change induced by radiatively active gases. *In*: The Greenhouse Effect Climatic Changes and Ecosystems, Warrick B., Jager D., (Eds). John Wiley & Sons.

Wigley T.M.L. (1992). Future climate of the Mediterranean basin with particular enphasis on changes in precipitation. *In*: Climatic Change and the Mediterranean, Jeftic L., Milliman J. D., Sestini G., (Eds). Edward Arnold.

Zimmermeyer G. (1990). Climatic risk and the role of energy use. *In*: Climate and Development, Karpe H.-J., Otten D., Trinidade S. C. (Eds). Springer-Verlag, Berlin.

CHAPTER 10

REGIONAL CLIMATE CHANGE IMPACTS ON HYDROLOGY AND WATER RESOURCES: A CASE STUDY FROM GREECE

MARIA A. MIMIKOU, YANNIS S. KOUVOPOULOS, AND PANAGIOTA S. HADJISSAVVA

Division of Water Resources, Hydraulic and Maritime Engineering,
Department of Civil Engineering, National Technical University of Athens,
5 Iroon Polytechniou, 157 73 Athens, GREECE

10-1. INTRODUCTION

It is widely known that the atmospheric concentration of radiatively active "greenhouse" gases, principally carbon dioxide (CO_2), nitrous oxide (N_2O), methane (CH_4), and chlorofluorocarbons (CFCs), has increased steadily over the past century. There is considerable consensus among the international scientific community that we are facing global warming due to this growing concentration of greenhouse gases in the atmosphere. This effect is known colloquially as "the greenhouse effect". It has been estimated that the average global temperature has increased 0.5 - 0.7 °C since the turn of the century, whereas current estimates from models indicate that a doubling of atmospheric CO_2 and other gases may increase the annual temperature by 3.0±1.5 °C during the next 50-100 years. Such predictions are, of course, far from being certain and can, at best, be only considered as possible scenarios. Although the properties of the radiative greenhouse gases are rather well-understood, and despite the improvements in our understanding of atmospheric dynamics and large-scale climatic processes, the climatic change due to greenhouse gases is still not entirely understood, and a clear consensus has yet to emerge as to the overall direction that global climate is evolving toward.

At our present state of knowledge, even if it was assumed that the global warming will take place on the scale and over the period predicted, the mechanism to translate climatic changes to specific changes in temperature and precipitation over a watershed for the planning, management and operation of specific water projects, would still be poorly understood.

The regional climatic change effects on hydrology and water resources have gained widespread research interest since the early '80s. Hydrological research suggests that plausible climatic changes will strongly influence various forms of water resources, such as mean and extreme surface runoff, soil moisture and groundwater, change sea-level,

NATO ASI Series, Vol. I 36
Diachronic Climatic Impacts on Water Resources
Edited by Andreas N. Angelakis and Arie S. Issar

and affect water quality. These climatic change impacts entail important implications for future water resources planning and management associated with serious environmental and socio-economic dislocations (Gleick, 1990; Mimikou, 1995).

The research work presented herein originates from a major research programme that addresses the regional impacts of greenhouse warming on water resources and water management works, over a mountainous region of central Greece. The work and findings presented here are restricted within the context of regional climatic impacts on different forms of water resources such as surface runoff and soil moisture. The paper is divided in two parts: The first part deals with the regional hydrological effects of climate change, whereas the second one with the subsequent effects on water management works.

A conceptual model of the water balance type has been used for the assessment of the regional hydrological effects of climate change. The model has been calibrated on a monthly basis and its predictability under changing climatic conditions (Klemes, 1985) has been successfully tested. Implementation of the model along with a series of future hypothetical climatic scenarios constructed on the assumption of uniform temperature increases by 1°, 2° and 4 °C and precipitation changes by -20%, -10%, 0%, 10%, 20%, has allowed an insight into the subject of basin response - as well as into the morphoclimatic mechanisms that modify it - by observing the model estimates for surface runoff, snow-melting runoff and soil moisture. Moreover, an attempt is made to evaluate the regional climate change impacts on some critical water management issues, such as reservoir storage and hydroelectric production, by analyzing their sensitivity to plausible climatic change scenarios. The study region is that of the Acheloos river diversion to the Thessalia Plain in central Greece comprising four drainage basins with four multi-purpose reservoirs at the respective outlets.

Results, showing a considerable distortion of water resources availability and distribution both in space and time, which is also shown to depend on specific regional characteristics, necessitate an extended regional approach to the problem of future water resource availability under a changing climate, in order alleviating actions to be taken.

10-2. THE STUDY REGION

The study area lies in the central mountainous region of Greece, about the 39° 30′ N parallel. It comprises four drainage basins with the characteristics given in Table 10-1:

Table 10-1. Characteristics of the study basins

Drainage basin	River	Area (km)	Mean elevation (m a.m.s.l.)
Mesohora	Upper Acheloos	633.0	1390
Sykia	Upper Acheloos	1173.0	1299
Pyli	Portaikos	134.5	800
Mouzaki	Pliouris	140.5	575

A major agricultural development plan for the Thessalia Plain has been initiated by the Greek Government. The main purpose is to transfer water from the humid parts of Greece to irrigate the fertile but semiarid Thessalia Plain. This will be accomplished be diverting a significant portion of the Upper Acheloos water to the Mouzaki reservoir. Hypsometric differences en route will be exploited for power production. The basic scheme comprises four multipurpose reservoirs (for irrigation and power generation) at the outlets of the respective basins, a diversion tunnel from the Sykia reservoir to the Mouzaki reservoir, a tunnel connecting the Pyli reservoir with the Mouzaki reservoir, and four power plants. The study area along with the associated hydrometeorological stations and the water development scheme are depicted in Figure 10-1.

The Mesohora reservoir, receiving runoff from the upstream drainage basin, is designed to guarantee annual energy production (E=231 GW h). The Mesohora power plant is situated at the end of the Sykia reservoir. The latter receives runoff from the intermediate drainage basin between the Sykia and Mesohora dam sites as well as the outflow from the Mesohora reservoir. A minor quantity of water from the Sykia reservoir is released through the Sykia power plant situated at the toe of the dam to produce primary energy (E=48.8 GW h annually).

The respective firm water flow downstream into the Acheloos River is W= 5 m^3 s^{-1}) is diverted from the Sykia reservoir through a tunnel into the Mouzaki reservoir on the western mountain sides of the Thessalia Plain. The diverted flow just before entering the Mouzaki reservoir passes through the Pefkofyto power plant. The primary energy generated there amounts to E=300 GW h per annum. The Mouzaki reservoir is also enriched from the Pyli drainage basin through a tunnel connecting it with the Pyli reservoir situated at the outlet of the Pyli basin. Finally, the Mouzaki reservoir is also enriched from the Pyli drainage basin through a tunnel connecting it with the Pyli reservoir situated at the outlet of the Pyli basin. Finally, the Mouzaki reservoir, also receiving runoff from the upstream basin, is designed to produce primary energy (E=370 GW h) annually and a firm water supply to the downstream Thessalia Plain of W=41 m^3 s^{-1}. The basic operation and design characteristics are given in Table 10-2.

Table 10-2. Design characteristics of the Acheloos diversion scheme

Reservoir	Storage capacity (10^6 m^3)	Hydroelectric power plant	Design head (m)	Primary energy (GW h $year^{-1}$)	Firm water supply (m^3 s^{-1})
Mesohora	228	Mesohora	200	231	-
Sykia	500	Sykia	137	48.8	5
		Pefkofyto	155.5	300	32
Pyli	47	-			Variable to Mouzaki reservoir
Mouzaki	530	Mouzaki	138.4	370	41

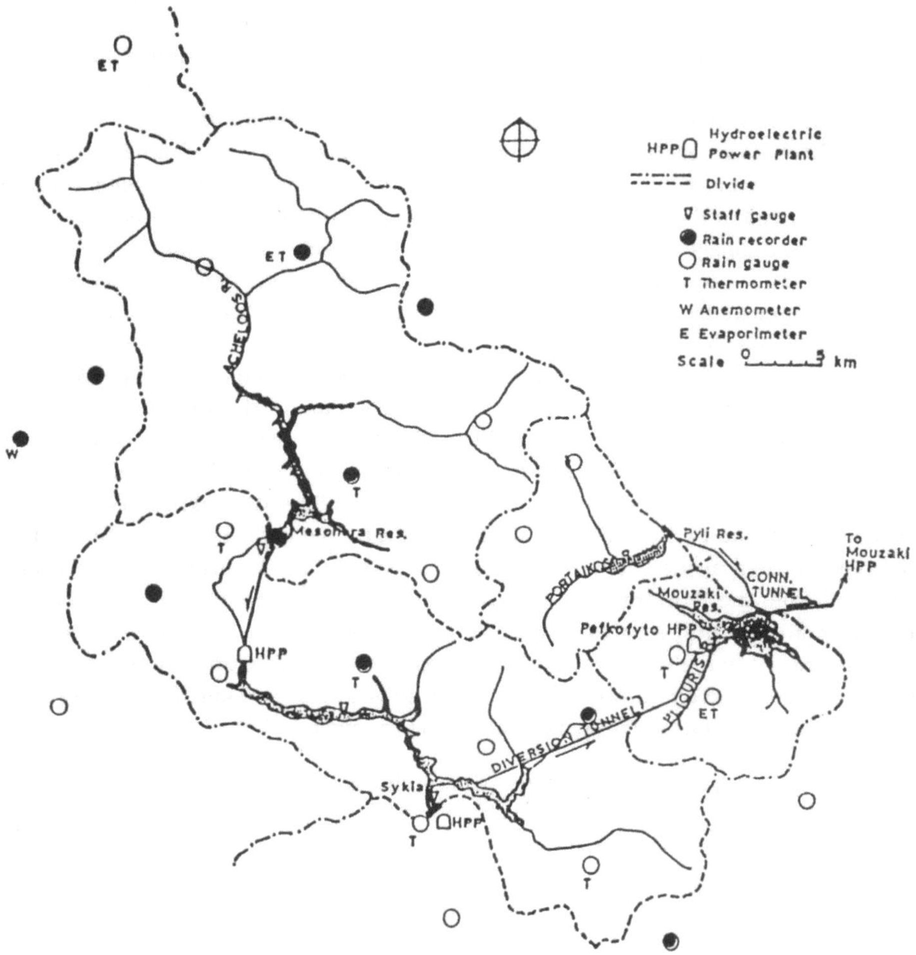

Figure 10-1. General plan of the study area

10-3. REGIONAL HYDROLOGICAL EFFECTS OF CLIMATIC CHANGE

Hydrological System Response to Climatic Changes

A very simple view of the hydrologic cycle in the greenhouse effect is illustrated in Figure 10-2. Of course, the full role played by the hydrologic cycle in the greenhouse effect cannot be inferred from this figure that highlights only consequences and oversimplifies the processes and feedbacks involved. The hydrologic cycle is instrumental in controlling and mediating greenhouse processes through its agency in other primary cycles that make up the global environmental system, such as momentum and energy transfer, carbon, etc. Pathways representing the influence of human activities through changes to the exploitation of water resources and catchment management on growing emission and atmospheric concentration of CO_2, would effectively complete the circle in Figure 10-2.

The hydrological response to the global greenhouse warming will be alterations in major climatic variables, such as precipitation and evapotranspiration, and consequent changes in hydrological variables, such as surface runoff, soil moisture, groundwater and sea-level. These climatic change impacts on various forms of water resources entail serious implications for future water resource planning and management, such as water supply, agricultural water use, reservoir reliability, energy production, land drainage, flood protection, etc., and will also affect water quality and treatment. As previously described, the expected global hydrologic changes include an overall increase in the intensity of the hydrologic cycle due to higher average temperatures and increased evapotranspiration, which, in turn, will mean higher global-average precipitation rates. But such general predictions are of little use for hydrologic studies which usually look at small-scale hydrologic processes or at watersheds that range in size from tens of hectares to hundreds of square kilometers. The essential problem here is to evaluate the regional hydrological effects of such global climatic changes since hydrologic patterns are often characterized by local physical phenomena as a result of which the scale of the surface grids of global climate models is too small to provide the appropriate level of details to evaluate regional impacts. Therefore, other different methods must be used to study in more detail the climatic change effects on small-scale hydrologic phenomena.

An assessment of the impact of climatic changes on hydrology and water resources involves the following steps (Gleick, 1989 and Klemes, 1985):

a) A quantitative estimate of changes in the long term indices of the major climatic variables, such as temperature, precipitation and evapotranspiration, by using quantitative climatic scenarios.
b) A simulation of the hydrologic cycle for a basin of interest, using the scenarios developed in the previous step, as well as estimation of the climatically induced variations in various hydrological variables of interest.
c) An assessment of the implications of the previously identified hydrologic variations

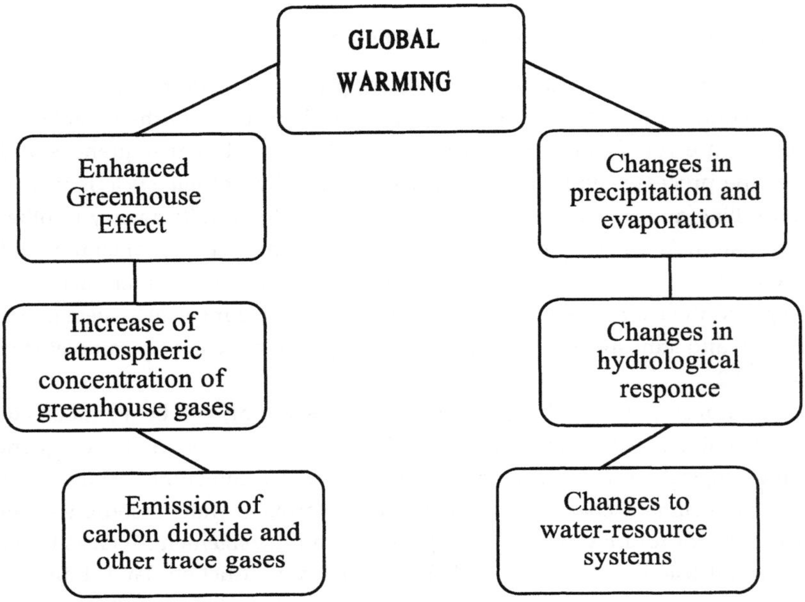

Figure 10-2. The hydrologic cycle in the greenhouse effect

for the performance and the reliability of water resource systems, such as aqueducts, dams, reservoirs, groundwater aquifers, etc.

The first two steps satisfy the needs of hydrologists, whereas the third one is necessary for water-resource planners, which are more interested in the effects of climatic changes on water supply, irrigation, reservoir reliability, energy production, and other water management issues.

Case Study

Hydrological simulation The impacts of climate change on major hydrological variables, such as surface runoff and soil moisture, are estimated by a monthly water balance model developed to this purpose and presented in other previous publications (Mimikou and Kouvopoulos, 1991 and Mimikou *et al.*, 1991a).

The underlying idea, as one can see in Figure 10-3, is that the variation in the key climatic parameters, precipitation P and temperature T, produces significant changes on the hydrological cycle of a basin. Scenarios of climatically changed P and T due to greenhouse warming can either be deduced from running GCMs or from reasonable hypotheses.

A regionally and temporally robust monthly water-balance model would then

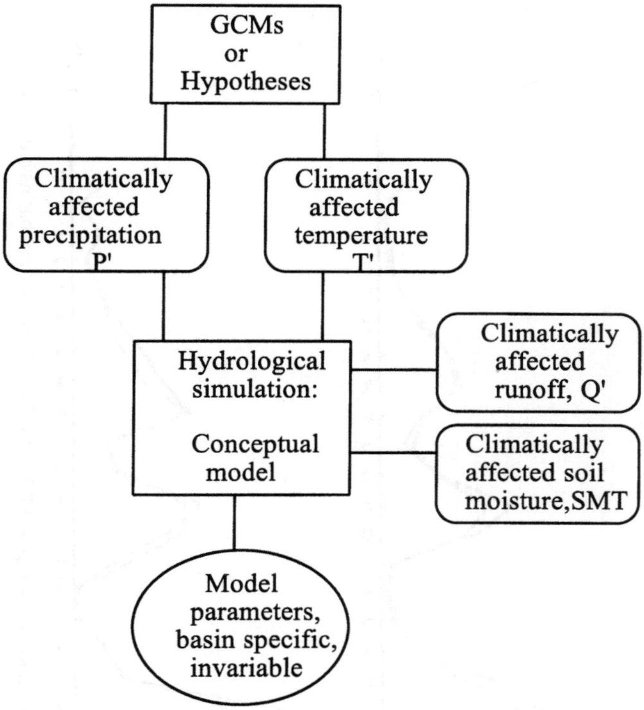

Figure 10-3. Hydrological simulation

provide the integrated simulation of hydrological processes in the basin and, by accepting as inputs the climatically affected time-series of precipitation and temperature, would produce respective estimates of surface runoff and soil moisture. The use of such models for estimating the hydrological impacts of the expected climate change is already prominent (Gleick, 1986 and Nemec and Schaake, 1982).

The model was calibrated and tested for its spatial and temporal robustness for the three basins separately in the 15-year long period of available hydrometeorological records. Regarding its accuracy, one can see in Figure 10-4 the very good fitting of actual and estimated monthly runoff series, for all three basins of the study area.

The Mouzaki basin was excluded from the calibration due to lack of sufficient runoff data and does not appear in the subsequent sections concerning basin hydrological response. However, since the Mouzaki reservoir plays an important role in the water management project, estimates of climatically affected runoff were also made for this catchment, based on a profound basin similarity to the well-gauged Pyli basin (Mimikou *et al.*, 1991b).

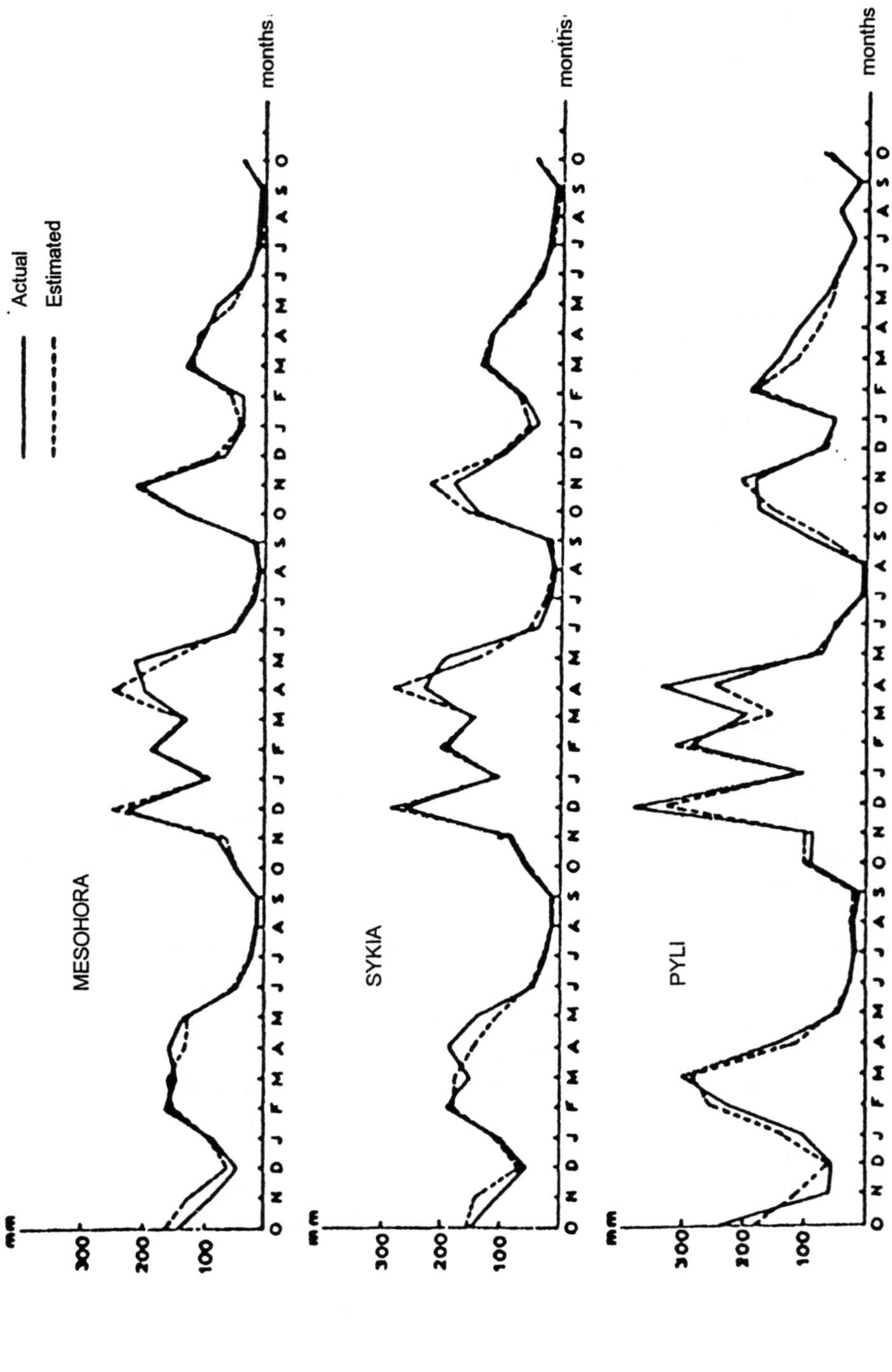

Figure 10-4. Model performance: actual and estimated hydrographs (1972-73 to 1974-75) for the three basins indicated

Hypothetical Climatic Scenarios The most widely accepted prediction regarding the effects of equivalent CO_2 doubling on temperature is that the mean annual global temperature will rise by 3±1.5 °C. The amount of rise is found to increase with increasing latitudes. However, neither the regional nor the temporal distribution of this rise are predictable yet. Therefore, for reasons of compatibility with the above estimated value and comparability with similar efforts, three scenarios were chosen with 1°, 2°, and 4 °C increases. Further, the simplest, yet plausible, assumption of uniform increase over the months of the year was made, taking as a basis the historical temperature time-series of each basin. Although increased rates of global evapotranspiration and precipitation are anticipated after greenhouse warming, little is known on the regional distribution of precipitation changes, regarding not only the quantities, but the direction of changes as well. To overcome this disadvantage, workers in the field usually assume a series of hypothetical scenarios of precipitation changes ranging from -25% to 25% (Gleick, 1986; Nemec and Schaake, 1982) in order to conduct rather a sensitivity analysis of the pertaining hydrologic variables and water-management systems. This approach was also maintained here. Five hypothetical scenarios were considered, with changes -20%, -10%, 0%, 10%, 20%. Furthermore, the simplest, yet plausible at any rate, assumption of uniform change over the months of the year was again made, taking as a basis the historical precipitation time-series of each basin. All temperature and precipitation perturbations mentioned previously (1972-73 to 1974-75) for the three basins combine into 3x5=15 climatic scenarios. Climatically affected series of runoff and soil moisture were obtained as output of the water-balance model. Especially, the estimated runoff series for zero change, referred to as the base run, was used for comparison purposes. The selection of the base runs as basis of comparison was made for uniformity reason and it was accepted since each base run was found to be similar to the historical record for all study basins.

Regional effects on runoff. To better appreciate the results, it is useful here to underline some essential differences between the basins considered, regarding their morpho-climatic characteristics:

a) Mesohora and Sykia basins are purely mountainous, cool and snowy (mean annual temp. 7.9 °C and 8.8 °C). A significant retention of precipitation occurs in the form of snow pack storage. On the other hand, Pyli basin is considerably lower and warmer (11.2 °C). Although it receives some snow, its snowpack storage is almost negligible.
b) Pyli basin is characterised by very high monthly coefficients of direct runoff, having to do with local physiographic characteristics (i.e. steep slopes and a thin soil layer).
c) Pyli basin is more humid than the other two. Although lower, it receives almost the same amount of yearly precipitation, approximately 2000 mm, and its specific discharge exceeds those of the other two by approximately 20%.

Some results regarding annual, and seasonal surface runoff characteristics have been tabulated in Table 10-3 (adapted from Mimikou, 1991a). In this table, base run results are presented in dimensional form (mm of equivalent depth), whereas perturbed climatic scenario outputs are presented as percentages of change of the base run values.

In Table 10-3, it can be seen that the effects of temperature rise alone vary in magnitude from basin to basin, depending on its specific morpho-climatic characteristics. Basins which, due to these characteristics, retain water for a longer period (such as Mesohora and Sykia as opposed to Pyli basin), display more sensitivity to temperature increases. In general, percentage reductions are more dramatic in dry years (minimum annual runoff in Table 10-3) and dry seasons, where the largest percentages of reduction appear, since evapotranspiration does not deplete water proportionally to the precipitation, as Gleick (1986) has also pointed out explaining similar results.

The effects of precipitation changes upon the annual quantities of runoff are characterised by a magnification with respect to the change of the causative factor (precipitation change). The magnification factor is almost independent of temperature increase and mainly depends on the local characteristics. This can be visualised in Figure 10-5, where the slopes of the parallel lines represent the constant magnification factor (Mimikou and Kouvopoulos, 1991).

The two seasonal quantities tabulated in Table 10-3 (winter season is from December to February and summer season is from June to August) which have been considered here, are marked by a contrasting behaviour, apparent in the example of Figure 10-6, which shows percent changes of monthly mean winter and mean summer runoff for Mesohora basin as a function of temperature and precipitation changes.

So while a temperature rise with no change of precipitation increases winter runoff, it reduces summer runoff. Also, summer runoff is more sensitive to temperature changes. The reason underlying this behaviour is common: it is the snow-cover that controls how temperature changes will affect seasonal runoff. So, the warm Pyli basin where snow-cover is almost absent, barely follows the above rule regarding winter runoff increase, whereas the cool Mesohora basin with a significant snow-cover exhibits increased winter runoff (up to 23%) and very large reduction in summer runoff (reduction by 48% for $f' = +4$ °C). The latter is attributed to the earlier onset of spring snow-melt as well as to increased evapotranspiration.

Table 10-3. Mean annual and seasonal runoff. Percent changes

	Base run (mm)	ΔP = 0			ΔP = 10%			ΔP = 20%			ΔP = -10%			ΔP = -20%		
		ΔT=1	ΔT=2	ΔT=4	ΔT=1	ΔT=2	ΔT=4	ΔT=1	ΔT=2	ΔT=4	ΔT=1	ΔT=2	ΔT=4	ΔT=1	ΔT=2	ΔT=4
Mesohora basin																
Annual																
Maximum	1611	-2	-4	- 9	12	10	5	26	24	19	-16	-18	-22	-30	-32	-36
Minimum	707	-3	-6	-13	13	10	3	30	27	20	-20	-23	-29	-36	-39	-45
Mean	1165	-1	-3	- 7	14	12	7	29	27	22	-16	-17	-21	-30	-31	-35

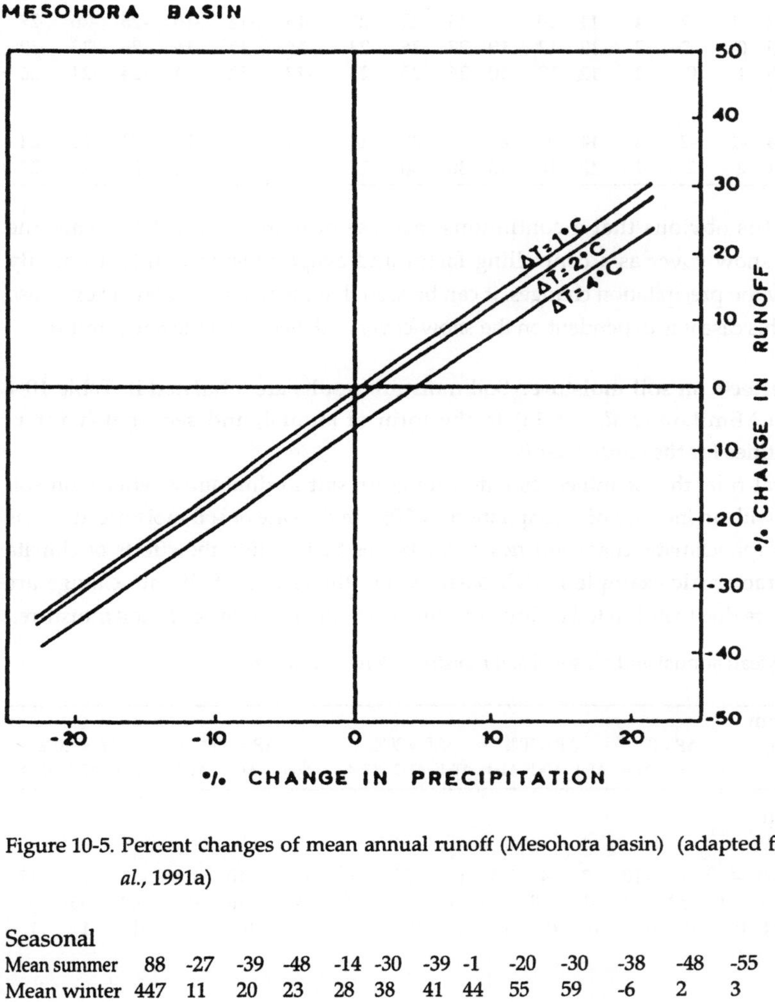

Figure 10-5. Percent changes of mean annual runoff (Mesohora basin) (adapted from Mimikou *et al.*, 1991a)

Seasonal																
Mean summer	88	-27	-39	-48	-14	-30	-39	-1	-20	-30	-38	-48	-55	-49	-57	-63
Mean winter	447	11	20	23	28	38	41	44	55	59	-6	2	3	-24	-17	-6
Sykia basin																
Annual																
Maximum	1882	-2	-4	-7	12	10	6	25	24	20	-15	-17	-20	-29	-30	-33
Minimum	711	-3	-7	-14	14	10	3	31	27	20	-20	-23	-30	-37	-40	-46
Mean	1277	0	-2	-6	14	12	8	29	27	23	-15	-16	-20	-29	-30	-34
Seasonal																
Mean summer	101	-21	-32	-40	-9	-22	-30	4	-11	-20	-33	-42	-49	-44	-52	-58
Mean winter	508	10	17	17	26	33	34	41	49	51	-6	0	0	-22	-18	-19

Pyli basin																
Annual																
Maximum	1808	-1	-2	-4	11	10	9	23	23	21	-13	-14	-16	-25	-26	-27
Minimum	1035	0	-2	-3	12	12	10	27	26	23	-14	-15	-16	-26	-27	-30
Mean	1355	1	0	-2	12	13	10	26	25	23	-12	-12	-14	-24	-24	-26
Seasonal																
Mean summer	84	-1	-2	-3	11	9	8	21	20	18	-11	-12	-13	-22	-23	-24
Mean winter	621	2	2	-1	13	16	13	30	30	27	-12	-12	-14	-25	-25	-27

However it is obvious that a continuous increase of temperature diminishes the importance of snow-cover as a controlling factor and evapotranspiration takes on. By accounting now the precipitation changes, it can be seen that the magnification factor is also present but in this case it is dependent on the snow-cover and, hence, on the temperature.

Regional effects on soil moisture. Soil moisture results are tabulated in Table 10-4 (adapted from Mimikou *et al.*, 1991a) in the form of annual, and seasonal (winter-summer) quantities for the study basins.

As a general rule, the examined climate changes result in diminutive effects on soil moisture, even with an increase of precipitation by 20%. In the case of soil moisture also, it is the regional morphoclimatic characteristics of the basins that modify the effects of climate change. A characteristic example is Pyli basin where the effects of climate change are particularly severe due to its limited ability to retain water either as snow or as soil moisture.

Table 10-4. Mean annual and seasonal soil moisture. Percent changes

	Base run (mm)	ΔP = 0			ΔP = 10%			ΔP = 20%			ΔP = -10%			ΔP = -20%		
		ΔT=1	ΔT=2	ΔT=4	ΔT=1	ΔT=2	ΔT=4	ΔT=1	ΔT=2	ΔT=4	ΔT=1	ΔT=2	ΔT=4	ΔT=1	ΔT=2	ΔT=4
Mesohora basin																
Annual																
Annual Mean	208	-4	-7	-10	-2	-4	-8	1	-2	-5	-7	-10	-13	-11	-14	-17
Summer Mean	91	-24	-38	-54'	-15	-28	-46	-5	-19	-38	-33	-46	-60	-42	-55	-67
Winter Mean	298	0	0	0	0	0	0	0	0	0	0	0	0	-1	-1	-1
Sykia basin																
Annual																
Annual Mean	207	-5	-7	-10	-2	-5	-8	1	-2	-5	-8	-10	-13	-11	-13	-16
Summer Mean	89	-29	-41	-56	-20	-32	-48	-10	-23	-41	-38	-50	-63	-46	-58	-69
Winter Mean	298	0	0	0	0	0	0	0	0	0	0	0	0	0	0	-1
Pyli basin																
Annual																
Annual Mean	116	-6	-10	-17	0	-6	-12	2	-2	-8	-11	-16	-23	-18	-23	-32
Summer Mean	1	-93	-100	-66	-96	-100	-40	-69	-100	-100	-100	-100	-100	-100	-100	-100
Winter Mean	217	-1	-1	-4	1	0	-1	2	2	0	-3	-5	-8	-8	-10	-15

Snow cover always plays an important role: temperature rise reduces spring snowmelt runoff and makes it shift towards winter when soil is near saturation and potential evapotranspiration is low. This results in a reduction (mainly in May) of the surface water available for evapotranspiration and soil moisture saturation. For this reason the

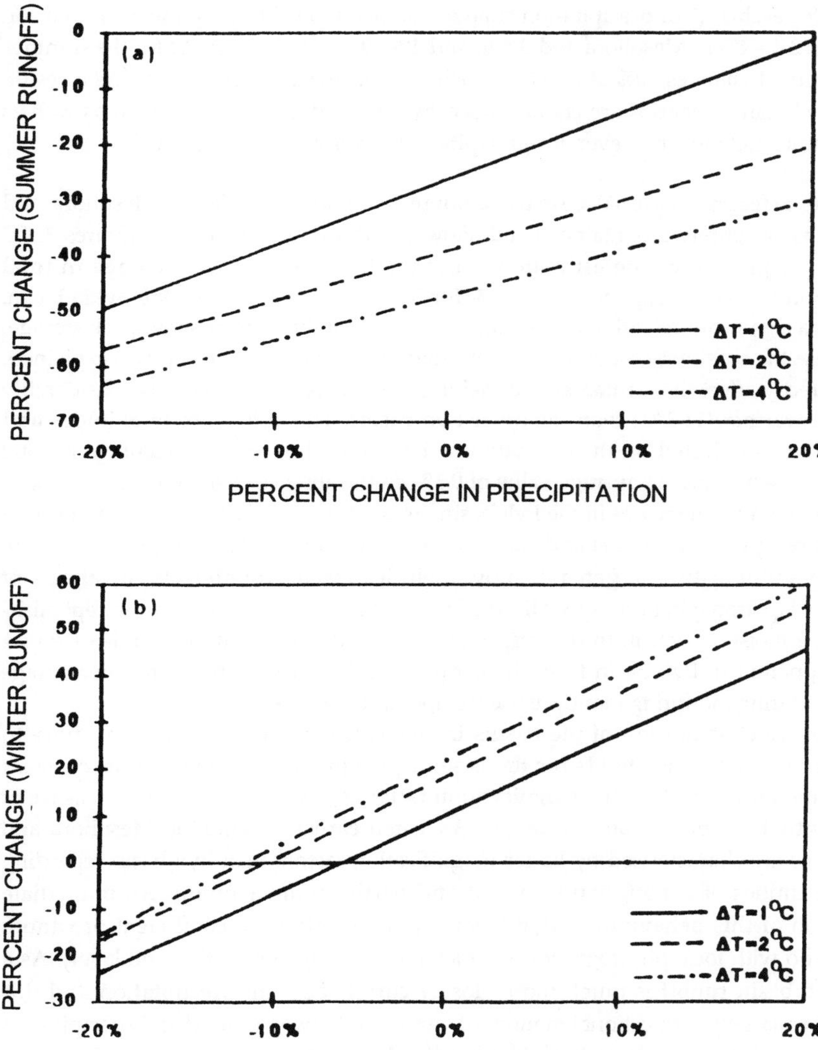

Figure 10-6. Percent changes of mean winter and summer runoff (Mesohora basin) (Mimikou *et al.*, 1991a and Mimikou, 1995)

warm season's significant soil moisture deficits start to appear in May in the resembling Mesohora and Sykia basins and earlier in Pyli basin which is characterised by the absence of snow pack and of the associated spring runoff in the base case.

Of particular interest are the effects of precipitation changes. Here, contrary to the magnification effect associated with runoff, a diminishing effect is observed. For example, for each 10% of precipitation change, the resulting change of mean annual soil moisture is 2% - 4% in Mesohora and Sykia and 4% - 9% in Pyli basin. As for the summer soil moisture, it changes 10% at most for each 10% of precipitation change. As a result, almost all climatic scenarios are characterised by a reduction of mean annual as well as mean summer soil moisture, even for precipitation increases as much as 20%.

Factors affecting regional climatic change hydrological effects. Mesohora and Sykia basins are purely mountainous and snowy, with mean annual temperatures 7.9°C and 8.8°C respectively. The estimate for the winter-time snow percentage in total precipitation has an average value of 27% for Mesohora and 23% for the Sykia basin. Therefore, a significant retention of precipitation occurs in the form of snow-pack storage. In this sense the sensitivity these two basins exhibit to temperature increases is explained, as they directly affect the formation and melting of snow-pack. Pyli basin is considerably lower and warmer (11.2°C) than the other two basins. This difference in altitudes and temperatures is reflected on the estimate for the winter-time snow percentage in total precipitation, which has an average value of 0.3%. In the absence of significant snow-pack storage, temperature increases in the Pyli basin are restricted to affect runoff only through the evapotranspiration process and the basin is shown to be not as sensitive to them. Regarding soil moisture in general, snow-melt, if present, counteracts the effects of increased evapotranspiration especially during spring, by contributing to the replenishment of soil moisture deficit. In this respect, the absence of the controlling role of snow-cover is apparent in Pyli basin from the comparatively much more intense percentage reductions during the spring months under temperature increases.

The above classification of the basins in two categories in terms of sensitivity to temperature changes also holds for the case of precipitation changes under constant temperature. In this context, the magnification factor can be considered as a measure of basin sensitivity to precipitation changes. As stated earlier, resembling Mesohora and Sykia basins are characterised by larger magnification factors than Pyli basin regarding annual quantities of runoff, winter runoff and partly summer runoff. An immediate explanation of this behaviour is Pyli basin's particularly high coefficient of runoff, having to do with local physiographic characteristics (steep slopes, thin soil layer). As a result, Pyli basin runoff is much more closely correlated with precipitation and this accounts in a largely proportional manner of variations between runoff and precipitation (magnification factor close to unity). Climatically, the observation is made that Pyli basin is more humid than the other two. In this respect basin aridity appears to be positively associated with basin sensitivity (as measured by the magnification factor) to precipitation changes (Dooge, 1989 and Mimikou and Kouvopoulos, 1991).

10-4. REGIONAL CLIMATE CHANGE IMPACTS ON WATER RESOURCES SYSTEMS

Basic Considerations

The necessity of an assesment of the climatic change impacts on water resources becomes apparent when one realizes that, for the majority of the water resource systems which are to start operation early in next century, the planning has already started. Such an assessment would show the changes in the efficiency of the water resource systems under the new climatic conditions and indicate the necessary measures needed to mitigate the impacts of adverse changes or to take advantage of favorable changes. An assessment of the climate change impacts on water resources presupposes that quantitative estimates of changes in the long-term indices of the major climatic and hydrological variables can be converted into estimates of how the availability and the quality of freshwater resources may change. As discussed in the preceding sections, the state-of-the-art real time climate forecasting does not permit any specific estimate of future changes in temperature and precipitation to be considered as a deterministic forecast. For this reason, though it cerainly would be preferable for water managers to have assessments based on reliable real-time forecasts, the assessment of climatic change impacts on water resources is usually made by using plausible climatic change scenarios. Therefore, the results of such assessments should be interpreted as a sensitivity analysis of water resource systems to alternative feasible climatic conditions rather than as predictions (Klemes, 1985).

The assessment of water-resource impacts is not only based on the perturbations to the hydrological variables, but also takes a use-oriented view focusing on water that is available for human exploitation; i.e. water at the right time and place and in the right quantity and quality (WMO, 1987). Changes in major hydrological variables will have serious impacts on the potential for water supply, entailing implications for domestic water use, irrigation, industrial water use and energy production. Since the key element of most large water supply projects is a storage reservoir, most of the studies are concentrated on the impacts on characteristics of a reservoir, and particularly, on the reliability of its performance under changed climatic and hydrologic conditions. Moreover, the effects on the occurrence frequency of extreme flow events, such as floods and droughts, will have potentially devastating results, and cause problems mainly related to physical security and food production. A possible sea-level rise would affect navigation and the quantity and quality of freshwater resources in coastal zones. With respect to flooding, a combination of a sea-level rise and an increased upstream runoff could prove particularly dangerous for coastal lowlands. In addition to these primary water resource management activities, there are other critical issues that would be seriously affected by a climate change, such as a wide range of environmental services, fisheries, wetlands preservation, wildlife, conservation and recreation.

In the earliest literature most of the studies focused on more direct hydrological impacts, whereas there have been relatively few attempts to determine the sensitivity of water management activities to climatic changes. Even fewer were the studies that considered further socio-economic impacts of altered water resources. The very unusual climatic conditions and the severe drought and flooding problems that have characterized much of the post-1987 period in many places of the world, stimulated considerable scientific and public interest on the possible impacts of the greenhouse warming on the availability and quality of freshwater resources. Research efforts have been accelerated, and attempts are being made towards an integrated approach to study all aspects of the problem from basic physical issues to questions of environmental, political and socio-economic nature.

Despite the increasing research efforts that focus on the water-resource impact area, the coverage of this multiplex problem is not satisfactory yet since several issues must be further investigated. Such issues include: impacts on groundwater withdrawal and recharge, effects of sea-level rise on quantity and quality of coastal freshwater, and the establishment of strategies to mitigate impacts on water resource systems of national, and especially of shared international rivers. Many other water resource problems will also arise related to impacts on environmental services, international-political security, and socio-economic development, which consequently require a lot more research.

Case study

For each of the 15 pairs of temperature and precipitation scenarios, climatically affected runoff time series I_t for the basins of the study area were produced (Mimikou *et al.*, 1991a). The 15 monthly time series I_t were finally used as inputs to the reservoirs at the outlet of each basin in order to check their sensitivity to the expected climate changes.

The operation of the system of four reservoirs shown in Figure 10-1 is simulated by a monthly water balance equation under various constraints concerning storage volume, outflow from the reservoir and energy production. Details on the reservoir operation simulating model and its performance are presented in other previous publication (Mimikou *et al.*, 1991b).

The impacts on firm water and energy supply levels were estimated by a probabilistic assessment of the risk involved under the various climatic scenarios. The monthly flow I_t, historical or climatically affected, were used as inputs to a stochastic model of the AR(2) type to generate synthetic series of inflows spanning 50 years. For each climatic scenario (current or hypothetical) a multitude of synthetic inflow series were thus fed into the system operation simulating model and the relative frequencies of monthly failures, or risk values, were estimated for various combinations of primary energy E and firm water supply W. A failure was considered to occur when either of the produced quantities failed to meet the preselected demand.

The risk analysis using the historical series revealed that the probability of failure at the examined reservoirs, associated with the design values for W and E, was well below 1% for all of them.

The results of the analysis on the probability of failure associated with the design values of E for the Mesohora reservoir and W and E for the Sykia and Mouzaki reservoirs under the 15 climatic scenarios are presented in Table 10-5 (Mimikou *et al.*, 1991b).

From this table it is deduced that, under the changed climatic conditions considered, reservoir failure is affected rather by precipitation reduction than by temperature increases. Under the same precipitation reduction, downstream reservoirs exhibit larger risk levels than upstream ones. This is explained by the cumulative effect of upstream failures transferred downstream.

Regardless of the temperature increase, precipitation increases do not significantly improve risk levels, while precipitation reductions significantly and disproportionately raise the risk levels. Regarding temperature increases alone, the Mesohora reservoir appears to be relatively more sensitive than the other two, owing to the intensification of the already high seasonality of runoff in conjunction with its comparatively smaller storage capacity.

Table 10-5. Risk (%) for the design values of energy and water yield under climate change

Climatic scenarios	Mesohora HPP	Pefkofyto HPP	Mouzaki HPP
Current-climatic conditions	0.35	0.04	0.33
ΔT= 1 °C			
ΔP= 20%	21.43	26.99	51.40
-10%	7.02	5.92	13.77
0%	1.5	0.88	1.35
10%	0.56	0.03	0.16
20%	0.33	0.03	0.04
ΔT= 2 °C			
ΔP=- 20%	21.29	27.28	49.37
-10%	9.01	6.95	15.12
0%	3.77	1.05	1.56
10%	1.74	0.31	0.38
20%	0.61	0.07	0.03
ΔT= 4 °C			
ΔP= -20%	25.29	32.99	53.65
-10%	11.15	9.90	20.54
0%	4.43	1.38	2.47
10%	2.10	0.30	0.28
20%	11.07	0.02	0.04

When temperature increases combine with precipitation reductions then the risk levels for water and energy supply were shown to be unacceptably high.

However, increasing net storage of the reservoirs compensates for this effect of precipitation reduction. To obtain an idea of how much should the net storage volume be increased in order the design (W,E) values be maintained at tolerable risk levels, the upstream Mesohora reservoir was examined under precipitation reduction scenarios. Results are shown in Figure 10-7. As can be seen in Figure 10-7, radical increases of net storage volume sometimes greater than 100% are required for the design energy production to be met with tolerable risk levels. From this aspect, precipitation reductions play again the important role, the effect of warming appearing to be marginal as compared to that of precipitation reductions.

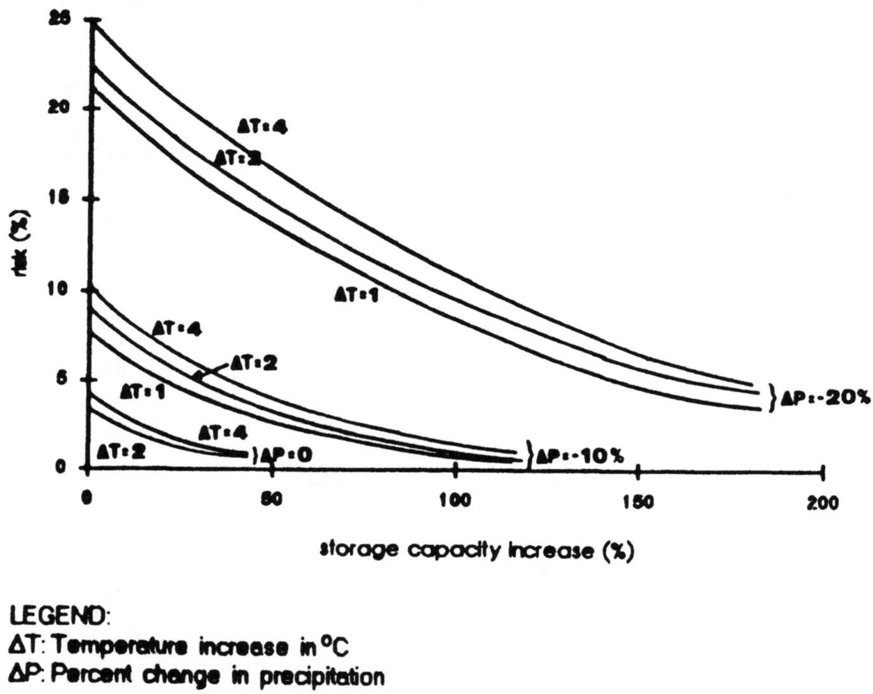

Figure 10-7. Risk of guaranteed annual production as a function of storage volume increase for the Mesohora reservoir (Mimikou *et al.*, 1991b and Mimikou, 1993)

10-5. CONCLUSIONS

The main conclusions drawn from this research are the following:

a) Mountainous, snow-covered Mediterranean basins with effective water-retentive characteristics under temperature increase exhibit reductions of soil moisture in general and more severely so in summer, serious reductions of mean annual runoff, even more serious reduction of mean summer runoff and increases of mean winter runoff accompanied by a shift of spring runoff. A warmer humid basin with local characteristics that unfavour water retention, exhibits minimal sensitivity of runoff to temperature increase.

b) The effects of precipitation change upon the quantities of runoff are characterised by a magnification factor. Basin aridity appears to be positively associated with the sensitivity of runoff to precipitation changes as measured by the magnification factor. The latter, on an annual basis, seems to be independent of the temperature increase, whereas on a seasonal basis it depends on the snow-cover and hence on the temperature.

c) Comparison renders snow (accumulation and melting) as most significant and determining factor of basin response to climatic change. This factor in turn depends on the orographic characteristics of the basin (altitude, area-elevation relationship, orientation) besides the general climatic ones.

d) The climatically affected runoff causes serious increases of the risk associated with the annual firm water and energy production from the system of reservoirs in the study area, when precipitation decreases are involved. The effect of temperature rise on risk levels is marginal. Downstream reservoirs appear to be more sensitive than upstream ones. Radical increases of reservoir storage volume are required to compensate for the effect of precipitation reduction on risk levels.

Acknowledgements

The authors wish to thank the Public Power Corporation of Greece for providing the hydrometeorological and other data used in the study.

REFERENCES

Dooge J.C.I. (1989). Effects of CO_2 increases on hydrology and water resources. *In*: Carbon Dioxide and Other Greenhouse Gases: Climatic and Associated Impacts. CEC - Kluwer Academic Publishers, Dordrecht, pp 204-213.

Gleick P.H. (1986). Regional Water Availability and Global Climatic Change: The Hydrologic Consequences of Increases in Atmospheric CO_2 and Other Trace Gases. Ph.D. Dissertation, Univ. of California, Berkeley, California, pp. 615.

Gleick P.H. (1990). Vulnerability of water systems. *In*: Waggoner P. (Ed), Climate Change and U.S. Water Resources, John Wiley and Sons, New York, pp. 223-240.

Klemes V. (1985). Sensitivity of Water Resource Systems to Climate Variations. World Clim. Appl. Programme WCP-98, World Meteorol. Organ. Geneva, pp. 17.

Mimikou M.A. (1995). Climatic change. *In*: Singh V.P. (Ed.), Environmental Hydrology. Kluwer Academic Publishers, pp. 69-106.

Mimikou M.A. and Kouvopoulos Y.S. (1991). Regional climate change impacts: I. Impacts on water resources. Hydrol. Sc. J., 36,3,6: 247-258.

Mimikou M.A., Y.S. Kouvopoulos, G. Cavadias, and Vayiannos N. (1991). Regional hydrological effects of climate change. J. Hydrol., 123: 119-146.

Mimikou M.A., P.S. Hadjissavva, Y.S. Kouvopoulos, and H. Afrateos. (1991b). Regional climate change impacts: II. Impacts on water management works. Hydrol. Sc. J., 36,3,6: 259-270.

Nemec J. and J. Schaake (1982). Sensitivity of water resource systems to climate variations Hydrol. Sc. J., 27(3): 327-343.

World Meteorological Organisation. (1987). Water Resources and Climatic Change: Sensitivity of Water Resources Systems to Climate Change and Variability. Rep. WCAP-4, WMO/TD 247, Geneva, pp. 50.

<u>*CHAPTER 11*</u>

CATCHMENT HYDROLOGICAL SENSITIVITIES TO CLIMATE CHANGES

DIONYSIA G. PANAGOULIA

Division of Water Resources, Hydraulic and Maritime Engineering,
Department of Civil Engineering, National Technical University of Athens,
5 Iroon Polytechniou, 15780 Athens, GREECE

11-1. INTRODUCTION

While climate changes over periods of thousands of years are well documented, hydrologists were rather reluctant to agree for as long as a decade (1960 to 1970) as to whether changes (signals) within typical water resources systems design periods (100 years or less) can indeed be distinguished from random variations (noise) in a physical hydrological time series (US. National Academy of Sciences, 1977). The advent of general circulation models (GCMs) over the last decade and consensus about the direction of future global climate change threw abundant light on the controversial theme, thus making acceptable the aspect that climate change does exist.

The continued burning of fossil fuels, plus deforestation and changes in use of land (converting it to buildup and industrial areas), let alone the increased use of nitrogen fertilizers, as well as large nuclear eruptions and primarily atmospheric pollution, have increased the concentrations of carbon dioxide and other trace gases (e.g. CH_4, N_2O, CFC-11, CFC-12) by about 20% over the last one 100 years. It is expected during the next 50-80 years (until 2070) that these activities will result in doubling the concentration of CO_2 in the atmosphere (US National Academy of Science, 1983; MacCracken and Luther, 1986; Manabe and Wetherald, 1985; Mitchell and Qingcum 1991) and cause through green house effect a gradual warming up of the Earth. By varying the level of carbon dioxide concentration in GCMs simulations, the models can provide quantitative estimates (predictions) of climate and hydrology variables for any level of CO_2 and its possible doubling (experiment of CO_2 doubling).

A GCM solves numerically the equations of mass, energy, momentum and state on a global grid of cells (Hansen *et al.*, 1983; 1988; Mitchell and Qingcuì, 1991). But, this spatial scale of the GCM outputs is inadequate for catchment hydrological simulation. The GCM predictions are provided as spatial averages over areas of the order of 10^4-10^5 km^2 (macroscale), where the fundamental differential equations of the continuum hydro- and

NATO ASI Series, Vol. I 36
Diachronic Climatic Impacts on Water Resources
Edited by Andreas N. Angelakis and Arie S. Issar

thermo- dynamics can conserve a real-world validity in contrast to the validity of these equations when they are applied to the hydrological modelling of a catchment (meso-micro scale simulation) (Becker and Nemec, 1987). Furthermore, it is doubtful whether GCM predictions on time scales shorter than one month do reflect the natural variability of field data, because these predictions represent grid-cell averages. On the other hand, all the climate models, including the better parameterized ones (GCMs), give different values of climate variable changes and not a single reliable estimate that could be advanced as a deterministic forecast for hydrological modelling.

The study of the hydrological system sensitivities to changes in meteorological inputs requires these inputs to be specified on a time step appropriate for conceptual catchment simulation to event storms. This is because the transfer function of rainfall to runoff is not linear, also because the infiltration and evapotranspiration processes, that play a major role in catchment runoff determination, are highly dependent on the storages and movement of water within the soil during storm occurences, as well as on the soil moisture state at the beginning of storm (determination of soil moisture capacities).

The operation of soil moisture system requires daily, hourly or even shorter time scales, depending on the size of the river basin. The GCM outputs can be represented on daily or shorter time steps, but it is not clear whether the results with these short scales properly reflect the short term dynamics of the atmospheric circulation process (Lettenmaier and Gan, 1990).

Since we have yet to achieve an appropriate space-time coupling between GCMs outputs (e.g. temperature, precipitation, reference evaporation, etc.) and hydrological models, we thought, for hydrological purposes, of adjusting the present day surface meteorological data to account for scenarios of meteorological variable long-term changes based on GCM predictions.

Yet, any calibration for climate studies demands accurate and physical climatic information (Hutchinson, 1990, and Wallis *et al.*, 1991). This information is not so easy to obtain in mountainous regions where high slopes and strong winds do affect the measurement of true precipitation. In addition, harsh conditions cause instruments to malfunction more frequently, thereby leading to gaps in the records, as well as to nonhomogeneity when instruments have to be replaced and/or recalibrated.

The sensitivity of water resource systems to climate variations was first studied by Nemec and Schaake (1982) for two river basins in USA. Nemec and Schaake introduced the global climate change (climate variations) by hypothetical scenarios of changes in precipitation (±10%, ±25%) and temperature (±1°C, +3°C), the latter converted into changes in reference evaporation (±4%, +12% Budyko- relation). They did not study snowcovered basins and deal with other catchment hydrological processes than runoff.

Subsequently, Lettenmaier and Gan (1990) investigated the hydrological effects of four subbasins in the Sacramento-San Joaquin river basin (California) for increases in temperature and reference evapotranspiration, as well as changes in precipitation that resulted from GFDL (Fluid Dynamics Laboratory), GISS (Goddard Institute for Space Studies), and OSU (Oregon State University) model scenarios.

The studies of Nemec and Schaake (1982) and Lettenmaier and Gan (1990) did not actually face the problem of incomplete meteorological point data integration for climate change adjustment. This difficult problem was dealt with in studies of Panagoulia.

So, the present chapter has been produced from a series of papers (Panagoulia, 1990, 1991a,b, 1992a,b, and 1993a,b) which deal with the long-term hydrological sensitivities (snow water equivalent, runoff, actual evapotranspiration and soil moisture) of a medium-sized mountainous catchment (in our case the Mesochora catchment in Central Greece) to hypothetical (given as range of changes in long- term average values of meteorological variables) and GISS-modelled climate changes. The used meteorological data included incomplete point values of daily precipitation and min-max temperature. In order to preserve the physical nature of climatic information and avoid the errors caused by the interpolation techniques (Georgakakos and Krajewski, 1991; Wallis *et al.*, 1991), we would rather not estimate the unavailable values, but integrate instead the existing ones for areal variation and change pro rata with elevation.

11-2. CATCHMENT FEATURES AND CLIMATE CHANGE SCENARIOS

The Mesochora catchment (632.8 km^2) lies in central mountain region of Greece and extends nearly 32 km from north (39° 42′) to south (39° 25′) with an average width of about 20 km (Fig. 11-1). The Pindus Mountains, with peak elevations of about 2300m, form the western boundary of the catchment, while the eastern one is formed by Koziakas mountains with peaks of about 2000m. Inside, the catchment presents intense topography with strong interchanges of lower and higher elevations. The mean elevation of catchment is 1390 m. The network of meteorological stations installed in and around the catchment is relatively dense, but 3.5% of daily precipitation values and 15.5% of daily min-max temperature values were missing for the 15-year period used in this study.

The climate in the Mesochora catchment is elevation-dependent, with hot summers and mild winters at low elevations and mild summers and cold winters higher up. Because of its high mean elevation, its hydrology is controlled by snowfall and snowmelt. The mean annual precipitation of catchment (weighted average over elevation bands) is 1898mm, and most of precipitation falls as snow at the higher elevations. The mean annual runoff of the catchment is about 1170 mm (or 23.5 m^3/s). The mean January and July daily temperatures (weighted average over elevation bands) are:

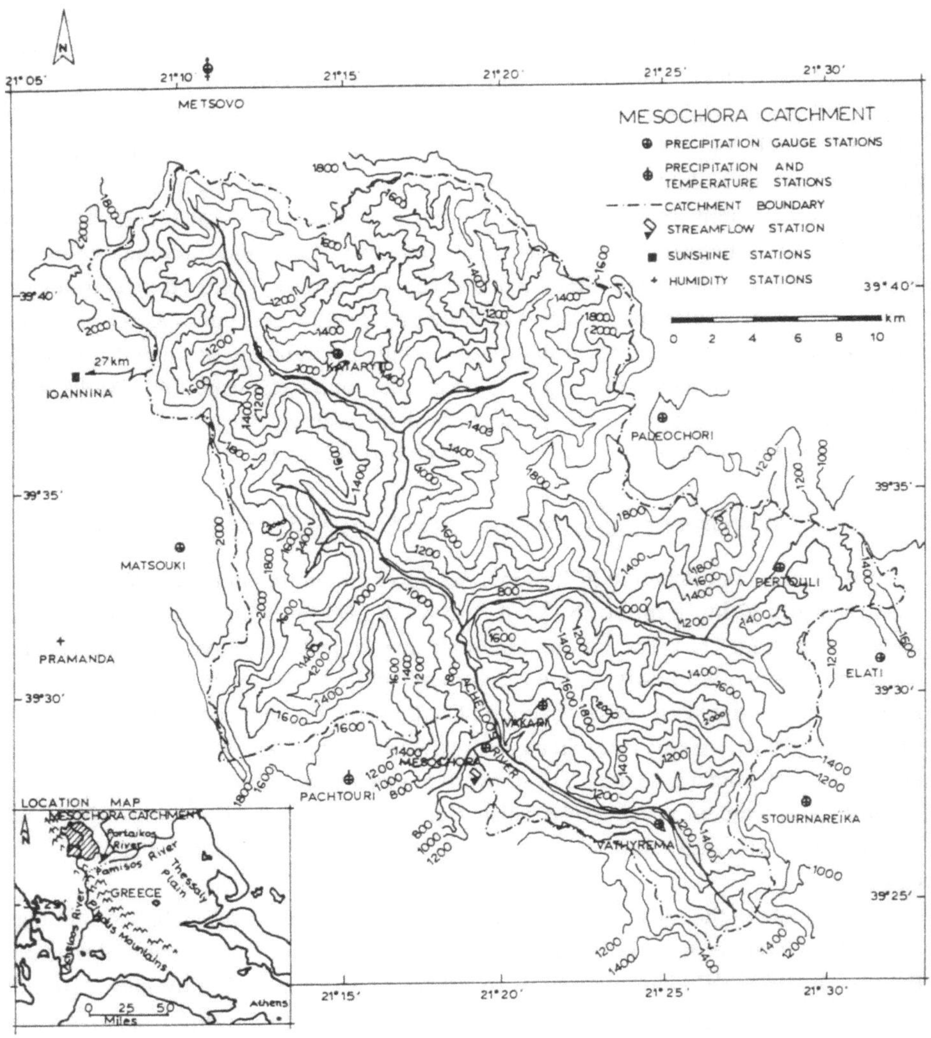

Figure 11-1. The Mesochora catchment, Greece. Topography and hydrometeorological stations

MEAN DAILY TEMPERATURE [^{0}C]

	JANUARY	JULY
Daily Average	0.100	17.53
Minimum	-3.511	11.43
Maximum	3.711	23.63

The soils of the catchment have been formed from decayed hard limestone and flysch. They are varied, but generally permeable. Mesochora, which constitutes the upper drainage catchment of Acheloos river, has a great singificance for Greece because the river will be partially diverted at the outfall of the Mesochora catchment through Pindus mountains to irrigate the arid Thessaly Plain. It is the largest construction project in Greece including five dams (one is Mesochora's), 40 miles of large tunnels and about 8000 miles of buried irrigation piping.

In order to analyze the sensitivities of catchment hydrological regime to climate changes we used:

a) Fifteen hypothetical scenarios denoted as HYPO(Δ,T, ΔP), where ΔT is temperature increase by 1, 2, 4°C and ΔP is precipitation change by 0, ±10, ±20%, (Panagoulia 1991a)
b) Two GISS-predicted scenarios (with both monthly precipitation and temperature changes GISS(t,p), and with monthly temperature changes alone GISS(t,0)) (Panagoulia, 1992b).

The range of HYPO precipitation changes was selected from the recent climatological literature (U.S. National Academy of Sciences, 1983; Manabe and Wetherland, 1985 and MacCracken and Luther, 1986). Such scenarios have been previously used by Nemec and Schaake (1982), Revelle and Waggoner (1983) and Gleick (1986), for streamflow modelling under climate change conditions.

The HYPO temperature increases were taken from the range of 1.5 and 4.5°C temperature increase according to climate predictions for CO_2 doubling by the year 2070 (US. National Academy of Sciences, 1983; Dickinson, 1982, and Mac Cracken and Luther, 1986).

The GISS scenarios resulted from the GISS general circulation model which was developed by Hansen *et al.* (1983) of the Goddard Institute for Space Studies (NASA) in USA and keeps on ever since. This global model has realistic topography, 8°x10° (latitude x longitude) resolution in the horizontal, and nine layers in the vertical. The model simulates climate by solving the fundamental equations for conservation of mass, energy, momentum and water. The source terms in these equations incorporate numerical representations of the physical processes of radiation, turbulent transfers at the ground-atmosphere boundary, cloud formation, condensation of rain, and transport of heat by ocean boundary currents. A complete description of the model appears in Hansen *et al.* (1983).

However, there are many uncertainties in the modelling and parameterization of terms. Various processes, such as cloud feedbacks, are only crudely modelled. Initial conditions of temperatures and precipitation, i.e. control runs, are rather difficult to determine at all points. As a result, the model is unable to completely simulate the present climate, especially precipitation. When the initial parameterizations are modified by doubling atmospheric CO_2 concentration, any errors inherent in the initial model runs would probably occur in the $2xCO_2$ simulations. Another problem has to do with the relatively coarse resolution of the model outputs (8°x10°).

Despite the limitations on spatial resolution and hydrological parameterizations of the model, the outputs of GISS model, as well as certain other relative models, such as GFDL, OSU, and NCAR, are used for the formation of climate change scenarios for several reasons:

a) To evaluate the strengths and weaknesses of the hydrological mechanisms used by GCMs
b) To study the sensitivity of a watershed hydrological system to changes in meteorological variables.
c) To throw light on the required information about an appropriate coupling of physically-based hydrological models and climate models.

The model outputs or grid-cell data used in the study were obtained from the Goddard Institute for Space Studies and include average monthly air temperature and precipitation for a control run ($1xCO_2$) and the GISS $2xCO_2$ experiment. The runs were integrated for 35 years and the data provided are the average figures for each month over the last 10 years.

The data are the average values over an 8°X10° cell, with the data-point references located at the center of each cell. Data for all cells of the global grid were also obtained as above. Three cells were determined: one lies within Greece, and two are adjacent. The center data of the cell that belongs to Greece were interpolated, using the center data of the other two cells, to latitude 39°34'N and longitude 21°19'E, which is approximately the centroid of the Mesochora catchment. The temperature difference between the $2xCO_2$ and control run on a monthly basis yielded the monthly temperature change scenarios (ranging from 3.37 to 4.98°C), while the ratio of the $2xCO_2$ monthly precipitation to the control run monthly precipitation provided the scenario of relative climate change (ranging from 0.925 to 1.487).

11-3. METHODOLOGY

The methodology of conceptual catchment simulation was adopted in this study for reasons of detailed representation of a medium-sized catchment (Panagoulia and Dimou, 1994). Actually, there were used the Snow Accumulation and Ablation model and the Soil Moisture Accounting model of US National Weather Service River Forecast System. Each of the models is described briefly in this section.

Snow Accumulation and Ablation Model

This model was developed by Eric Anderson within the US National Weather Service Hydrologic Research Laboratory (Anderson, 1973). This is a deterministic, conceptual model consisting of a set of equations which describe the accumulation and ablation of a snowpack. The model inputs are air temperature and precipitation at a six-hourly time step. In this study, daily precipitation was interpolated to six-hourly increments and six-hourly temperatures were estimated from daily temperature maxima and minima using equations furnished by Anderson.

The model can be summarized as follows. Accumulation of snowpack occurs when air air temperature, T_a, is less than the delineation temperature which can be 0°C or other. In the opposite case (T_a> delineation temperature) the model assumes that the precipitation is rain. The ablation of snowpack is controlled by the heat exchange at the air-snow interface. For heat exchange computations there are two basic conditions, namely:

- when the air is warm (T_a>0°C) in which case melt takes place at the snow surface, and
- when the air is too cold (T_a<0°C) for melt to occur. Furthermore, the melt is computed for rain or non-rain periods. For melt during rain periods the following assumptions are made:
  * there is no solar radiation
  * incoming longwave radiation is equivalent to blackbody longwave radiation at T_a
  * snow surface temperature is 0°C
  * the dew point is T_a
  * the rain temperature is T_a.

Under these assumptions, the amount of melting snowpack expressed as heat losses, ΔQ is:

$$\Delta Q = Q_n + Q_e + Q_h + Q_{px}$$

where

Q_n long wave radiation

Q_e latent heat transfer due to condensation

Q_h sensible heat transfer

Q_{px} heat transfer by rain water.

For melt during non-rain periods, the model checks whether the snowpack is isothermal at 0°C. If the snowpack is not isothermal, no melt occurs and the net heat flux is added to the heat content of the snowpack. If the snowpack is isothermal, and the air temperature is T_a>0°C, melt occurs at a rate proportionate to a seasonally varying melt factor and the difference between the air temperature and 0°C.

During non-melt periods (the model assumes T_a<0 °C), an antecedent temperature index, *ATI*, is used as an index to the temperature of the surface layer of snowpack. The heat exchange is assumed proportional to the temperature gradient defined by current air temperature and the andecedent temperature index. The proportionality constant is a parameter "called the negative melt factor" which varies seasonally in the same way as does the melt factor used during non-rain periods.The model accounts for the areal extent of snow cover. During the periods of snow accumulation, this is assumed to be 100%. During periods of depletion, the model uses an areal depletion curve of snow, that is a function of the areal extent of snow cover versus the ratio of mean areal water equivalent to an index value, which is the smaller of the maximum water equivalent (since snow began to accumulate), or a preset maximum. The six major and six minor parameters of the snow model are described in the section of parameter estimation.

Soil Moisture Accounting Model

The model was developed by Burnash *et al.* (1973) and forms the basis of the US. National Weather Service's basic catchment hydrologic response model for operational forecasting. It is a deterministic, continuous, lumped-parameter conceptual model. The original model was designed for daily precipitation input but later versions allows finer time increments (6 hours or less). Input to the model is pseudoprecipitation (rain plus melt model output) and reference evaporation (actual or long-term average). The model is based on a system of percolation, soil moisture storage, drainage and evapotranspiration characteristics to represent the significant hydrologic process in a rational manner. The components of the soil moisture hydrological model is shown in Figure 11.2. As seen from this figure, the model is represented by an upper and lower zone. The upper zone is divided into tension water storage and free water storage for the permeable portion of the catchment. Tension water is considered as that water which is closely bound to soil particles. This water is available for evapotranspiration based on the upper zone soil moisture. Tension water storage should be filled up before moisture becomes available to enter the free water storage. Free water can descend to lower zone by percolation or move laterally to produce interflow. Percolation is controlled by the contents of the upper zone free water and the deficiency of lower zone moisture volume.

When the precipitation rate exceeds the percolation rate and the maximum interflow drainage capacity, then the upper zone free water capacity is filled completely, whereupon the excess raifall will result in surface runoff.

Lower zone consists of tension water storage and two free water storages. Again, the tension water is available for evapotranspiration. The two free water storages fill up simultaneously from percolated water and drain independently at different rate, giving a variable ground water recession. Direct runoff from impervious areas, surface runoff, interflow and base flow from lower zone contribute to generate the channel inflow. The model employs about 21 parameters. These are soil moisture storage parameters for

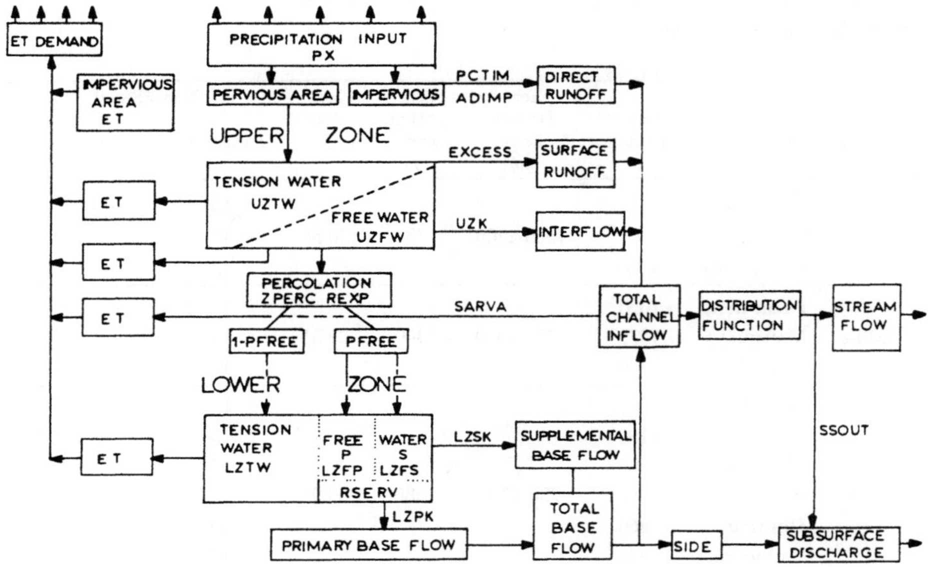

Figure 11-2. Soil moisture accounting model components

upper and lower zone, percolation parameters, catchment characteristics. These parameters are described in the section of parameter estimation. Some parameters can be estimated from semi-log plot of discharge or geographic maps of the study area.

11-4. HISTORICAL INPUT DATA

Three categories of stations and data, according to their use by the models, are hereby described. These are:

a) Precipitation and temperature stations with daily data
b) Temperature, sunshine, and humidity stations with monthly data
c) Streamflow station with daily data.

In addition, the methods used to average the station measurements data over area and elevation are described as well.

Precipitation and Temperature Stations Data

Eleven precipitation stations are installed within and around the Mesochora catchment, with the greater density at the lower part of the catchment (Fig. 11-1). The general characteristics of the stations for the study period (1972-1986) are presented in Table 11-1.

Table 11-1. Precipitation gauge stations by elevation zone, and temperature stations for the entire Mesochora catchment

Zone	Station	Station Elevation (m)	Zone Elevation Range (m)	Zone Median Elevation (m)	Zone Area %	Years of Record Precipitation	Temperature max	Temperature min
	Metsovo	1157				15	14	15
Upper	Katafyto	980	1580-2200	1830	30.54	15		
	Palaiochori	1050				15		
	Palaiochori	1050				15		
Middle	Matsouki	1079	1280-1580	1400	29.51	15		
	Pertouli	1160				15		
	Elati (Tyrna)	900				15		
	Vakari	1150				15	15	15
	Mesochora	780				15		
Lower	Pachtouri	950	780-1280	1080	39.95	15	11	11
	Stournareika	860				15		
	Vathyrema	920				15		

The precipitation stations are consistent with and representative of the catchment, but in some precipitation records daily data are missing. We would rather not interpolate them for three reasons in the main:

a) In order to preserve the real nature of precipitation-temperature series elements.
b) In order to avoid the computing errors that are introduced by the estimation techniques of daily missing data (Linsley *et al.*, 1988).
c) Because these last techniques are applied to with difficulty and especialy so the multisite stochastic models.

Thus, the technique used to estimate the mean areal daily precipitation was a combinatorial one consisting of a similar Thiessen method plus station daily availability, and including elevation correction (Panagoulia, 1991a). The correction areal precipitation factor for a given elevation (e.g. midpoint catchment elevation) is obtained from the following algorithm.

$$p_e = [P + (e - e_{ws}) p_f] / P \qquad (11\text{-}1)$$

where

p_e = corrective precipitation factor for elevation e,
P = mean areal precipitation,
e_{ws} = weighted mean station elevation based on station daily availability,
p_f = variation rate of precipitation with elevation.

As regards the network of temperature stations there are three stations. One is installed inside the catchment, while the other two are installed outside it and present significant deficiencies of daily maximum and minimum data. The consistency of the data was checked by the double-mass curve on monthly basis (Anderson, 1973), and some deviations from straight line were observed for minimum data. The inconsistent data were corrected by applying an appropriate corrective factor of the order -0.6 to -1.0°C. For the implementation of consistency checking only, in the case of missing monthly minimum temperatures, these were interpolated by the long-term average values of their existing monthly data. The consistency checking of the corrected actual data did not show an absolute straight line in the double-mass curve diagram, but merely two or three hardly distinguishable parallel straight-lines. The technique used to estimate the mean areal maximum and minimum daily temperature was, also, a combinatorial one of a look like Thiessen method and station daily availability. The correction areal temperature factor for a given elevation is obtained from the following algorithm:

$$t_e = (e - e_{ws})\, t_f \qquad (11\text{-}2)$$

where

t_e = additive corrective temperature factor for elevation e
e_{ws} = weighted mean station elevation based on station daily availability
t_f = rate of temperature decrease with elevation (lapse rate).

Reference Evapotranspiration

In order to obtain the catchment reference evapotranspiration (RET), (Allen *et al.*, 1994; Doorenbos and Pruitt, 1977, and Jensen *et al.*, 1990) the sunshine, temperature and humidity data were considered. Their measurement stations are described in Table 11-2. The sunshine stations are outside the catchment, the Ioannina is at the north-west boundary of the catchment, while the M.Kerasia station is at the eastern boundary. The catchment sunshine is computed from the long-term monthly arithmetic average of the sunshine of the two stations. With the above manner was also calculated the catchment humidity from the two stations. The Pramanta station is installed outside the catchment near the western boundary.

The catchment temperature was estimated by the aforesaid Thiessen method combined with the monthly station availability. The areal temperature was corrected for catchment mean elevation by applying a monthly dependent lapse rate according to Equation 11-2. The long-term monthly mean catchment sunshine, temperature, and humidity were used as inputs to Penman equation which is given in Veihmeyer (1964) to estimate the catchment reference evapotranspiration.

Other inputs to Penman equation were the average wind speed, (200 miles/day), the monthly percent of reflecting surface and the solar radiation for the midpoint catchment latitude. The sunshine was entered to Penman equation as monthly ratio of duration of bright sunshine to maximum possible duration of bright sunshine.

Table 11-2. Sunshine temperature and humidity gauge stations of Mesochora catchment

Meteorological Variable	Station	Station Elevation (m)	Period of Records
Sunshine	Ioannina	483	1972-86
	M. Kerasia	560	1972-86
Temperature	Metsovo	1156	1961-86
	Vakari	1150	1972-86
	Pachtouri	950	1972-81
Humidity	Metsovo	1156	1970-86
	Pramanta	835	1970-86

Streamflow

The daily streamflow data of the Mesochora gauge station for the period 1972-86 were used in this study. Most of the streamflow records were complete, but some missing daily data were computed and included. For estimating the missing data, the Avlaki station was used as a backup station, while the average monthly streamflow for both stations was being computed. The Mesochora missing data were arrived at by multiplying the complete daily data of Avlaki station by the ratio of the monthly average streamflow (Mesochora/Avlaki).

11-5. MODEL CALIBRATION

The procedures used for models calibration (parameter estimation) and soil moisture accounting model verification are elaborated on as follows.

Snow Accumulation and Ablation Model Parameter Estimation

The mountainous catchment was divided into elevation zones, and the snowmelt model was applied to each zone separately, since low elevations are likely to receive rain, while higher elevations receive snow from the same storm. The weighted mean value of the pseudo-precipitation from all zones was treated as the mean areal precipitation which is the input to the Soil-Moisture Accounting model. In general, the weighting factors used were equal to the ratios of the elevation zone subareas for the total catchment area.

The elevation bands were delineated as follows: First hypsometric curve (elevation versus area fraction) was developed. The catchment was then divided into three zones of area, depending on the elevation range, and the elevation of the midpoint of each band was identified (Table 11-1).

The snowmelt model was manually calibrated for the three elevation zones. According to available station data on that particular day, for every elevation zone, the precipitation-elevation correction factors p_e, were estimated through a trial and error approach, which was carried out concurrently with the calibration of Soil Moisture Accounting model. As with the function p_e, the lapse rate is usually nonlinear. It was found that average lapse rates for four periods (each 6 hours long) of each day is - 0.80°C/100 m. The lapse rates, t_f and the temperature correction factors, t_e, were also estimated concurrently with the calibration of Soil model.

Seasonal melt factors were interpolated between *MFMAX* and *MFMIN* and estimated to 0.9 and 0.4 mm/°C/6 hours, respectively. *PXTEMP*, the temperature [°C] above, where precipitation was assumed to be rain was assumed to be 1°C for the first six months of calendar year, and 0°C for the rest ones. SI, the mean areal water-equivalent above which 100% areal snow cover always exist, was assumed to be 100mm. (See Table 11-3 for description and calibrated values of the snow-melt model parameters).

Soil Moisture Accounting Model Parameter Estimation

Parameter estimation for the soil moisture accounting model was based on a process of initial parameter estimation, as suggested by Peck (1976). The final values of parameters were obtained through manual model calibration on the Mesochora catchment. The description of model parameters and their final values are listed in Table 11-4. The model was calibrated for all the study period of from 1972 to 1986, which included dry, medium and wet years, so that the model could be subjected to a broad range of changes in conceptual storages.

The results of error analysis of daily flow, the three-day volume including peaks, as well as the comparison between simulated and observed flow components (Panagoulia, 1992a) showed that both hydrological models are capable of reproducing the observed streamflow (the snowmelt model indirectly, and the rainfall-runoff model directly). Also, the typical monthly simulation errors (monthly differences between simulated and observed streamflows), expressed in per cents of observed flows, were of the order of 10-15%, higher in low runoff months (August and September) and lower in high runoff months.

Table 11-3. Snow melt model parameter description and parameter calibrated values

Parameter	Description	Calibrated values
SCF	A multiplying factor to correct for gauge catch deficiency in the case of snowfall	1.10
MFMAX	Maximum melt factor during non-rain periods which occurs on June 21	0.90 mm/°C/6hr
MFMIN	Minimum melt factor during non-rain periods which occurs on December 21	0.40 mm/°C/6hr
UADJ	Average wind function during rain on snow periods	0.10 mm/mb/6hr
SI	Mean areal water-equivalent above which there is always 100% areal snow cover	100 mm
NMF	Maximum negative melt factor	0.12 mm_e/°C/6hr
TIMP	Antecedent temperature index parameter	0.30
PXTEMP	Temperature which delineates rain from snow	1.0 - 0°C
MBASE	Base temperature for snow melt computation during non-rain periods	0°C
PLWHC	Percent liquid-water holding capacity of ripe snow	0.05
PAYGM	Average daily ground at the snow-soil interface	0.020 mm
EFC	Percent area over which evapotranspiration occurs when there is 100% snow cover	0.61

Table 11-4. Soil moisture accounting model parameter description and parameter final values

Soil Moisture Phase	Parameters	Description	Final values
Direct runoff			
	PCTIM	Minimum impervious catchment	0.01
	ADIMP	Additional impervious catchment	0.01
	SARVA	Catchment covered by streams, lakes and riparian vegetation	0.0
Upper zone			
	UZTWM	Upper zone tension under capacity	4.50 cm
	UZFWM	Upper zone free water capacity	4.61 cm
	UZK	Daily upper zone free water drainage rate	0.57

Percolation			
	ZPERC	Proportional increase in percolation from saturated to dry condition	6.00
	REXP	Exponent affecting rate of change of percolation between wet and dry conditions.	1.80
Lower zone			
	LZTWM	Lower zone tension water capacity	25.00 cm
	LZFSM	Lower zone supplementary free water capacity	9.00 cm
	LZFPM	Lower zone primary freewater capacity	30.00
	LZSK	Daily lower zone supplementary free water drainage rate	0.15
	LZPK	Daily lower zone primary free water drainage rate	0.015
	PFREE	Percolation water fraction passing directly to lower zones free water	0.20
	RSERV	Fraction of lower zone free water unavailable for transpiration	0.10
	SIDE	Ratio of non-channel baseflow to channel baseflow	0.00
Initial water			
	UZTWC	Upper zone tension water content	4.50 cm
	UZFWC	Upper zone free water content	0.11 cm
	LZTWC	Lower zone tension water content	21.41 cm
	LZFSC	Lower zone supplementary free water content	0.088 cm
	LZFPC	Lower zone primary free water content	2.26 cm
	ADIMC	Tension water content of the additional impervious catchment	25.91 cm

Model Verification (robustness) for Climate Change Studies

The plot of the long-term annual mean catchment pseudo-precipitation (rain plus melt) (Fig. 11-3) reflected three distinct periods with different climate conditions. A modified differential split sample test was implemented in order to verify the ability of the model to respond, without significant deviation, to the three different climate periods. The model was run for each period separately, and the statistical variables: long-term annual mean runoff, standard deviation of annual runoff, standard error and correlation coefficient of monthly runoff, were accordingly computed (Table 11-5). The null hypothesis H_0 of the variable difference between two climate periods and any climate period and calibration period was also tested. The results for all variables fell within 95% of the critical region.

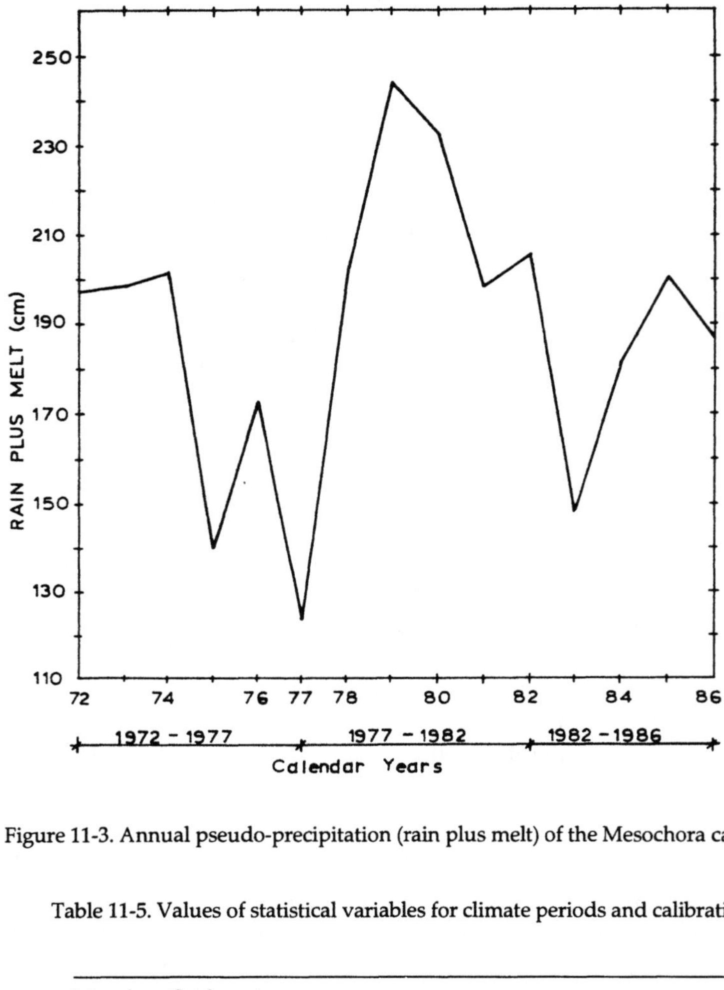

Figure 11-3. Annual pseudo-precipitation (rain plus melt) of the Mesochora catchment

Table 11-5. Values of statistical variables for climate periods and calibration period

Mesochora Catchment Pseudo-precipitation	Annual Runoff		Monthly Runoff	
Period	Average Value (cm)	Standard Deviation (cm)	Standard Error (cm)	Correlation Coefficient
Intense Descending	99.67	27.80	2.213	0.944
Intense Rising	128.32	45.63	2.940	0.961
Mild Descending	113.43	21.16	2.989	0.935
Calibration Period	119.23	30.16	2.581	0.954

11-6. ALTERED CLIMATE INPUTS

The historical input data were adjusted to reflect the altered climates simulated by each of fifteen HYPO and each of two GISS-conceived climate change scenarios. Particularly, the climate change applied to the historical time series: (a) daily precipitation, (b) daily minimum and maximum temperature, and (c) monthly temperature in order to be converted into changes of reference evapotranspiration. The description of the climate change and its implementation on each of the (a), (b), and (c) time series runs as follows.

a) Precipitation : All the daily values of the historical input precipitation series were multiplied for the HYPO changes by a uniform factor and for the GISS cases by the monthly precipitation ratio for CO_2 doubling and control runs.Thus, there resulted five altered daily precipitation series (time series scenarios) corresponding to the five climate change scenarios, and one series with zero precipitation change.

b) Temperature : The HYPO temperature increase scenarios were applied uniformly to both minimum and maximum daily values of the historical input temperature series, while the GISS-predicted monthly temperature differences between the CO_2 doubling and control run were added to the min/max daily values of the input historical temperature data as well. Thus, we obtained four altered daily temperature time series of minima and four of maxima, corresponding to the four temperature increase scenarios.

c) Reference Evapotranspiration : RET was computed on a monthly basis using the Penman equation, as a function of the meteorological variables: temperature, relative humidity, wind speed, relative sunshine, and mean solar radiation. Of these input variables, only temperature can be associated with climate changes, while the other variables are indirectly influenced by the climate change through a not well known mechanism that does not permit their change assessment. Therefore, in Penman's equation the only variable which was linked to climate change was the monthly temperature, while the rest were assumed not to be affected by it.

The monthly temperature increase scenarios were the same with those applied to daily time series because the mean monthly temperature is computed by averaging either the mean minimum and maximum daily temperature or three mean daily temperature values (8hr observations) by doubling the third temperature value (night's). Therefore, if the minimum and maximum daily temperature goes up by the same amount, this implies that during all hours of the day there will prevail the same temperature increase, a fact that identifies this increase with the mean monthly temperature, this regardless of the way it was actually computed. The monthly temperature increase scenarios were 1°C, 2°C, 4°C (HYPO case) and GISS-predicted for CO_2 doubling and control run.

The HYPO scenarios were applied uniformly to monthly values of the historical time series of catchment temperature and three altered monthly temperature time series (time series scenarios) were obtained. These altered temperature time series were used as inputs to Penman equation which in its turn yielded three monthly time series of *RET* (time series reference evapotranspiration scenarios) corresponding to the three principal temperature increases. For the GISS cases the *RET* was also computed with the same equation for the monthly temperature data for the CO_2-doubling and control run.The monthly differences in *RET* were computed and these differences were added to the historical *RET* data. The long- term average of each monthly reference evapotranspiration series was used as input to soil moisture accounting model.

11-7. HYDROLOGICAL RESPONSE ANALYSIS

As described previously, the long-term hydrological response of the Mesochora catchment was simulated for climate regimes associated with a base case (nominally, present climate conditions, solid line in graphs), as well as fifteen hypothetical and two GISS-predicted climate change scenarios.

Because the snow accumulation and ablation model, as well as the soil moisture accounting model operate on daily or shorter time steps and all scenarios involved running both models for fifteen years, large amounts of computer output were generated. To simplify the analysis of the results that are reported as averages over the 15year simulation period, we selected the following simulated variables to describe the alternative hydrologies:

a) Monthly average snow water equivalent over catchment.
b) Monthly average catchment runoff.
c) Monthly average catchment evapotranspiration.
d) Monthly average catchment soil moisture storage in simulated zones.

For the seventeen climate change scenarios the (a) simulated variable yielded seventeen monthly snow water equivalent scenarios, the (b) plus seventeen monthly runoff scenarios, the (c) plus seventeen monthly evapotranspiration scenarios, and (d) seventeen monthly soil moisture storage for each of the five model moisture zones, resulting totally to one hundred and thirty six catchment hydrological response scenarios plus the eight ones of the base case (model outputs for present climate conditions, solid line in graphs). The hydrological scenarios of the above variables are plotted in Figures 4 to 11 and a brief interpretation of these figures follows.

Snow Water Equivalent

The long-term monthly snow water equivalent over the study catchment for all alternative climates is presented in Figure 11-4. There was a marked reduction in average snow water equivalent for all alternative scenarios. The GISS scenarios and the HYPO(4, all) produced the maximum reduction in snow water equivalent. They generally generated similar annual snow water equivalent hydrographs in the same

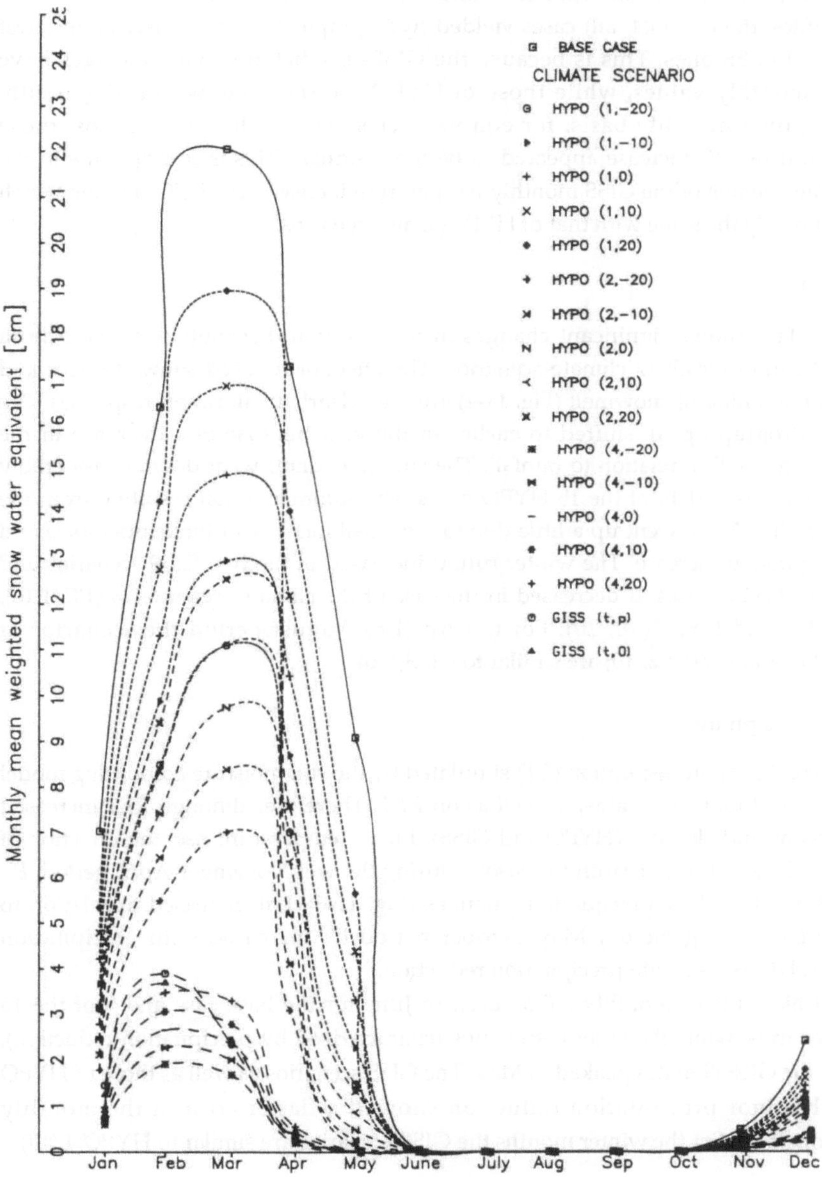

Figure 11-4. Mesochora catchment monthly mean snow water equivalent for the HYPO, GISS and base case climate scenarios

month of snow maximization, extinction and return. But there is a difference in snow water profiles: the HYPO(4, all) cases yielded hydrographs with obviously flatter crest than that of GISS ones. This is because the GISS- predicted climate changes have different monthly values, while those of HYPO are the same within the month. Searching, on a monthly basis, for equivalencies among all the scenarios, those characterized by 4°C increase appeared to be more similar. This is perhaps due to the fact that the average of the GISS monthly temperature increases is 3.94°C, a value which is approximately the same with that of HYPO(4, all) scenarios.

Runoff

Figure 11-5 shows significant changes in the seasonal distribution of Mesochora catchment runoff for all 17 climate scenarios. The effect of reduced snow storages and change in the timing of snowmelt (Fig. 11-4) are seen clearly in all runoff responses. The annual hydrograph peak shifted to earlier in the year because of a decrease in the amount of snowfall in relation to rainfall. The summer runoff went down considerably in GISS scenarios and 14 of the 15 HYPO cases. The summer runoff resulting from the scenarios HYPO(1, 20) went up a little due to the small increase of the temperature and large precipitation increase. The winter runoff increased in the two GISS scenarios and 10 of the 15 HYPO cases. It decreased in the case of the climate scenarios HYPO(1,10), HYPO(2,10) and HYPO(all, 20). For the April to August period the scenarios of HYPO(1,-20) and HYPO(2,-10) are similar to GISS(t, p).

Evapotranspiration

The actual evapotranspiration (*ET*) simulated by the soil moisture accounting model depends on soil moisture status, as well as on *RET*. Therefore, although *RET* increased for all months and climates (HYPO and GISS) due to temperature rise, the direction of change in *ET* varied from season to season. During the wet November-April period, *ET* remained unaffected by precipitation changes (Fig. 11-6), but increased in relation to base case *ET*. During the dry May-October period *ET* increased with precipitation increase and decreased with precipitation reduction.

The peak value of monthly *ET* occured in June for the base case and 9 of the 15 HYPO scenarios, while the other 6 scenarios (characterized by precipitation reduction), as well as the GISS climates peaked in May. The GISS scenarios as well as those of HYPO ones with minor precipitation reduction showed a flatter crest in the monthly distribution of *ET*. For the winter months the GISS scenarios are similar to HYPO(4, all).

Upper Zone Free Water

The free water contents in all three zones of the model are strongly and erratically influenced by HYPO and GISS scenarios, as well as from month to month under the

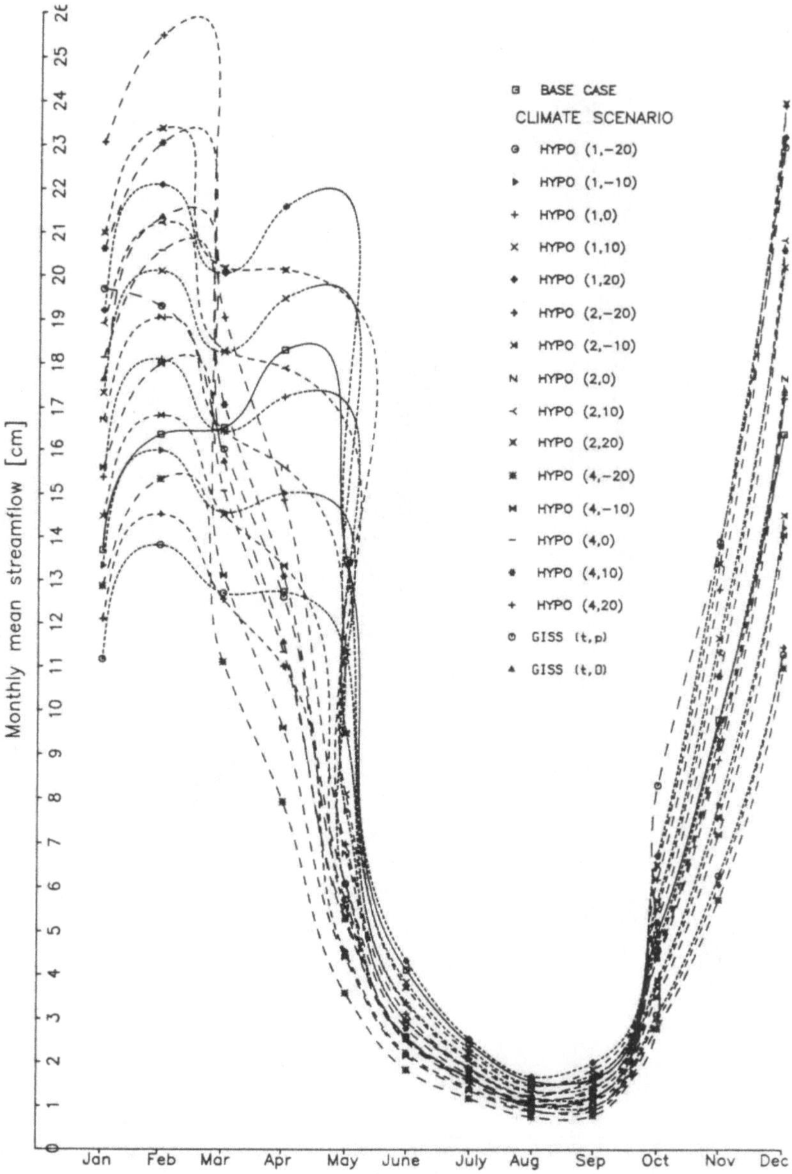

Figure 11-5. Mesochora catchment monthly mean runoff for the HYPO, GISS and base case climate scenarios

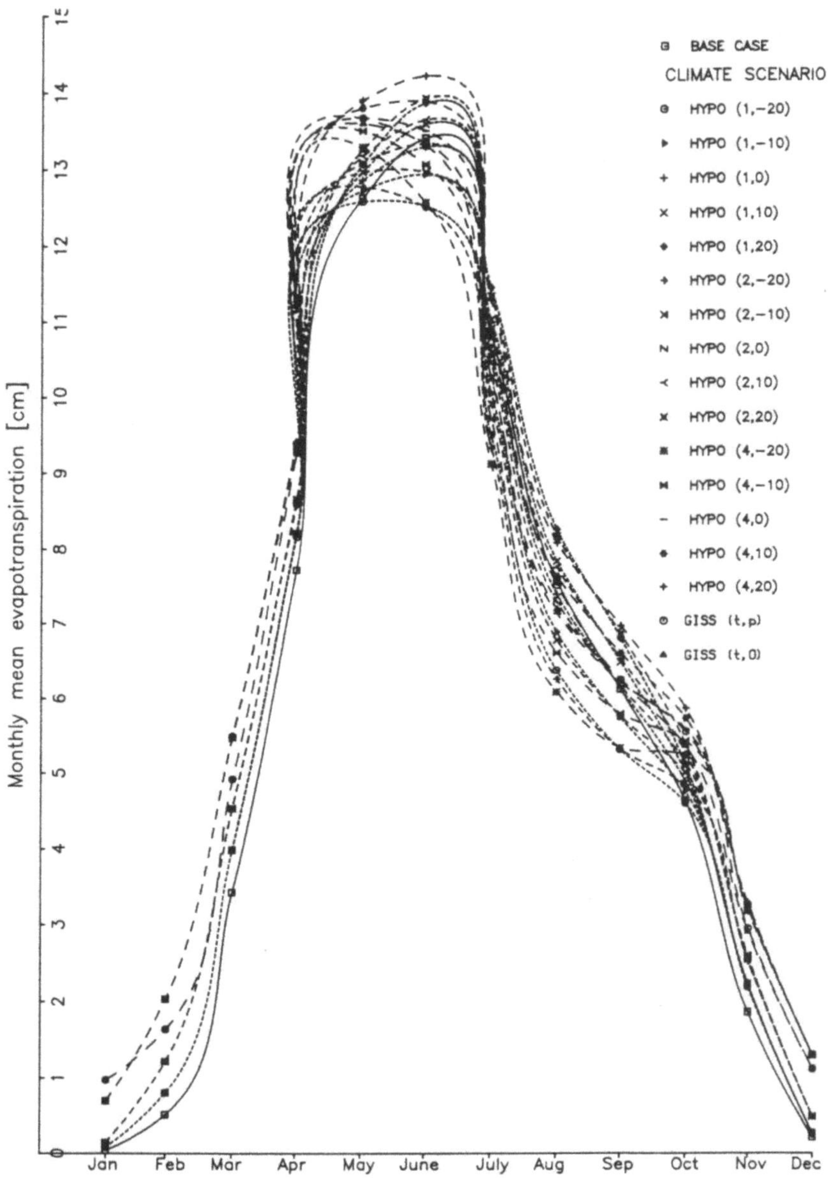

Figure 11-6. Mesochora catchment monthly mean evapotranspiration for the HYPO, GISS and base case climate scenarios

same scenario. Notwithstanding that, the free moisture content of the upper zone (Fig. 11-7) posted larger fluctuations than those of the lower zones (Figs 11-8 and 11-9). Indeed, it peaked in January for the drier HYPO climate scenarios and in December for the rest HYPO, GISS and the base case, while during the summer dry period (July-October), the upper zone free water went down for all HYPO and GISS cases.

Lower Zone Primary Free Water

The free moisture content of the lower primary zone (Fig 11-10) supplies the baseflow with larger amounts than those of lower supplement zone. It peaked in March for 13 of the 15 HYPO and GISS cases, while that of the base case reached its maximum in May. The other two scenarios HYPO(1,-20) and HYPO(1,-10) caused the primary free moisture content to come to a maximum in April. For all HYPO and GISS scenarios the primary free moisture was minimized in October.

Lower Zone Supplemental Free Water

For 12 of the 15 HYPO scenarios and GISS(t, 0) the supplemental moisture content (Fig. 11-11) peaked in March, while for the GISS(t, p), HYPO(2,-10), HYPO(4,-20) and HYPO(4,-10) the peak shifted to February. The GISS(t, 0) case appeared shifted earlier by one month from the peak month (April) for the base case. This content is minimized in August for HYPO and July-September period for GISS climate cases.

Upper Zone Tension Water

The moisture content of the upper tension zone (Fig. 11-8) is little affected by HYPO and GISS climate changes in winter period (December to February), while all the other months are affected much more. The tension moisture content of this zone is reduced for all months and its maximum fall occurs in May for both HYPO and GISS scenarios. The seasonal tension moisture profiles are very similar for both cases of HYPO and GISS scenarios.The fact of the considerable tension moisture reduction in spring and early summer months is to be attributed to snowmelt reduction that contributes to soil moisture.

Lower Zone Tension Water

The moisture content of the lower tension zone (Fig. 11-9) shows the same monthly distribution profile for both HYPO and GISS scenarios. The largest reduction in lower tension moisture occurs in October for the two GISS and HYPO(4,-20) scenarios. Generally, the larger lower zone tension moisture reductions appear shifted forward by two or three months compared with these of upper zone tension moisture storages.

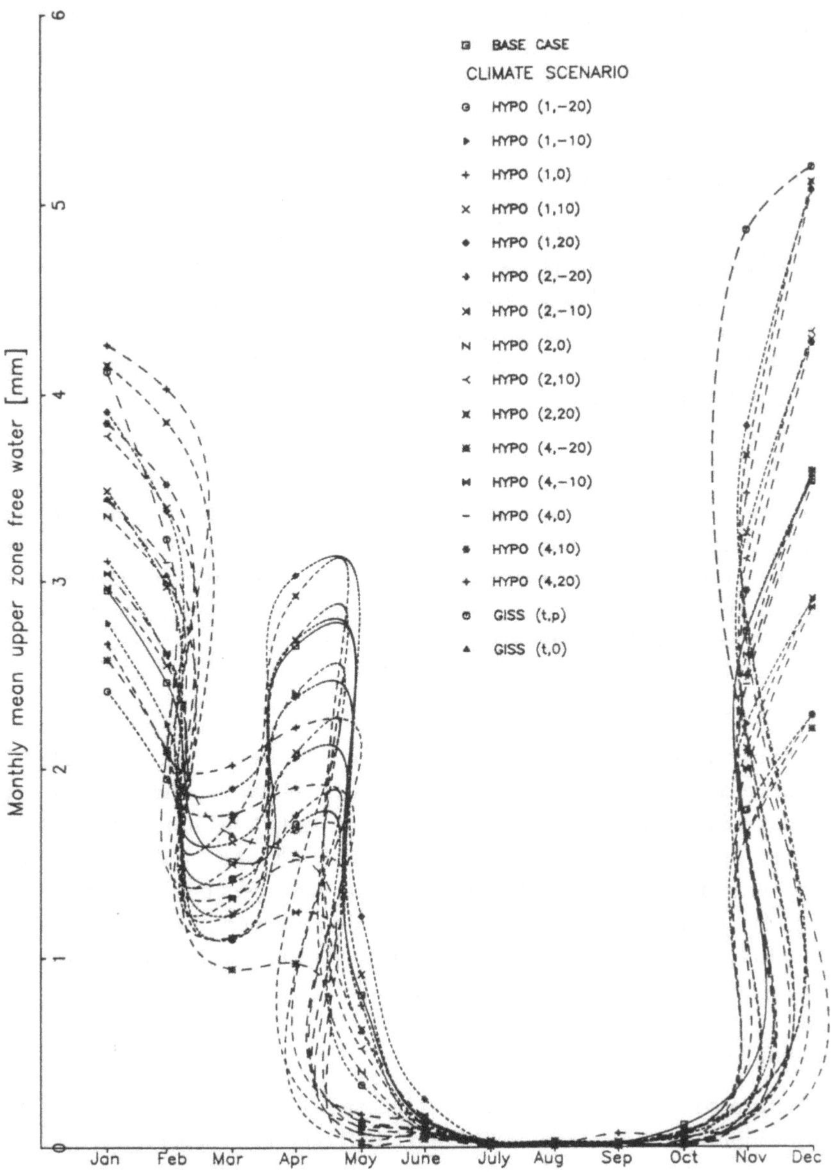

Figure 11-7. Mesochora catchment monthly mean upper zone free water for the HYPO, GISS and base case climate scenarios

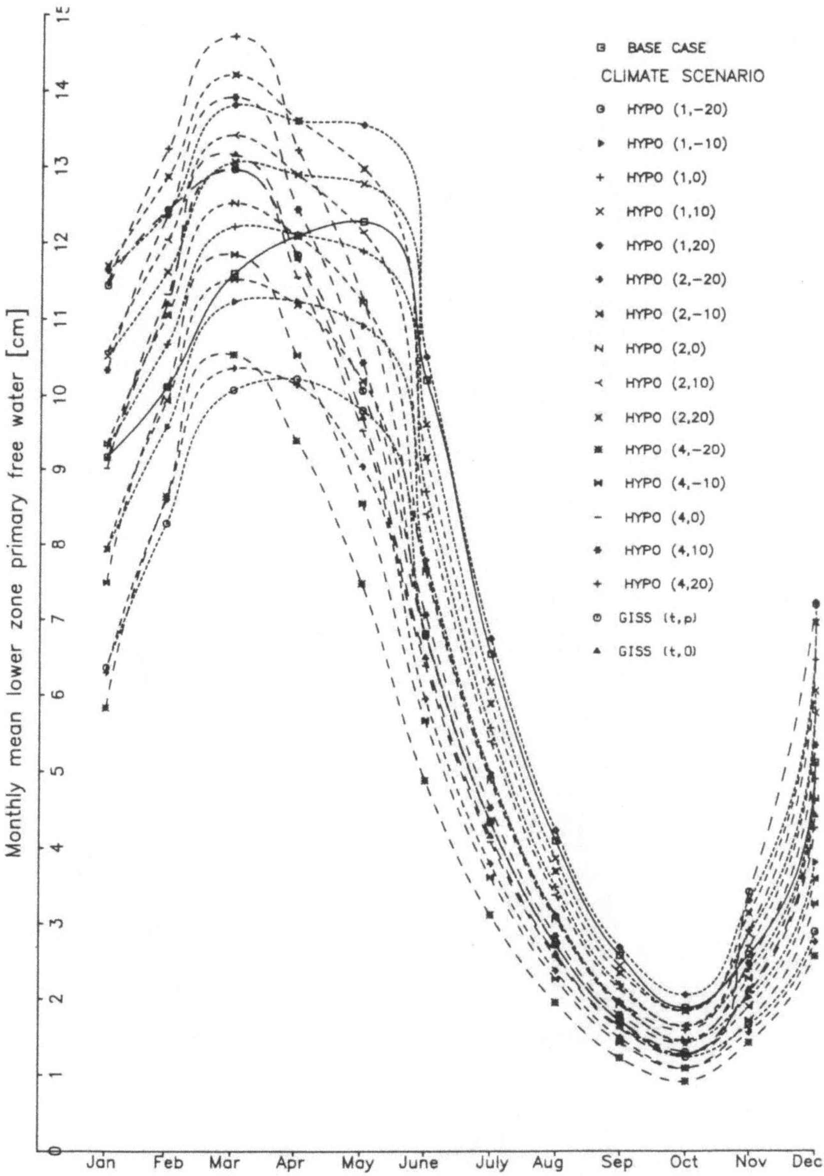

Figure 11-8. Mesochora catchment monthly mean lower zone primary free water for the HYPO, GISS and base case climate scenarios

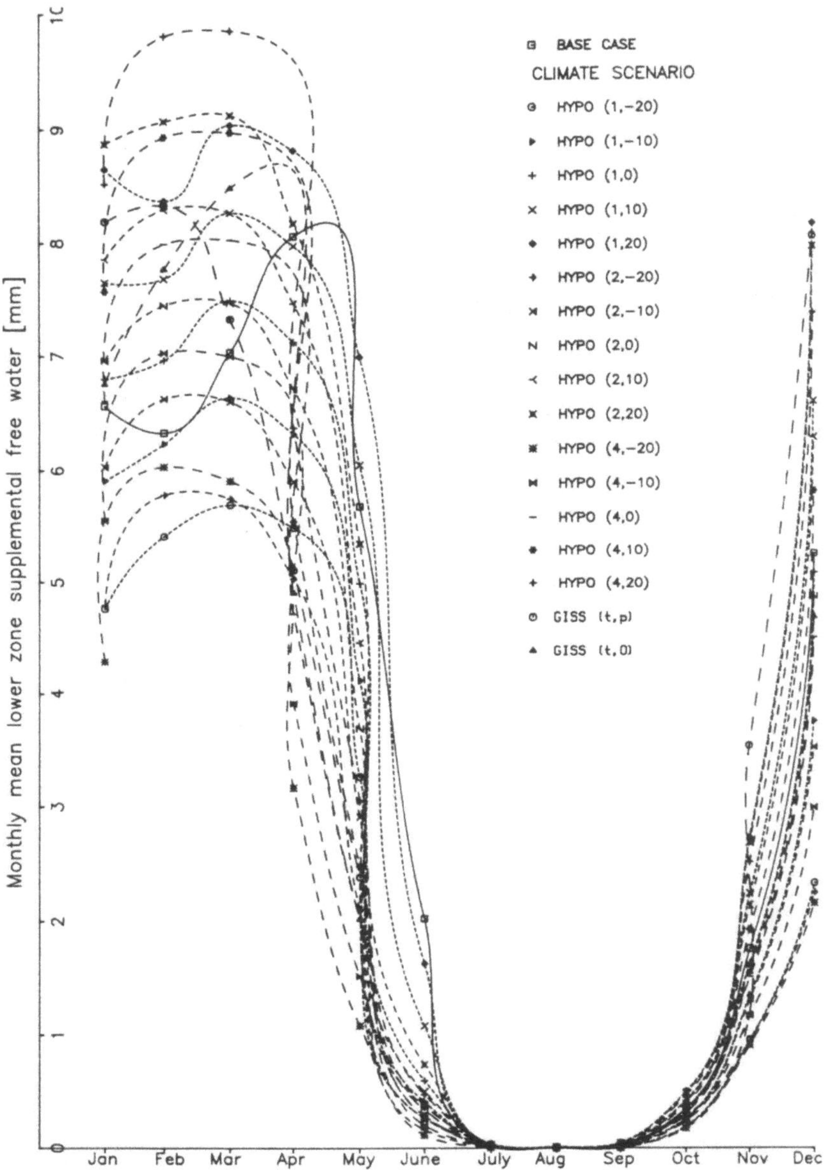

Figure 11-9. Mesochora catchment monthly mean lower zone supplemental free water for the HYPO, GISS and base case climate scenarios

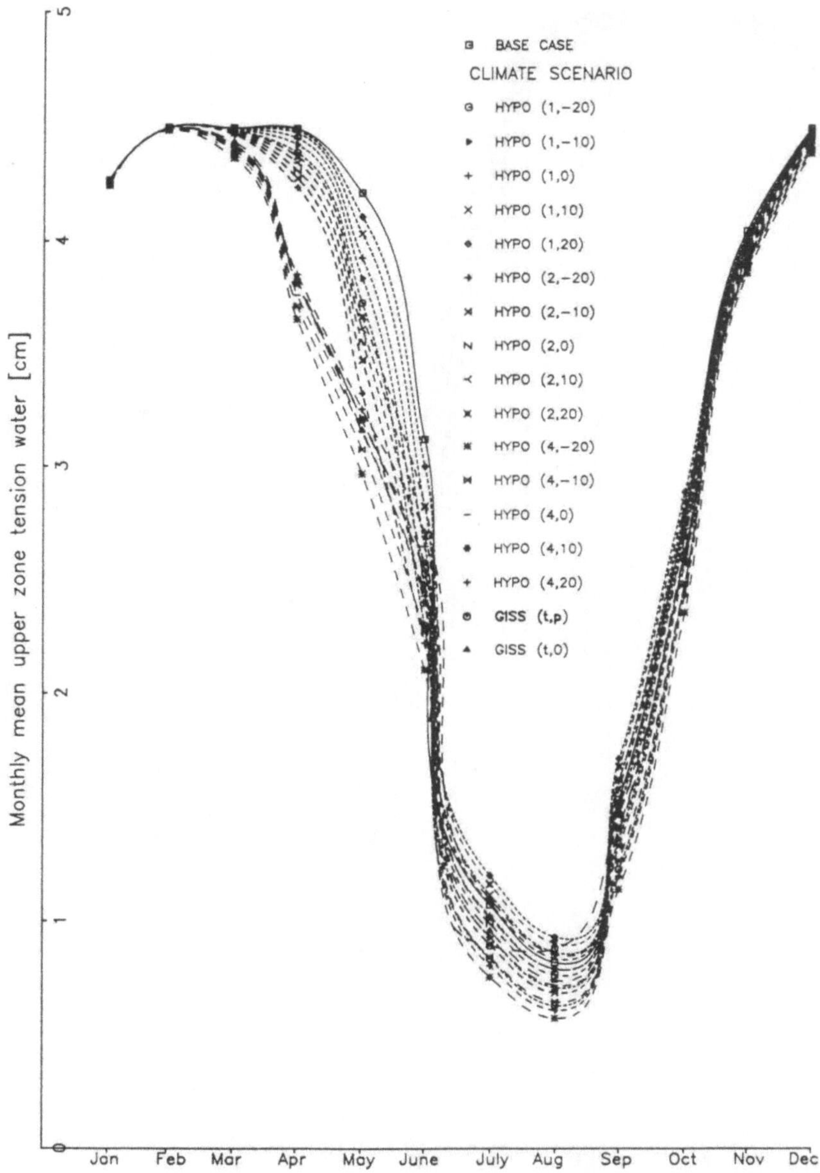

Figure 11-10. Mesochora catchment monthly mean upper zone tension water for the the HYPO, GISS and base case climate scenarios

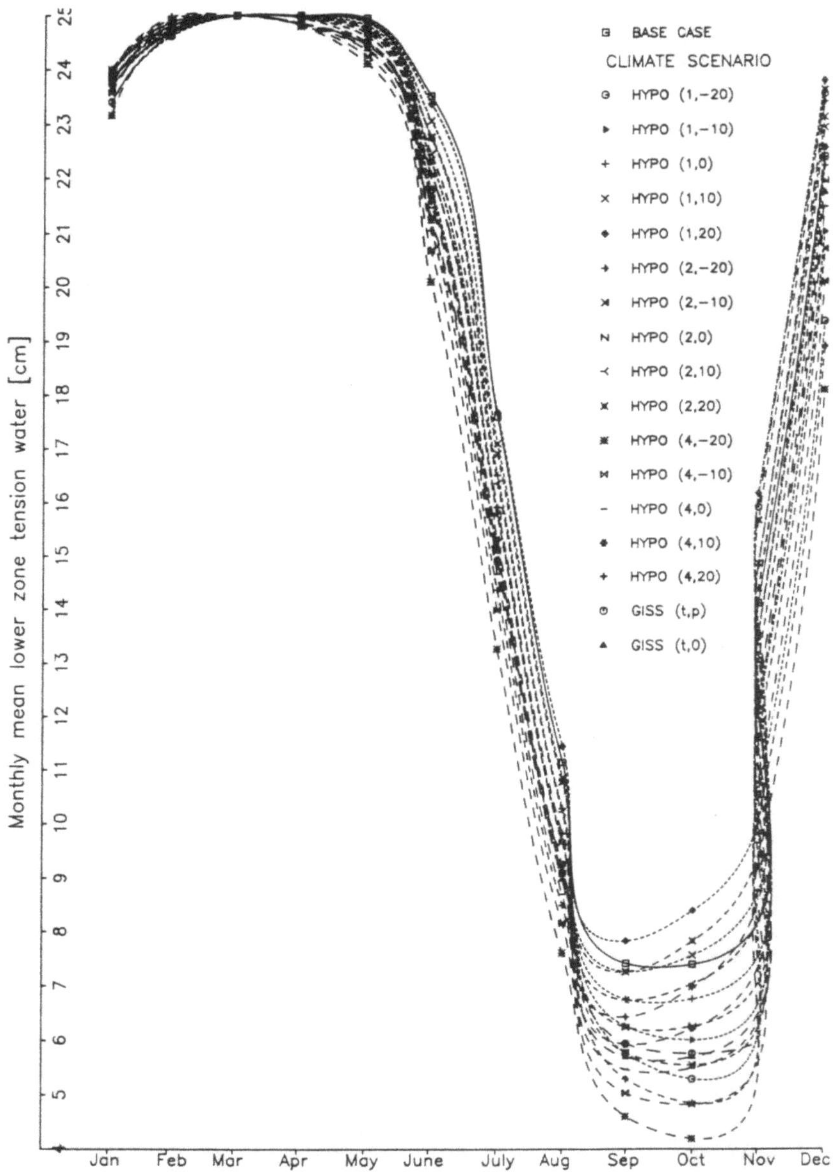

Figure 11-11. Mesochora catchment monthly mean lower zone tension water for the the HYPO, GISS and base case climate scenarios

11-8. CONCLUSIONS

Before jumping to the final conclusions of this research, it is necessary to make clear at this juncture that (a) the simulation results constitute alternative scenarios and not deterministic predictions and (b) because the assumptions and simplification incorporated in the models are reflected in the results, we thought if advisable to elaborate on these assumptions, namely:

a) The application of uniform climate change: Because the precipitation was adjusted for climate change with a fixed factor, this implies that the coefficient of variation (standard deviation divided by the mean) is the same for the altered climate scenarios as for the base case. For precipitation factors greater than one this means that the precipitation and its variability will increase, a fact that could affect the streamflow variability and hence the operation of a given reservoir or the design of new facilities.
b) The capability of hydrological models to provide a sufficient description of the catchment dynamics to altered climates: Although the NWS models contain the appropriate level of detailed dynamics for medium-sized catchments and can also capture the basic elements of the long-term hydrological response of the catchment, their adaptability to altered climates is rather difficult to determine. The parameters of the soil moisture accounting model are climate- dependent and hence their yields for altered inputs are strongly model- dependent. Another major problem of this model is that it cannot possibly be relied on long-term changes in vegetation (Panagoulia, 1994).

Having considered the limitations imposed by the various assumptions, the following general conclusions can be reached, viz:

a) The monthly distribution patterns of the catchment hydrological variables for HYPO and GISS climate scenarios are very alike.
b) Both the HYPO and GISS temperature increases associated with all precipitation changes, could possibly cause substantial decreases in average snow accumulations.
c) Reduction in the amount of precipitation that is falling like snow could increase the winter runoff volumes and decrease the summer and spring runoff ones, thus causing the floods to rise in winter and water shortages to occur in summer.
d) Increased precipitation falling like rain in the winter could possibly increase the winter soil moisture storages, thereby making much more moisture available for actual evapotranspiration in the early spring. Increased temperatures could increase the spring's actual evapotranspiration.
e) The reduction in moisture supplied in the form of snowmelt in spring combined with increased spring actual evapotranspiration could reduce the soil moisture of late spring, summer and fall, which could in turn reduce runoff during these periods.
f) Significant differences were noted in hydrological numerical results among the two GISS scenarios and HYPO scenarios due to the wide range of the climate variables changes (e.g. the GISS precipitation increase in October was up 50%).

Acknowledgment

I thank sincerely the civil engineer George Dimou who helped me enormously in computer themes.

REFERENCES

Allen R.G., Smith M., Perrier A., and Pereira L.S. (1994). Updated Reference Evapotranspirartion Definition and Calculation Procedures. ICIC Bulletin .

Anderson E.A. (1973). National Weather Service river forecast system. Snow accumulation and ablation model NOAA Technical Memorandum NWS HYDRO 17.

Becker A. and Nemec J.(1987). Macroscale hydrologic models in support to climate research. The Influence of Climate Change and Climatic Variability on the Hydrologic Regime and Water Resources, Proceedings of the Vancouver Symposium, August 1987. IAHS Publ. no. 168: 431-445.

Budyko M.I.(1989). Climatic conditions of the future. Conference on Climate and Water, Helsinki, National IHP Committee Finland, 9-30.

Burnash R.J.C., Ferral R.L., and Mcquire R.A. (1973). A generalized streamflow simulation system conceptual modelling for digital computers. U.S. National Weather Service, Sacramento, CA.

Dickinson R.E. (1982). Modelling climate change due to carbon dioxide increases. *In*: W.C. Clark, (ed.), Carbon Dioxide Review. Oxford University Press, New York.

Doorenbos, J. and Pruitt W.O. (1977). Guidelines for predicting crop water requirements. Irrigation and Drainage Paper 24, 2nd Ed. FAO, Rome, pp. 156.

Georgakakos K.P. and Krajewski W.F. (1991). Worth of radar data in the real- time prediction of mean areal rainfall by nonadvective physically models Water Resour. Res., 27(2): 185-197.

Gleick P.H.(1986). Regional water availability and global climatic changes: The hydrologic consequences of increases in atmospheric CO_2 and other trace gases. Energy and Resources Group, PhD Thesis ERG-DS 86-1, University of California, Berkeley, California, USA.

Hansen J., Russell G., Rind D., Stone P., Laus A., Lebedeff S., Ruedy R., and Travis L. (1983). Efficient three-dimensional global models for climate studies: Models I and II. Monthly Weather Review, 111(4): 609-662.

Hansen J., Fung I., Laus A., Rind D., Lebedeff S. Rueddy R. Russel G., and Stone P. (1988). Global climate changes as forecast by Goddard Institute for space studies three-dimensional model. J. Geophys. Res., 93 (D8): 9341-9364.

Hutchinson M.F. (1990). A point rainfall model based on a three-state continuous Markov occurence process J. Hydrology, 114: 125-148.

Jensen M.E., Burman R.D., and Allen (1990). Evapotranspiration and Irrigation Water Requirements. ASCE Manuals and Reports on Engineering Practices No 70. American Society of Civil Engineers. New York, pp. 442.

Lettenmaier D.P. and Gan T.Y. (1990). Hydrologic sensitivities of the Sacramento-San Joaquin River Basin, California, to global warming. Water Resour. Res. 26(1): 69-86.

Linsley R.K., Kohler Max A., and Paulhus J.L.H. (1988). Hydrology for Engineers, McGraw-Hill, N.Y.

MacCracken M. and Luther F. (Ed.). (1986). Projecting the climatic effects of increasing carbon dioxide. U.S. Office of Energy Research, Office of Basic Energy Sciences. Carbon Dioxide Research Division Washington, D.C.

Manabe S. and Wetherald R. (1985). CO_2 and hydrology. *In*: Advances in Geophysics. 28(a). Academic Press, Washington, D.C.

Mitchell J.F.B. and Qingcum Z. (1991). Climate change prediction. *In*: Climate Change: Science, Impacts and Policy, Proceedings of the Second World Climate Conference, Geneva, October-November 1990. J.Jager and H.L. Ferguson (Eds), pp. 59-70.

Nemec J. and Schaake J. (1982). Sensitivity of water resources systems to climate variation. Hydrological Sciences, 27(3): 327-343.

Panagoulia D. and Dimou G. (1994). Temporal scale effects on modelled catchment hydrological processes in respect of global climate change. Western Pacific Geophysics Meeting, AGV, Book of Abstracts 30, Hong Kong, July 25-29, 1994.

Panagoulia D. (1994). Global climate changes and water resources decision- making, Second European Conference on Advances in Water Resources Technology and Management, Lisbon Portugal, 14-18 June 1994.

Panagoulia D. (1993a). Catchment hydrological responses to climate changes calculated from incomplete climatological data, Proceedings of Exchange Processes at the Land Surface for a Range of Space and Time Scales, IAHS Publications, No 212, Yokohama Japan, 13-16 July, pp. 461-468.

Panagoulia D. (1993b). Documentation of global climate changes through catchment hydrological modelling. *In*: NATO Proceedings of Engineering Risk and Reliability in a Changing Physical Environment, May 24-June 4, Deauville, France.

Panagoulia D. (1992a). Hydrological modelling of a medium-sized mountainous catchment from incomplete meteorological data. J. Hydrol., 137: 1-4, 279-310.

Panagoulia D. (1992b). Impacts of GISS-modelled climate changes on catchment hydrology. Hydrol. Sci. J., 37(2): 141-163.

Panagoulia D. (1991a). Hydrological response of a medium-sized mountainous catchment to climate changes. Hydrol. Sci. J., 36(6): 525-547.

Panagoulia D. (1991b). A technique estimating daily catchment precipitation with elevation correction for conceptual simulation. *In:* Proc. Advances in Water Resources Technology, Published for ECOWARM by A.A.Balkema, Athens, Greece, pp. 89- 101.

Panagoulia D. (1990). Sensitivity Analysis of Catchment Hydrological Response to Climate Changes. PhD Thesis, National Technical University of Athens, Greece.

Peck E.L. (1976). Catchment modelling and initial parameter estimation for the National Weather Service Forecast System. NOAA Techn. Memorandum NWS HYDRO 31.

Revelle R.R. and Waggoner, R.E. (1983). Effects of a carbon dioxide-induced climatic change on water supplies in the Western United States. *In*: Changing Climate, National Academy of Sciences. National Academy Press, Washington D.C.

US. National Academy of Sciences (1983). Changing climate. National Academy of Sciences, National Academy Press, Washington D.C., USA.

US. National Academy of Sciences (1977). Climate, Climatic Change and Water Supply - Overview and Recommendations, 1-23, National Academy of Sciences, Washington D.C., USA.

Wallis J.R., Lettenmaier D.P., and Wood, E.F. (1991). A daily hydroclimatological data set for the Continental United States. Water Resour. Res., 27(7): 1657-1663.

Veihmeyer F.J. (1964). Evapotranspiration. *In*: Handbook of Applied Hydrology, V.T. Chow. (Ed.), McGraw Hill, New York.

CHAPTER 12

THE IMPACT OF CLIMATIC CHANGE ON AVAILABLE MOISTURE IN ARID LANDS: EXAMPLES FROM THE AMERICAN SOUTHWEST

OWEN K. DAVIS

Department of Geosciences, University of Arizona, Tucson, Arizona 85721, USA

12-1. INTRODUCTION

About one-third of the Earth's land surface is classified as arid to extremely arid (Fig-1), receiving insufficient precipitation to support continuous ground cover, or to permit cultivation without irrigation (Meigs, 1952). The bulk of these deserts lie between 15° and 30° North and South latitude (Fig. 12-1). In these tropical and temperate deserts, precipitation primarily arrives during the summer via. monsoonal circulation. However, in western North America and central Asia, where arid lands may extend to 55° North, winter precipitation dominates. Figure 12-2 compares the temperature and precipitation regimes in the northern (winter precipitation) and southern (summer-precipitation) Great Basin.

Winter precipitation in arid lands generally occurs in the form of frontal storms of several days duration. Plant utilization of this moisture, stored in the soil column, ordinarily is delayed until air temperatures rise. Summer drought in this "Mediterranean" climatic regime is accentuated by heightened evaporation resulting from elevated air temperatures (Fig. 12-2).

Monsoonal circulation is associated with summer precipitation and winter drought. It is driven primarily by the thermal contrast in the lower atmosphere between land and sea (Webster, 1987). Since air temperatures are high during the rainey season (Fig. 12-2), evapotranspiration is relatively great, and the effectiveness of the precipitation is reduced. However, utilization of water in the soil column is more immediate than for cold-season precipitation.

Winter-rainfall and summer-rainfall deserts on the same continent may be separated by gradual, but persistent climatic boundaries. Frequently, these boundaries correspond to topographic barriers such as mountain ranges. In arid western U.S.A., these climatic boundaries have persisted through time (Davis, 1989). For example, on the decadal time scale, the principal geographical variation in western U.S.A. tree-ring series is between

NATO ASI Series, Vol. I 36
Diachronic Climatic Impacts on Water Resources
Edited by Andreas N. Angelakis and Arie S. Issar

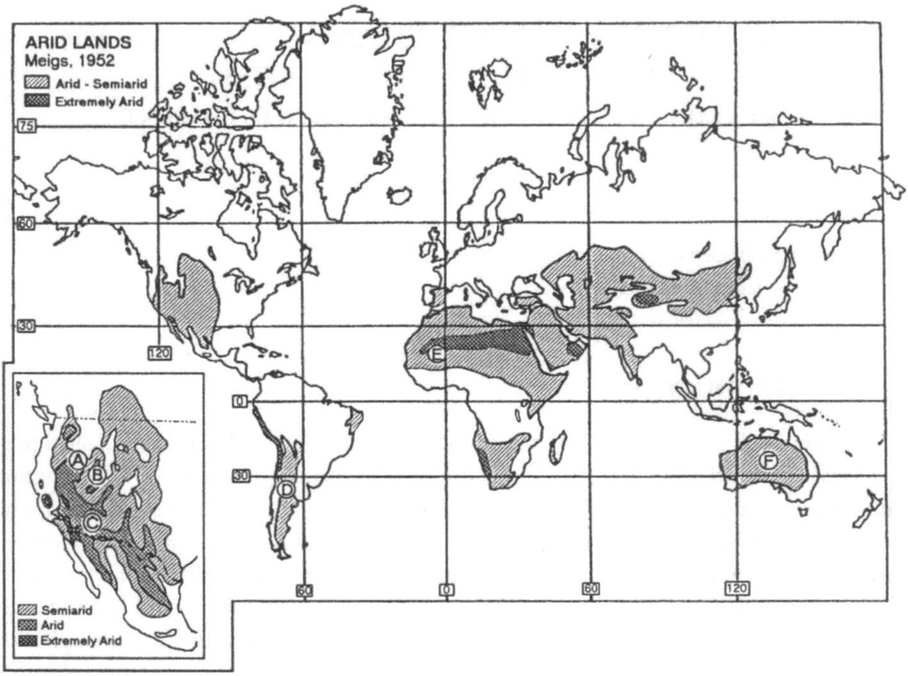

Figure 12-1. Global arid lands and site localities. Location of semiarid, arid, and extremely arid regions based on Meigs, 1952. A. Rattlesnake Cave, B. Great Salt Lake (pollen site) and Salt Lake City (meteorological station), C. Tucson, Arizona (meteorological station), Montezuma Well (pollen site), Pecks Lake (pollen site), Head Lake (pollen site), Beef Pasture (pollen site), and San Pedro Valley (isotope study), D. Argentina, E. Mali , F. Australia - regions with monsoonal precipitation

the Pacific Northwest (winter-rainfall) and American Southwest (summer-rainfall) regions (Fritts, 1991). On the millennial time scale, the northern Great Basin experienced an "early Holocene Xerothermic (Hebda and Mathewes, 1984)" whereas the southern Great Basin experienced an "early Holocene Pluvial (Davis, 1989)".

The Arizona monsoon (Bryson and Lowry, 1955) shares many characteristics with the Asian and African monsoons. In general, monsoonal precipitation occurs in summer when the contrast in air temperature between land and ocean is greatest. Tang and Reiter (1984) demonstrate that the plateau of western North America functions in the same way as the Tibetan highland in magnifying the land-sea thermal contrast. In summer, surface winds flow toward the low pressure area over the plateau, producing forced lifting of the air masses and convective precipitation (Tang and Reiter, 1984).

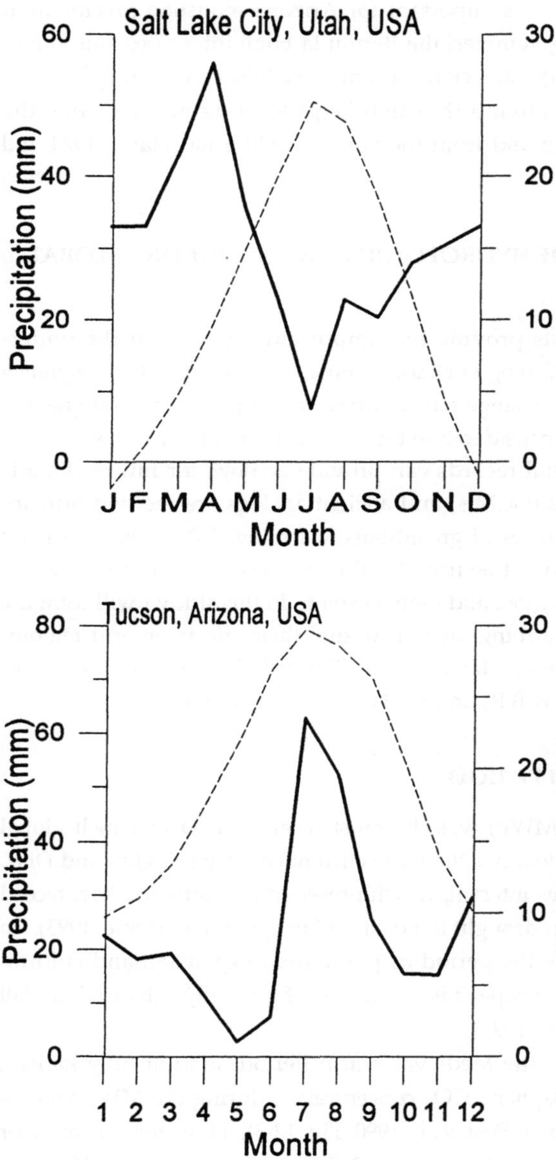

Figure 12-2. Northern and southern Great Basin precipitation and temperature. Mediterranean-type precipitation (Salt Lake City), maximum temperature in July (month 7) coincides with minimum precipitation. Monsoonal precipitation (Tucson), maximum temperature in July coincides with precipitation maximum

General atmospheric circulation also is important for Arizona monsoon precipitation (Bryson and Lowry, 1955). During summer, the Bermuda High intensifies and moves westward, and a high-level (5000 m) anticyclone becomes established over the Colorado Plateau. Atmospheric circulation around this high helps to bring moist air into the southwest from the Gulf of Mexico and from the Gulf of California (Hales, 1974 and Brenner, 1974).

12-2. ANCIENT ANALOGS OF HYDROLOGIC CHANGE DURING GLOBAL WARM PERIODS

Ancient environmental records provide two important services for the understanding of climatic (including hydrologic) change. First, they indicate the frequency and amplitude with which climatic change has occurred in the past. For example, the frequency of floods is an important consideration for construction in river valleys.

Secondly, ancient environmental records can provide analogs for future climatic change. Global circulation models (GCM's) and ancient analogs are the two primary means for predicting the consequences of greenhouse warming (EPA, 1989). Several warm periods in the Earth's past might be used for this purpose; but, these past warm periods vary in their causation, duration, and global extent. In this study I will compare three ancient analogs for global warming, demonstrating their effects on arid regions, primarily the western U.S.A. These are the Medieval Warm Period, *ca.* A.D. 1000-1200, the early Holocene, *ca.* 8,000-10,000 yr B.P.; and the late-Pliocene, *ca.* 3 Ma.

12-3. THE MEDIEVAL WARM PERIOD

The Medieval Warm Period (MWP) was the most recent period in which global temperatures were as warm as, or warmer than, the current century (Hughes and Diaz, 1993). Historic documents from this interval, which preceded the Little Ice Age, record longer growing seasons, and greater drought in northern Europe (Serre-Bachte, 1993). In the Southwestern U.S.A., this was the period of population expansion and cultural florescence (Cordell, 1984). Unlike Europe, this was a time of generally enhanced rainfall in the Southwestern U.S.A. (Petersen, 1993).

The climatic factors that led to the Medieval Warm Period are not fully known. Based on polar ice cores, the atmospheric CO_2 concentration during the MWP was no higher than the pre-industrial average (Post *et al.*, 1990, Fig. 12-3). However, the number of volcanic eruptions was low from A.D. 1100-1600 (Fig. 12-3, see Bryson, 1988), so reduced atmospheric aerosols may have permitted increased solar radiation at the earth's surface. In addition, increased solar activity during the MWP has been suggested by Eddy (1977) and by Hood and Jirikowic (1990) who use the concentration of ^{14}C in tree rings as an index of solar activity. When solar activity is low, global climate is cold

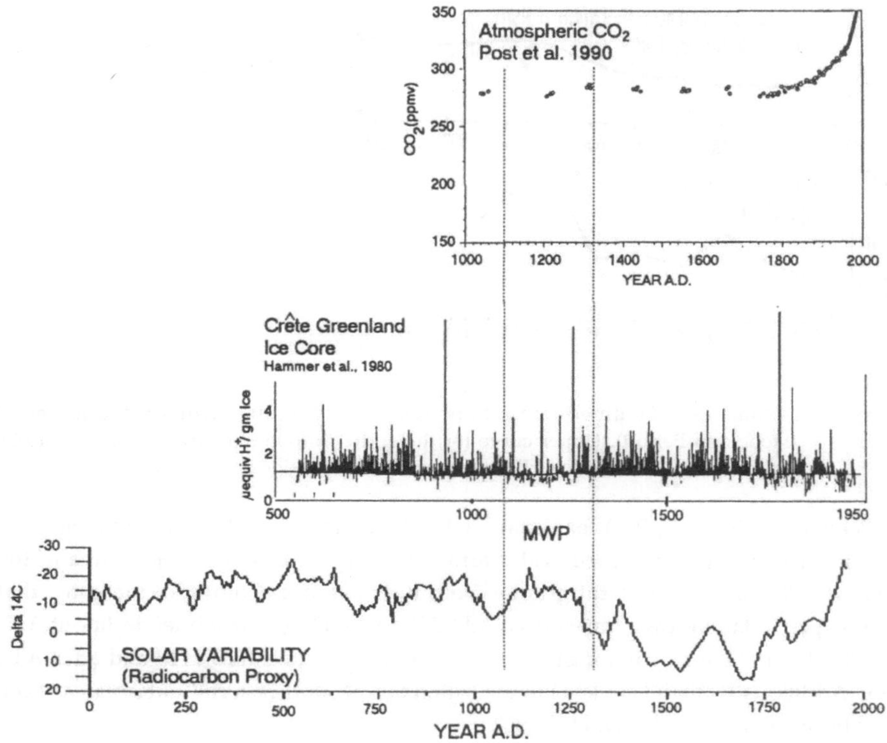

Figure 12-3. Climate forcing during the Medieval Warm Period. Ice Core acidity data reflect generally fewer volcanic eruptions during the MWP. Delta ^{14}C records, a proxy for solar activity, indicate greater solar radiation during the MWP, prior to the Little Ice Age

and ^{14}C production in the upper atmosphere is increased. Hood and Jirikowic (1990) refer to the interval of low ^{14}C production from A.D. 1120-1280 as the "Medieval" or "Grand" cosmogenic isotope anomaly (Fig. 12-3). Likewise, Schmidt and Gruhle (1988) note the occurrence of drought in western Europe during low ^{14}C production periods of the Medieval Warm Period.

The classical prehistoric cultural groups of the Southwest: Anasazi, Mogollon, Hohokam, and Patayan were agricultural people who cultivated corn, beans, squash, and cotton (Cordell, 1984). They all lived within the Arizona monsoonal boundary and practiced dryland cultivation as well as irrigation. Although each locality shows considerable variation, these cultural groups each began slow population growth at about 500 A.D., reaching a climax *ca.* 1100-1200 A.D. and had begun major decline before A.D. 1400 (Fig. 12-4).

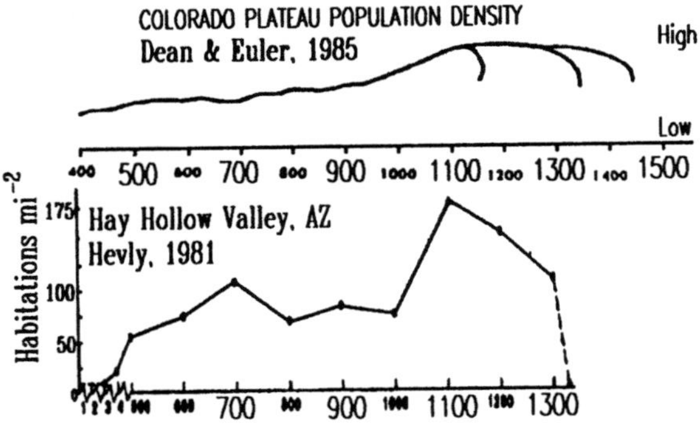

Figure 12-4. Estimates of Southwestern U.S. population during the Medieval Warm Period (Hohokam Period). Upper curve for the entire four-corners area (Utah, Colorado, Arizona, and New Mexico). Lower curve for small region in central Arizona

Petersen (1988 and 1993), has compiled palynological and tree ring evidence for climatic change during the Medieval Warm Period in the Dolores Project Area of the Colorado Plateau. Pollen stratigraphy from Beef Pasture, Colorado, indicates that summer precipitation was greatest from A.D. 780 - A.D. 1130, with a brief decline at A.D. 980 (Fig. 12-5). Overall soil moisture was greatest from A.D. 1025-1110 and after A.D. 1325. Additionally, high lake-level in the Salton Sea of southeastern California has been dated by Waters (1989) at A.D. 1187-1312.

Pollen analysis of other sites within the Arizona monsoon boundary, such as Pecks Lake, central Arizona (Fig. 12-5), indicate increased lake level from A.D. 700 - 1350, thereby supporting Petersen's (1988 and 1993) finding of summer precipitation during the MWP greater than that of today. These sites south of the Arizona Monsoon boundary indicate the coincidence of the florescence of the Anasazi, Mogollon, Hohokam, and Patayan cultures with increased monsoonal precipitation during the Medieval Warm Period.

In contrast, north of the Arizona monsoon boundary, climatic records generally indicate greater aridity during the MWP; e.g., the western Sierra Nevada (Stine, 1990). Head Lake (2300 m), 150 km northeast of Beef Pasture shows heightened aridity coincident with greater moisture at Beef Pasture and Pecks Lake (Fig. 12-5).

12-4. THE EARLY HOLOCENE WARM PERIOD

The climatic factors that led to Early Holocene global warming resulted from increased solar insolation in the Northern Hemisphere due to changes in the Earth's orbit

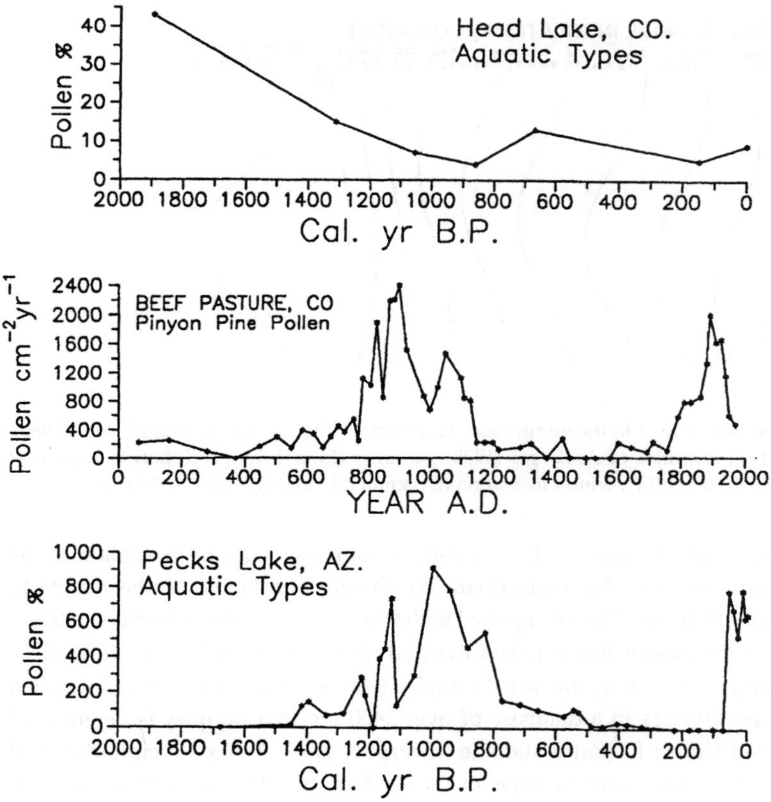

Figure 12-5. Climate reconstruction of the Southwest U.S.A. for the Medieval Warm Period. For Head Lake (southern Colorado) percentages of aquatic plants are low during the MWP, indicating greater aridity, which is the normal condition in Mediterranean-type precipitation regimes. Beef Pasture and Pecks Lake indicate greater summer precipitation and higher lake levels during the MWP, indicating strengthened monsoonal precipitation

(Fig. 12-6, see Kutzbach and Guetter, 1984 and 1986). Increased monsoonal precipitation at that time was driven by a 50% increase in the temperature difference between land and water during summer (Sellers, 1984). Short and Mengel (1986) show that the land-sea temperature difference was greatest 9,000 years ago for monsoonal Africa.

As with the Medieval Warm Period, the Early Holocene exhibits contrasting precipitation effects in winter- and summer-rainfall dominated regions. This is clearly shown by Rognon (1987) for northern and southern Africa (Fig. 12-7), and also has been shown for the American Southwest by Davis and Sellers (1993), and Barnosky (1987). In

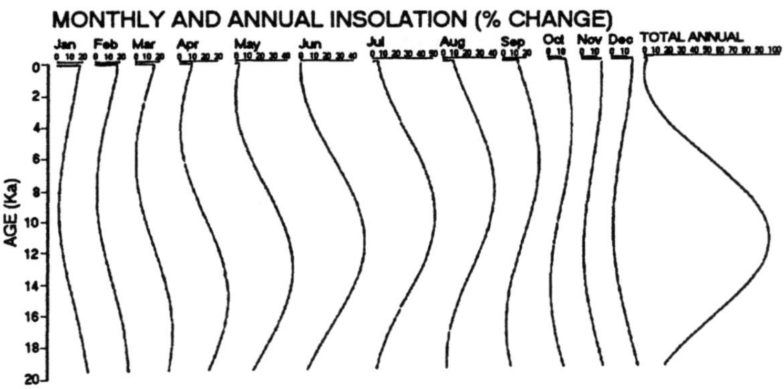

Figure 12-6. Percentage change in incoming solar radiation for the last 20,000 years (Davis, 1989), based on formulae of A. Berger (1978). Summer (June, July) precipitation is greatest 12,000 - 8,000 yr B.P., winter insolation (December, January) is least at this time

arid North America, the phrase "Early-Holocene Xerothermic" has been applied to the climate of the northern Great Basin and Pacific Northwest (Hebda and Mathewes, 1984), and the term "Early-Holocene Pluvial" applied to the Southwest (Davis and Sellers, 1993).

The climatic estimates for Rattlesnake Cave and Montezuma Well in Figure 12-7 are based on modern pollen analogs for fossil samples (Davis, 1994). For each fossil pollen sample, the closest analogs in a database of over 1400 modern samples is determined using the Squared Chord Distance statistic (Overpeck *et al.*, 1985). Precipitation and temperature are calculated as the average for the modern samples such that the dissimilarity is less than 0.14. If fewer than 2 modern analogs are found (SCD < 0.14), then no climate value is recorded.

In the northern Sahara and the northern Great Basin, the early Holocene is arid, but in the southern Sahara and southern Great Basin, precipitation is greater than today. Mosture reconstructions for the northern Sahara and the northern Great Basin differ in that moisture is greatest 3-5 *ka* in the Sahara, but is greatest in the late Holocene in the Great Basin (Fig. 12-7). Both the southern Sahara and southern Great Basin show a brief arid pulse about 8,000 yr B.P. They differ in that the monsoonal maximum in the southern Great Basin begins earlier, about 12,000 yr B.P. (Spaulding and Graumlich, 1986) compared to about 10,000 yr B.P. in the southern Sahara.

12-5. THE PLIOCENE WARM PERIOD

The mid-Pliocene is useful as an analog for elevated global temperatures because it is the most recent period in the earth's history when atmospheric CO_2 may have been higher than modern values (Fig. 12-8, see Zubakov and Borzenkova, 1988). In fact, the

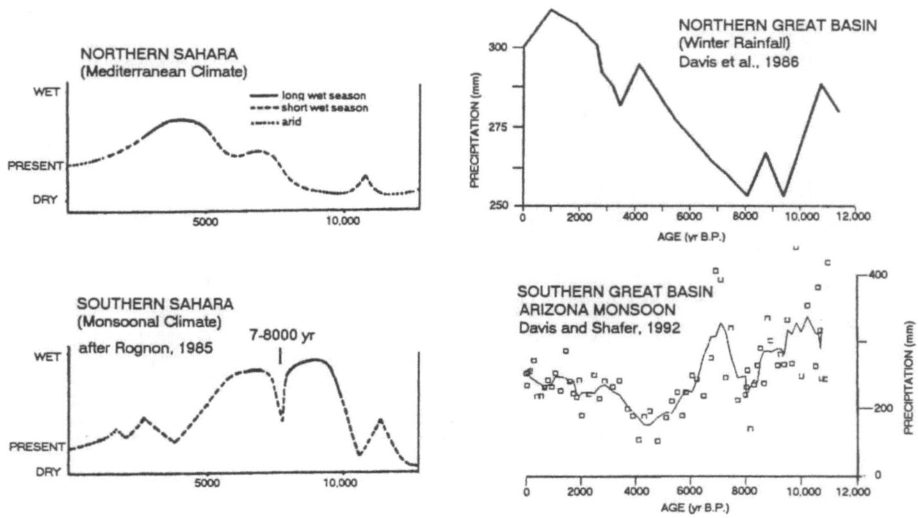

Figure 12-7. Comparison of early-Holocene climate of Mediterranean and monsoonal regions of Africa and western U.S.A. Both regions show greater aridity in the early Holocene in northern, Mediterranean-climate regions, but greater moisture during the same period for southern, monsoonal-climate regions

exact atmospheric concentration of CO_2 is not precisely known, and it might range from 1 to 3 times modern (Cerling, 1991 and Freeman and Hayes, 1992).

Climatic estimates for the Pliocene are fewer than for the MWP or early-Holocene. Floristic records of the western U.S. generally indicate a gradual reduction in temperature and increased aridity for the region from the Oligocene onward (Axelrod and Bailey, 1969 and Wolfe and Hopkins, 1967). Early to mid-Neogene pollen floras contain conifer assemblages with subtropical elements, and mesic broadleaved taxa now limited to the southeastern U.S.A. and southeastern Asia. Miocene pollen floras from the Columbia Plateau (Martin and Gray, 1962) contain over 50% bald cypress (*Taxodium*) and 5 - 20% sweet-gum (*Liquidambar*) and hickory (*Carya*). The Succor Creek pollen flora of southeastern Oregon (Cross and Taggart, 1982) contains 50% *Glyptostrobus* (a southeast China endemic) and bald cypress (*Taxodium*).

Beginning in the mid-Miocene, increasing summer drought led to the gradual elimination of exotic conifers and broadleaved trees in western North America (Leopold and Denton, 1987; Wolf, 1981). Bald cypress declined from > 70% to virtual absence during the late Miocene of south central Washington (Leopold and Denton, 1987).

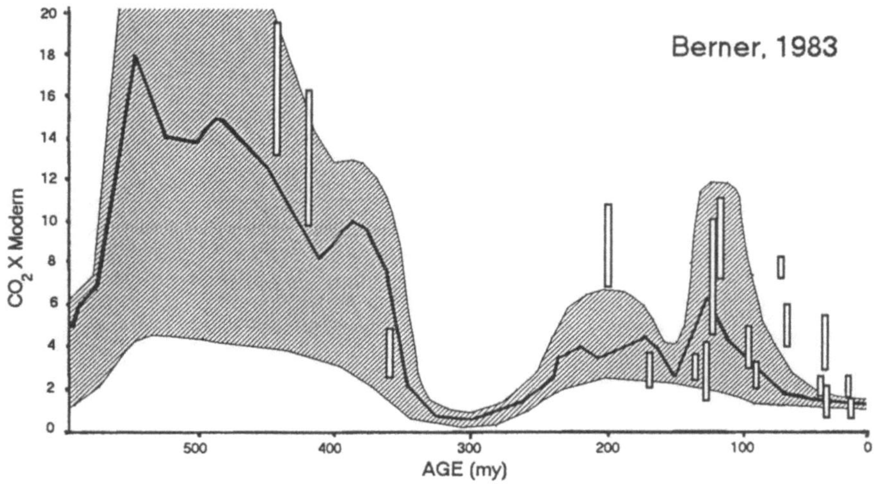

Figure 12-8. History of atmospheric CO_2 through Phanerozoic time. Vertical bars are geologic estimates of atmospheric CO_2. Solid line is model estimates (with shaded area of uncertainty) of atmospheric CO_2 (Berner, 1993)

Chenopod pollen (steppe or halophytic plants) increased to > 40% at the top of the Mio-Pliocene Wilmington, California, core (Martin and Gray, 1962). By the Plio-Pleistocene transition, steppe elements had become abundant, and the pollen of exotic trees was rare. For example, Pliocene pollen spectra from Jackson Hole, Wyoming, were dominated by types prevalent in the modern pollen rain such as fir (*Abies*), pine (*Pinus*), greasewood (*Sarcobatus*); but exotic deciduous trees also were present in small numbers; e.g., elm (*Ulmus-Zelkova*), hickory and wing nut (*Pterocarya*; Barnosky, 1984). Early Pleistocene pollen floras contained few or no exotic tree taxa and abundant pollen of steppe plants (Gray, 1961; Thompson, 1991).

Several climatic events interrupted this general trend of Neogene cooling and increasing aridity in the western U.S.A. An interval of cool winters is recorded before 3 Ma in northern California (Adam 1989) and southern Idaho (Thompson, 1991). This correlates with global cooling dated 3.4-3.2 Ma in the North Atlantic (Raymo *et al.*, 1986), and also recorded in terrestrial pollen records from Russia (Svetlitskaya, 1993 and Naidina, 1993), southern Europe (Suc *et al.*, 1993) and northern Africa (LeRoi and Dupont, 1993).

Following this cooling event, many floras indicate greater moisture and warmth than today (Wolf, 1981, Thompson, 1991). Ostracode morphology (Forester, 1991) generally indicates the presence of large, stable lakes 3.2 - 2.5 Ma. This correlates with global warming of 10° to 20° C in the Antarctic (Webb and Harwood, 1991) and sea levels perhaps 35 m higher than today (Krantz, 1991). In the American Southwest this event is reflected in high lake levels and increased winter precipitation (Smith *et al.*, 1993).

A transition to cool, dry conditions *ca.* 2.5 Ma is recorded in many floral and ostracode records from western North America, corresponding to the re-glaciation of Antarctica (Webb and Harwood, 1991) and the initiation of carbonate and isotopic oscillations in the marine record (Raymo *et al.*, 1986). Although several floral records indicate a return to warmer and wetter climate 2.0 - 1.8 Ma (Thompson, 1991), diatom stratigraphy for north central California records cooler summer temperature 2.0 - 1.7 Ma (Bradbury, 1991).

A major cooling event 0.8 Ma is recorded in southern Colorado (Rogers *et al.*, 1985) coincident with the initiation of deep-lake oscillations in the Northern Great Basin Pluvial Lakes (Morrison, 1991) and with the beginning of low-frequency (100 ka) oscillations in the marine record (Raymo *et al.*, 1986).

Herein I present preliminary results of pollen analysis of sediments of the Great Salt Lake. This lake is the largest body of water in western North America (4,700 km^2). Its surface elevation has fluctuated between 4192 and 4212 feet during the historic times primarily in response to changing discharge from its tributaries. At its Pluvial maximum (L. Bonneville, *ca.* 15 Ka) the lake was size of Lake Michigan, 51,690 km^2, in eastern North America. Pluvial oscillations of Lake Bonneville (and Lahontan to the west) began *ca.* 800 Ka (Morrison, 1991).

The Great Salt Lake pollen samples are provided as part of a cooperative research agreement between Amoco Production Company and the University of Arizona Department of Geosciences. This includes over 1,400 processed pollen samples, 15,000 cuttings, and 63 core samples from 15 wells drilled to bedrock in the Great Salt Lake. This represents the most complete continental record of the Pleistocene - Pliocene vegetation and climate in western North America, and provides a record of basin evolution in the eastern Basin and Range since at least the Miocene.

Age-control is provided by a date of 16.2 ± 1.2 Ma K-Ar at the base (12,050 ft) of the Bridge well, by the Huckleberry Ridge ash (2.01 Ma) at 2606 ft, and by the Rozel Basalts (3.1 Ma) at 3700 ft (Ron Calveret, personal communication, 9/93). Preliminary pollen analysis for the mid-lake Indian Cove core are shown in Figure 12-9. The counts are for the prepared samples provided by Amoco, each sample consists of 3 adjacent cutting samples (i.e., a 30-ft interval) taken at 100-foot intervals. Only the abundant types are shown, and only for the upper 5000 ft of the well (base = 12,050 ft.). Percentages are based on 300 pollen grains of upland plants per sample; pollen and spores of aquatic plants, freshwater algae, fungal spores, dinoflagelates, and charcoal are outside the sum.

Overall, the Indian Cove pollen diagram (Fig. 12-9) records vegetation similar to that can be found in the Great Basin today. Prior to 3.1 Ma, pine (*Pinus*), sagebrush (*Artemi-*

sia), and salt desert shrubs (*Chenopodiaceae-Amaranthus*) dominate the pollen spectrum; sedge (*Cyperaceae*) and cattail (*Typha*) are rare aquatic representatives; and deteriorated percentages exceed 50% at the base, declining to *ca.* 10% at 3.1 Ma. Degradation has altered this assemblage, but xeric steppe could produce this spectrum in the modern vegetation.

From *ca.* 3.0 - 3.2 Ma pine, sagebrush, and juniper (*Cupressaceae*) pollen percentages approach modern values, and the climate reconstruction indicates cooling (Fig. 12-10) From 3.1 Ma to *ca.* 800 Ka, salt desert shrubs (*Chenopodiaceae-Amaranthus* and *Sarcobatus*) dominate the upland pollen assemblage, and sedge and cattail are abundant (*ca.* 40%). The abundance of the wetland sedge and cattail indicates shallow water with relatively low salinity. The abundance of salt desert shrubs suggests hot, probably xeric steppe vegetation.

After 800 Ka, pine and sagebrush become abundant, comparable to modern values in the Great Salt Lake (Spencer *et al.*, 1984). Cattail and sedge become rare, and *Pediastrum*, a freshwater planktonic algae is common in a few samples (Fig. 12-9). This assemblage indicates the beginning of pluvial oscillations, confirming Morrison's (1991) findings based on exposed sediments north of the modern lake.

Temperature and precipitation reconstructions for the fossil pollen samples from the Great Salt Lake are reconstructed in the same way as for the Rattlesnake Cave and Montezuma Well ones (see above). Temperature and precipitation both show a gradual decline from *ca.* 5 Ma onward, confirming reconstructions based on disparate sites (Thompson, 1991), and consistent with gradual uplift of the Sierra Nevada and Cascade Ranges. Cooling 3.0 - 3.2 Ma is consistent with other global records. Wide fluctuations after 800 Ka correlate with the pluvial oscillations for Lake Bonneville. Milankovitch-band fluctuations cannot be detected due to the wide (100 ft.) sampling interval. Temperature and precipitation during the Pliocene are 3-4°C warmer and 50-100 mm drier than modern annual averages at Salt Lake City.

The Northwest-Southwest contrast, apparent in the MWP and early Holocene global-warming analogs, possibly is seen in the GSL cores (Fig. 12-10). If the late Pliocene (3.2 - 2.5 Ma) is wetter in the southwest (Thompson, 1991), then the aridity apparent in the GSL cores may indicate a NW-SW contrast. However, the $\delta\ ^{18}O$ data from the San Pedro Valley (Smith *et al.*, 1993, Fig. 12-10) show progressive lightening after 2.5 Ma, consistent with reduced winter precipitation, which is consistent with declining precipitation recorded at GSL (Fig. 12-10) and with gradual uplift of the Sierra Nevada and Cascade Ranges.

12-6. CONCLUSIONS

During each of three ancient periods of global warming, the climate of the Northern Great Basin was drier than today and the climate of the southern Great Basin was wetter.

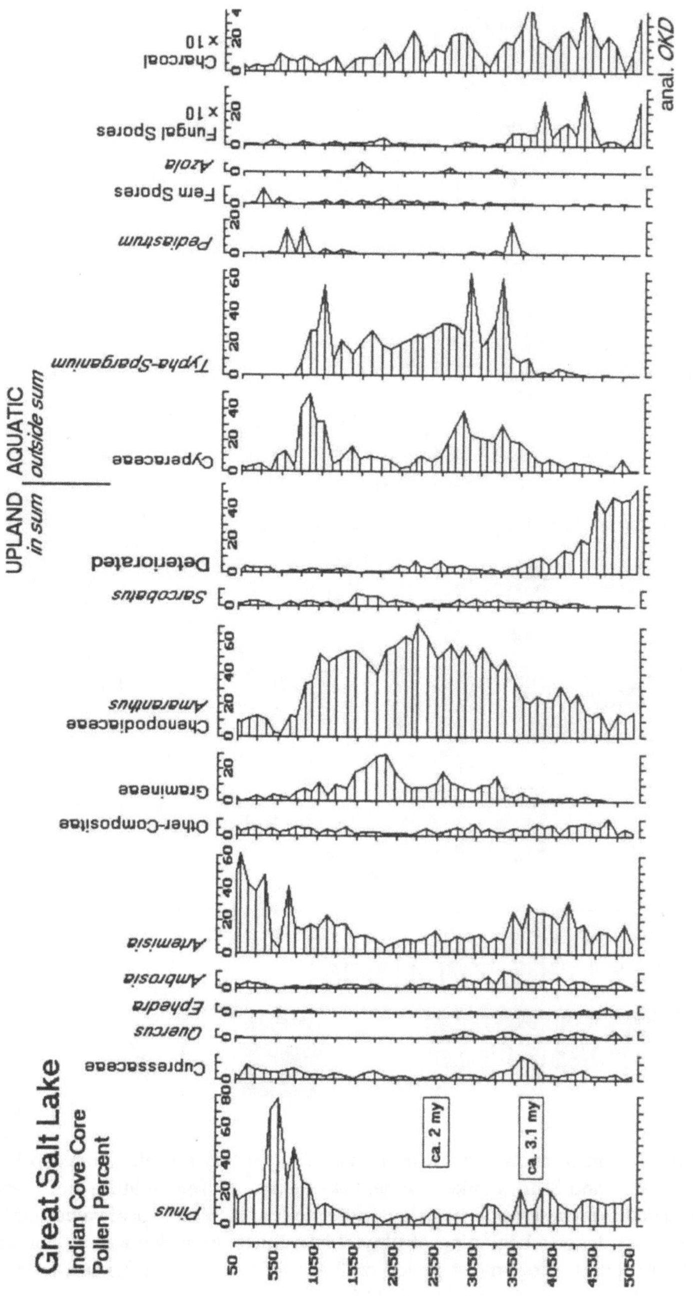

Figure 12-9. Percentage pollen diagram for Indian Cove well, Great Salt Lake. Only most abundant types shown. Note 1/10 scale for fungal spores and charcoal. Vertical scale is feet

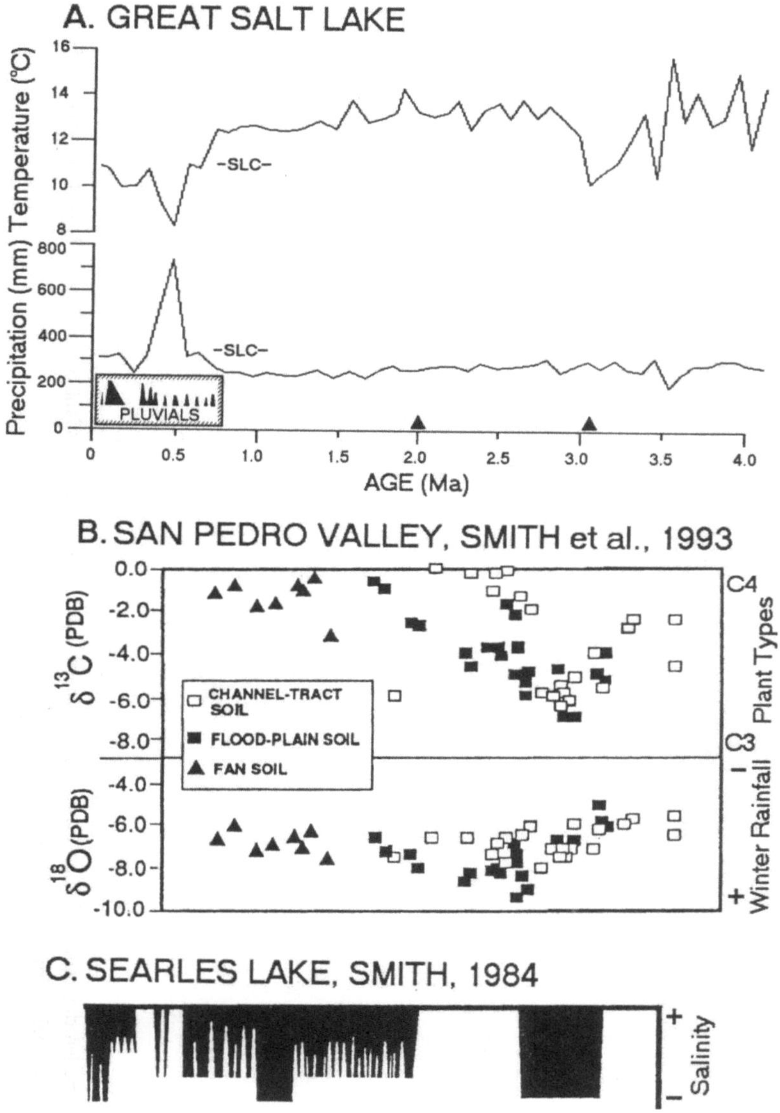

Figure 12-10. Climate reconstructions for Great Salt Lake, Indian Cove well, San Pedro Valley soil carbonates, and Searles Lake sediment chemistry. A. Temperature and precipitation estimated from best modern analogs, note modern meteorological values for Salt Lake City, inset denotes beginning of pluvial lake fluctuations 800 Ka. B. Soil carbonate isotopes reflect a cool moist period *ca.* 3 Ma. C. Sediment geochemistry for Searles Lake indicates a lake high-stand *ca.* 3 Ma

During the MWP, the population of indigenous people reached a maximum, demonstrating the correlation between increased moisture availability and these agricultural people. I conclude that effects of global warming have differed among summer-rainfall and winter-rainfall regions of the Great Basin, and have varied in the same fashion in the Sahara as recorded by Rognon (1987). These three periods differed in their causative factors, so the conclusion may be general: during periods of global warming, summer-rainfall deserts may become wetter and winter-rainfall deserts may become drier.

Acknowledgements

Pollen and climatic analysis of the Great Salt Lake cores was made possible my a cooperative research agreement between Amoco Production Company and the Department of Geosciences, University of Arizona, and by NSF grant EAR-9219505.

REFERENCES

Adam D.P., Sarna-Wojcicki A.M., Rieck H.J., Bradbury J.P., Dean W.E., and Forester R.M. (1989). Tulelake California: the last 3 million years. Palaeo Palaeo Palaeo, 73:89-103.

Axelrod D.I. and Bailey H.P. (1969). Paleotemperature analysis of tertiary floras. Palaeogeography, Palaeoclimatology, Palaeoecology, 6:163-195.

Barnosky C.W. (1987). The northwestern U.S. during deglaciation; vegetational history and paleoclimatic implications. *In*: Ruddiman WF, Wright HE Jr (Eds), North America and Adjacent Oceans During the Last Deglaciation. GSA Geology of North America, Vol. K-3.

Barnosky C.W. (1984). Late Miocene vegetational and climatic variations inferred from a pollen record in Northwest Wyoming. Science, 223:49-51.

Berger A.L. (1987). A simple algorithm to compute long term variations of daily or monthly insolation. Contrib 18, Inst D'Astronom Geophysiq Univ Catholique Louvain.

Berner R. (1993). Source for figure by T Keaton. *In:* Appenzeller T., (Ed.), Searching for Clues to Ancient Carbon Dioxide. Science, 259:908-909.

Brenner I.S. (1974). A surge of maritime tropical air - Gulf of California to the southwestern United States. Monthly Weather Review, 102:375-389.

Bryson R.A. (1988). Late Quaternary volcanic modulation of Milankovitch climate forcing. Theoretical and Applied Climatology, 39:115-125.

Bryson R.A. and Lowry W.P. (1955). Synoptic climatology of the Arizona summer precipitation singularity. Bulletin American Meteorological Society, 36:329-339.

Cerling T.E. (1991). Carbon dioxide in the atmosphere: Evidence from Cenozoic and Mesozoic Paleosols. Amer J Sci., 291:377.

Cordell L.S. (1984). Prehistory of the Southwest. Academic Press New York, pp 409.

Cross A.T. and Taggart R.E. (1982). Causes of short-term changes in fossil plant assemblages: some considerations based on a Miocene flora of the northwest United States. Ann. Missouri Bot. Gard., 69:676-734.

Davis O.K. (1984). Multiple Thermal Maxima During the Holocene. Science, 255:617-619.

Davis O.K., Sheppard J.C., and Robertson S. (1986). Contrasting climatic histories for the Snake River Plain resulting from multiple thermal maxima. Quaternary Research, 26:321-339.

Davis O.K. (1989). The regionalization of climatic change in western North America. *In*: Leinen F., Sarntein M. (Eds) Atmospheric Transport. Nato ASI Series.

Davis O.K. (1994). Climate and vegetation patterns in surface samples from arid western U.S.A.: Application to Holocene climatic reconstructions. Palynology, in review.

Davis O.K. and Sellers W.D. (1994). The effects of seasonal isolation on the history of the Arizona monsoon. Climate Change, in press.

Davis O.K. and Shafer D.S. (1992). An early-Holocene maximum for the Arizona monsoon recorded at Montezuma Well, central Arizona. Palaeo Palaeo Palaeo, 92:107-119.

Dean J.S., Euler R.C., Gumerman G.J., Plog F., Hevly R.H., and Karlstrom N.V. (1985). Human behavior, demography and paleoenvironment on the Colorado Plateaus. Amer Antiq., 50:537-554.

Eddy J.A. (1977). Climate and the changing sun. Climate Change 1:173-190.

EPA (U.S. Envir. Protection Agency). (1989). Policy Options for Stabilizing Global Climate. Report to Congress (Draff). D.A. Lashof and D.A.Tirpak (Eds), U.S. Printing Off., Washington D.C.

Freeman K.H. and Hayes J.H. (1992). Fractionation of carbon isotopes by phytoplankton and estimates of ancient CO_2 levels. Global Biogeochemical Cycles, 6:185.

Fritts H. (1991). Reconstructing Large Scale Climatic Patterns from Tree Ring Data, a Diagnostic Analysis. Univ. Arizona Press.

Gray J. (1961). Early Pleistocene paleoclimatic record from Sonoran desert Arizona. Science, 133:38-39.

Hammer C.U., Clausen H.B., and Dansgaard W. (1980). Greenland ice sheet evidence for post-glacial volcanism and its climatic impact. Nature, 288:230-235.

Hebda R.J. and Mathewes R.W. (1984). Holocene history of cedar and native Indian cultures of the North american Pacific coast. Science, 225:11-712.

Hevly R.H. (1981). Pollen production, transport and preservation: Potentials and limitation in archaeological palynology. J. Ethnobiology, 1:39-55.

Hood L.L. and Jirikowic J.L. (1990). Recurring variations of probable solar origin. Geophys. Res. Lett., 17:85-88.

Hughes M.K. and Diaz H.F. (Eds.). (1993). The Medieval Warm Period. Climate Change.

Krantz D.E. (1991). A chronology of Pliocene sea-level fluctuations, U.S. Atlantic Coastal Plain. Quaternary Science Reviews, 10:163-174.

Leopold E.B. and Denton M.F. (1987). Comparative age of grassland and steppe east and west of the northern Rocky Mountains. Ann. Missouri Bot. Gard., 74:841-867.

LeRoi S.A.G. and Dupont L. (1993). Upper Pliocene history of the Sahara by high resolution marine palynology: ODP site 658. Abstracts 26th Ann. Mtg. Amer. Assoc. Stratigr. Palynol. Oct. 23-29, 1993, pp. 50.

Martin P.S. and Gray J. (1962). Pollen analysis and the Cenozoic. Science, 137:103-111.

Meigs P. (1952). World distribution of arid and semi-arid homoclimates. Reviews of Research on Arid Zone Hydrology. UNESCO, Paris.

Naidina O.D. (1993). Climatostratigraphic interpretations of the upper Pliocene palynological data of the southeastern Russian Plain. Abstracts 26th Ann. Mtg. Amer. Assoc. Stratigr. Palynol., Oct. 23-29, 1993, pp. 60.

Overpeck *et al.*, (1985). Quantitative interpretation of fossil pollen spectra: Dissimilarity coefficients and the method of modern analogs. Quaternary Research, 23:87-108.

Petersen K.L. (1988). Climate and the Dolores River Anasazi. Univ. Utah Anthropolog Pap No 113. pp 152.

Petersen K.L. (1994). A warm and wet little climatic optimum in the southern Rocky Mountains, U.S.A. *In*: Hughes M.K., Diaz H.F. (Eds.), The Medieval Warm Period. Climate Change.

Post W.M., Peng T.-H., Emanuel W.R., King, A.W., Dale, V.H., and DeAngelis, D. (1990). The global carbon cycle. Amer. Sci., 78:310-32.

Raymo M.E., Ruddiman W.F., and Clement B.M. (1986). Pliocene-Pleistocene paleoceanography of the North Atlantic at Deep Sea Drilling Project site 609. *In*: Initial reports of the Deep Sea Drilling Project ,Vol. 94, US Gov. Printing Off, pp. 890-901.

Morrison R.B. (1991). Quaternary stratigraphic, hydrologic and climatic history of the Great Basin, with emphasis on lakes Lahontan, Bonneville, and Tecopa. *In*: Morrison RB (Ed), Quaternary nonglacial geology: Conterminous U.S. GSA Geology of North America, Vol. K-2.

Rogers K.L., Larson E.E., Smith G., Katzman D., Smith G.R., Cerling T.E., Wang-Yang and Baker R.G., Lohmann K.C., Repenning C.A., Patterson P., and Mackie G. (1992). Pliocene and Pleistocene geologic and climatic evolution in the San Luis Valley of south-central Colorado. Palaeogeography Palaeoclimatology Palaeoecology, 94:55-86.

Rognon P. (1987). Aridification and abrupt climatic events on the Saharan northern and southern margins, 20,000 Y B.P. to Present. *In*: Berger W.H., Labeyrie L.D. (Eds.), Abrupt Climatic Change Evidence and Implications. Nato ASI Series C, Vol 216.

Schmidt B. and Gruhle W. (1988). Klima, Radiokohlenstoffgehalt und Dendrochronologie. Naturwissenschafliche Rundschau, 41:177-182.

Serre-Bachte F. (1993). Middle Ages Temperature reconstructions in Europe: a focus on northeastern Italy. *In*: Hughes M.K., Diaz H.F. (Eds.), The Medieval Warm Period. Climate Change.

Smith G.A., Wang Y., Cerling T.E., and Geissman J.W. (1993). Comparison of a paleosol-carbonate isotope record to other records of Pliocene-early Pleistocene climate in the western United States. Geology, 21:691-694.

Smith G.I. (1984). Paleohydrologic regimes in the southwestern Great Basin, 0-3.2 my ago, compared with other long records of "global" climate. Quaternary Research, 22:1-17.

Spaulding W.G. and Graumlich L. (1986). The last pluvial climatic episodes in the deserts of southwestern North America. Nature, 320:441-444.

Spencer R.J., Baedecker M.J., Eugster J.P., Forester R.M., Goldhaber M.B., Jones B.F., Kelts K., McKenzie J., Madsen D.B., Rettig S.L., Rubin M., and Bowser C.J. (1984). Great Salt Lake and precursors Utah: The last 30,000 years. Contributions to Mineralogy and Petrology, 86:321-334.

Stine S. (1990). Late Holocene fluctuations of Mono Lake, eastern California. Palaeogeography, Palaeoclimatology, Palaeoecology, 78:333-381.

Suc J.-P., Drivaliari A., Bessais E., Bertini A., Zheng Z., Abdelmalek S.-M., Diniz F., Combourieu-Nebout N., Leroy S., Ferrier J., and Duzer D. (1993). Pliocene vegetation and climate in the west European and Mediterranean regions. Abstracts 26th Ann. Mtg. Amer. Assoc. Stratigr. Palynol. Oct 23-29, 1993, pp. 74-75.

Svetlitskaya T.V. (1993). The main stages of environmental evolution in the northern Black Sea coastal region during the Pliocene according to paleomagnetic data. Abstracts, 26th Ann. Mtg. Amer. Assoc. Stratigr. Palynol., Oct 23-29, 1993, pp. 77-78.

Tang M. and Reiter E.R. (1984). Plateau monsoons of the northern hemisphere: a comparison between North America and Tibet. Monthly Weather Review, 112:617-637.

Thompson R.S. (1991). Pliocene environments and climates in the western United States. Quaternary Science Reviews, 10:115-132.

Webster P.J. (1987). The elementary monsoon. *In*: Fein J.S., Stephens P., (Eds), Monsoons. John Wiley and Sons, N.Y., pp. 3-32.

Webb P.-N. and Harwood D.M. (1991). Late Cenozoic glacial history of the Ross embayment, Antarctica. Quaternary Science Reviews, 10:215-223.

Wolf J.A. (1981). Paleoclimatic significance of the Oligocene an Neogene floras of the northwestern United States. *In*: Niklas K.J. (Ed), Paleobotany, Paleoecology, and Evolution. Praeger, New York, 2:79-101.

Wolf J.A. and Hopkins D.M. (19670. Climatic changes recorded by Tertiary land floras in northwestern North America. *In*: Tertiary Correlations and Climatic Changes in the Pacific. 11th Pacific Sci. Congr. , Tokyo, Japan, pp. 67-76.

Zubakov V.A. and Borzenkova II. (1988). Pliocene palaeoclimates: past climates as possible analogues of mid-Twenty-First century climate. Palaeogeography, Palaeoclimatology, Palaeoecology, 65:35-49.

CHAPTER 13

PERSPECTIVES OF FUTURE CROP WATER REQUIREMENTS IN SPAIN:
THE CASE OF MAIZE AS A REFERENCE CROP

MAPLA I. MINGUEZ[1] AND ANA IGLESIAS[2]

[1]*Departamento de Producciõn Vegetal: Fitotecnia,*
Escuela Tecnica Superior de Ingenieros Agrõnomos,
Ciudad Universitaria, and [2]*Departamento de Sistemas Forestales, CIT-INIA,*
Carretera Coruna km 7, 28040 Madrid, SPAIN

13-1. INTRODUCTION

The demand of water for consumption in Spain is 30,500 hm^3 $year^{-1}$. Industrial and urban consumptions amount to 1,900 and 4,300 hm^3 $year^{-1}$ respectively and the rest, 24,200 hm^3 $year^{-1}$ is consumed in agriculture and represents 80% of the total consumption (MOPT, 1993a). One of the main objectives of the future law on water use ("Ley del Plan Hidrologico Nacional") is to increase use efficiency in order to increase water availability (MOPT, 1993b, pp. 22). The goal is to economize at least 1,000 hm^3 $year^{-1}$, that represents 4% of the volume of water currently consumed in agriculture. The specific water savings projected programs deal mainly with improvement of the irrigation systems and with an efficient adjustment between crop water demand and supply (MOPT, 1993a).

The recognized perspective of a changing climate would increase the present uncertainties of water availability in regions as southern Europe. Although this perspective is considered in the spanish law on water use ("Plan Hidrologico Nacional") as an additional reason for increasing the efficiency of water use (MOPT, 1993b), there are very few studies in Spain that have addressed this problem from a scientific point of view. This study pretends to be a contribution on the assessment of possible changes in irrigation water requirements under climate change, focussing on alterations of crop evapotranspiration (ET) and net irrigation requirement: the first as part of the hydrological equation and the second as preliminary data for future analysis of available supplies. The approach taken here is to use a water balance model embedded in a crop model in combination with scenarios of climate change generated by GCMs.

The growing competition for water and the importance of irrigated agriculture in Spain has provided additional stimulus for the study. Although there are a few case

NATO ASI Series, Vol. I 36
Diachronic Climatic Impacts on Water Resources
Edited by Andreas N. Angelakis and Arie S. Issar

studies that evaluate the impacts of climate change on the agriculture of Mediterranean regions (Carter *et al.*, 1991a, 1991b; Kenny and Harrisson, 1992; Iglesias and Minguez, 1994; Imerson *et al.*, 1987; Parry *et al.*, 1988, 1992; Santer, 1985), none of them have analysed the irrigation water demand.

Some climatic impact studies on water resources have used historical climate data. Following this approach, Revelle and Waggoner (1983) correlated past temperatures changes with river flows of the Colorado River, and with a 2°C increase they deduced a 30% decrease in runoff. Kenny and Harrison (1992) calculated changes in crop ET driven by an increase in 2°C across Europe. But most of the attempts to evaluate vegetation responses to climate have used output weather variables from General Circulation Models (GCMs) (Adams *et al.*, 1990; Goudriaan and Unsworth, 1990; Overpeck and Bartlein 1990; Rosenzweig, 1990; Rosenzweig *et al.*, 1992). The recent IPCC scientific and impact assessments (Tegart *et al.*, 1990) endorse this approach. Allen *et al.* (1991) evaluated possible changes in irrigation water demand in the Great Plains, using a water-balance-irrigation model in combination with climate scenarios generated by two GCMs, and analysed the economic consequences.

An advantage of using output from GCMs for vegetation-related impact studies is that they produce output climate variables (temperature and precipitation) that are internally consistent, since in a doubled-CO_2 climate past relations between seasonal temperatures and precipitations may not longer hold. In addition, the use of GCMs to develop scenarios allows for comparisons between and among regions. Nevertheless, since precipitation changes are not consistent from model to model, the resulting changes in soil moisture as calculated with output of different models are not consistent (for example, see Kellogg and Zhao, 1988). Therefore, the approach taken here is to compare output from three GCMs to produce a range of possible scenarios that should be considered only as a sensitivity analysis given the current knowledge on climate projections.

13-2. PROJECTED FUTURE CLIMATE IN SPAIN: A RANGE OF SCENARIOS

This study considers the effect of an equivalent doubling of atmospheric CO_2 on the climate of the region (a "2xCO_2" climate scenario). In order to characterize this climate, outputs from experiments with 3 GCMs were used (Goddard Institute for Space Studies (GISS) (Hansen *et al*, 1983); Geophysical Fluid Dynamics Laboratory (GFDL) (Manabe and Wetherald, 1987) and United Kingdom Meteorological Office (UKMO) (Wilson and Mitchell, 1987)). In addition the study has considered a "low-end" scenario created with the output of the transient run of the GISS model for the decade of the 2030s (Hansen *et al.*, 1989). The output from the GCMs is not accurate enough for direct use in crop model on a daily basis, so mean monthly changes in climate variables from doubled CO_2 simulations of three GCMs were applied to observed daily climate records (1974-1985) (obtained from the Instituto Nacional de Meteorologia) to create climate change scenarios for each site. The sites selected represent all gridboxes of the three GCMs in the Iberian peninsula. Because the GCMs compute mean climatic variables for large (~100,000 km^2) gridboxes and they do not account for variation of these quantities within the gridbox, no interpolation was made of the GCM output. The study is based on six sites located in main agricultural regions (Fig. 13-1) chosen to represent climates with differences in seasonal temperatures, precipitation amounts, and crop management practices.

In general, precipitation concentrates between October and April (Southern regions) or May (Northern regions) and average seasonal temperatures decrease with increasing altitude and latitude, and the weather at all sites is characterized by large inter-annual variability (Font-Tullot, 1983). The low precipitation especially in the summer months (June, July and August) makes irrigation a necessity (Bignon, 1990, Iglesias and

Figure 13-1. Schematic re-presentation of the Spanish agricultural regions and sites considered in this study. In these regions maize has to be grown under irrigation in contrast with region I where it can be grown as a rainfed crop. In region II, horticultural crops are mainly grown

Minguez, 1994) since the minimum summer (June, July, August) precipitation requirement for rainfed grain maize is 150 mm (Shaw, 1977). The summer precipitation is well under this threshold in all regions (Font-Tullot, 1983).

The three GCM models are in accord on the temperature change projections, with the GISS and GFDL models showing warming of some 4°C over the Iberian Peninsula, varying somewhat with season, and the UKMO showing larger increases (*ca.* 6° C)

(Iglesias and Minguez, 1994). Nevertheless, the models have large discrepancies in the precipitation changes projected for each region (Fig. 13-2). Since the current summer precipitation is low or non-existant, the model discrepancies do not imply large net precipitation variations. This should be taken into account when predicted irrigation needs of a summer crop are evaluated (see below).

13-3. EVAPOTRANSPIRATION

Cloudiness, windiness, humidity and temperature are factors that are affected by greenhouse warming so atmospheric demand for water will also be affected. Higher atmospheric temperatures asssociated with the future climate can have profound implications on hydrologic processes in general and in water availability in particular (Rosenberg, 1981; Rosenberg *et al.*, 1989). Evapotranspiration constitutes an important component of the hydrologic balance, therefore its accurate calculation is essential. Changes in ET are not only related to vegetative responses, but affect other processes of the water cycle such as runoff and water availability (McKenney and Rosenberg, 1993).

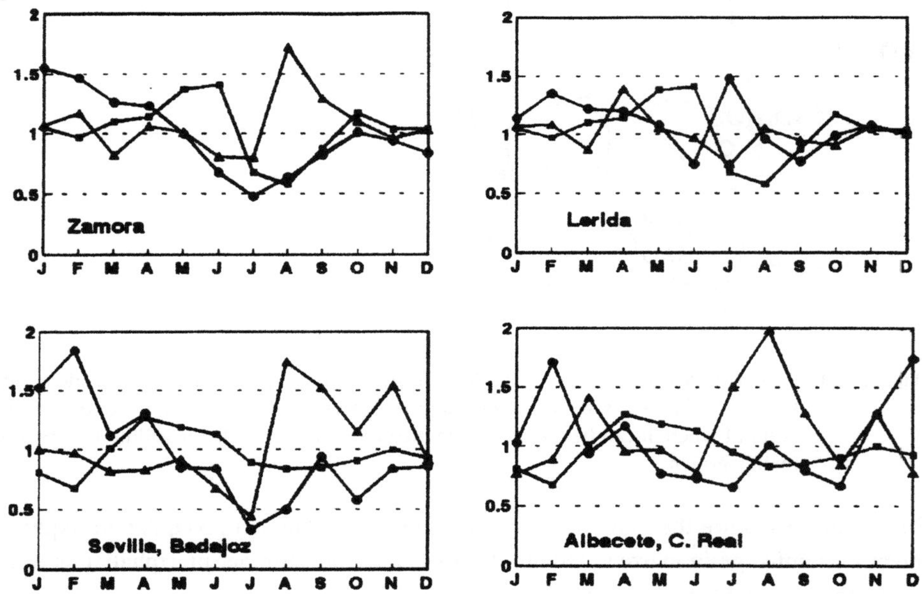

Figure 13-2. Monthly precipitation ratios ($1xCO_2/2xCO_2$) of the GCMs used for the climate change simulations. (▲) GISS (Hansen *et al.*, 1983); (●) GFDL (Manabe and Wetherald, 1987); (■) UKMO (Wilson and Mitchell, 1987)

Rind *et al.* (1992) examined the likelihood of future drought, discussing the importance of an accurate calculation of the potential (reference) evapotranspiration in the models for climate change assessments. The projected changes in climate imply higher actual evapotranspiration rates, leading to an increased rate of moisture loss from plants and soils.

In this study we use a crop model (CERES-maize; Jones and Kiniry, 1986) that has an embedded water balance model to simulate changes in hydrological variables. Potential evaporation is evaluated in the CERES model using the Priestley-Taylor relation (Priestley and Taylor, 1972). This relation is extensively used because it uses meteorological data that are generally available. Evapotranspiration (ET) is directly related to the equilibrium ET by a coefficent α. Priestley and Taylor (1972) obtained values for α ranging from 1.08 to 1.34 with 1.26 as an overall mean value.

Other processes integrant of the water balance of the CERES model are calculated as following: Precipitation is a daily input, either the observed (for the baseline runs) or the modified by the ratio $2xCO_2/1xCO_2$ for the climate change runs; Runoff is a function of soil type, soil moisture and precipitation; Infiltration is calculated as precipitation minus runoff; Total Evaporation is a function of potential evaporation, leaf area index and time; Drainage, when the soil moisture is greater than the soil water holding capacity of the bottom layer; Soil moisture, calculated as precipitation minus evaporation minus runoff minus drainage.

13-4. THE DIRECT EFFECTS OF CO_2 ON WATER USE EFFICIENCY

Increases in the atmospheric CO_2 concentrations affect plants direcly by influencing the plant physiological processes of photosynthesis and transpiration. Because the climate change is associated with higher levels of CO_2 than the current climate, the direct effects of CO_2 on crop yield and water use have to be included (see revision of experimental literature by Acock and Allen, 1985; Allen *et al.*, 1985; Cure, 1985; Kimball, 1983; and Rose, 1989). Increased concentrations of atmospheric CO_2 have been shown to allow plant stomata to remain partially closed and limit transpirative water loss per unit of leaf area. In any case, the crop water-use efficiency (a measure of the yield per unit amount of water transpired) tends to rise with increased CO_2 (Acock and Allen, 1985).

For maize the photosynthesis is reported to increase 10% (Cure, 1985) and the stomatal resistance is reported to increase from 56 (seconds/meter) in 330 ppm to 106 (s/m) in 555 ppm CO_2 (Rogers *et al.*, 1983). While these results have been demonstrated in experimental settings, their applicability outside of the enclosed systems in which they have been observed, for example, over wide geographical areas and long time periods, is currently being debated (Körner, 1990). Based in these experimental data Peart *et al.* (1989) modified the CERES model to account for the direct CO_2 effects on the crop.

13-5. THE CROP RESPONSES TO FUTURE CLIMATE

Potential changes on crop responses were estimated with the CERES-maize model (Jones and Kiniry, 1986). CERES is a dynamic crop growth model that simulates crop growth and development on a daily basis in response to climatic (daily solar radiation, maximum and minimum temperature, and precipitation), soils (albedo and a variety of measures related to water in the profile) and management (cultivar, planting date, plant population, row spacing and sowing depth, etc.) factors. Input soils and management variables were determined for each location according to information provided by the Agricultural Experimental Stations. The initial soil moisture conditions were specified for each site. The water demand at all sites was calculated assuming: 100% efficiency of the automatic irrigation system; irrigation management depth according to the local soil; automatic irrigation when the available water was 50% of soil water capacity and soil water for each layer was re-initialized to 100% capacity at the start of each growing season. Because of these assumptions, only relative changes on water requirements are considered in this study.

The uncertainties of irrigated maize production are large due to high production costs and the limited and unreliable water available for irrigation. Yield, evapotranspiration and irrigation demand of corn are determined in this study in order to evaluate potential changes in water use efficiency and irrigation demand under climate change scenarios. Strategies that optimize water use efficiency are emphasized. The choice of maize for this type of analysis was guided by several reasons: (a) it is a crop with a high water consumption that has the maximum requirements during the season with largest water shortages and rainfall variability in Spain; (b) the readily available CERES model has an embedded water balance model that simulates water requirements and it has been calibrated previously in Spain (Iglesias and Minguez, 1994); (c) the model permits changes in management to evaluate adaptation strategies to improve water use efficiency under different climate scenarios.

In most regions, current management strategies are based on the cultivation of long growing-season hybrids or varieties (following the FAO classification cycles FAO 700 and FAO 800 (over 2850 thermal units)) as first crop and are sown in spring. The changes in crop yield under the GCM climate change scenarios have been already described by Iglesias and Mînguez (1994) and were driven by two interacting effects, changes in climate and CO_2 enrichment. We will focuss here on the effects of climate change on water use and comparisons will be made mainly between the base (current climate under 330 ppm CO_2) and the GCM climate changes scenarios with increased CO_2 (555 ppm CO_2).

Evapotranspiration

Evapotranspiration was summed over the entire period of crop growth for the current and climate change simulations. The effects of climate change on maximum crop ET (Table

13-1) are mainly influenced by two opposing factors: (a) the shortening of vegetative period associated with higher temperatures (Table 13-2); and (b) the direct effect of temperature increases in the daily ET. The maximum ET over this period increased at all sites in the simulation under climate change alone, in relation to the baseline climate ET, except in the coldest site (Zamora). When the direct CO_2 effects on crop water use are included in the simulations, projected overall ET varied between -17% (Zamora, GISS) and +18% (Badajoz, UKMO). The reductions in overall season ET have to be evaluated in the view of the significant shortening of the crop growing season associated with the warmer climate. In all cases, daily evaporative rates increased for most of the growing season.

Table 13-1. Effects of GCM climate change on simulated maximum crop evapotranspiration (mm) with and without the direct effects of increased CO_2 on crop water use. BASE indicates are average ET values under current climate

Site	Climate Change Alone				Climate Change + CO_2 Effects		
	BASE	GISS	GFDL	UKMO	GISS	GFDL	UKMO
Sevilla	629	712	735	712	619	642	623
Badajoz	594	684	721	818	597	629	721
Albacete	624	637	659	720	553	571	630
C. Real	612	660	703	777	575	612	683
Zamora	561	540	579	586	467	499	508
Lerida	580	605	616	689	528	538	605

Table 13-2. Effect of GCM climate change on the length of the crop growing season. Values are changes from current (observed) season length values (in days)

Site	GISS	GFDL	UKMO
Sevilla	-13	-19	-22
Badajoz	-12	-13	-11
Albacete	-23	-21	-24
C. Real	-22	-21	-20
Zamora	-33	-32	-38
Lerida	-21	-23	-24

Crop responses to temperature changes are non-linear, therefore a sensitivity study was carried out in Albacete to characterize the increases in evaporative demands and lengths of the crop growth duration with increased temperatures. An increase of +1°C implied an average increase of 8% in the daily ET and a decrease of 2% in total crop ET, associated with a shortening of the crop duration of 12 days. Under current climate, crop growth duration of the FAO 700 and 800 hybrids range from four and a half to five months in the region considered. An increase of +3°C implied increases of average daily ET of about 20%, decrease of total crop ET of 2% and decrease of growth duration of 23 days.

An annual increase in potential (reference) evapotranspiration of the order of 55 to 70 mm for each 1°C rise in annual temperature has been estimated for the southern mediterranean (Le Houerou, 1990). A similar order of magnitude of change is indicated by Rowntree (1990), who estimates an increase of 50 mm per 1°C temperature increase over a 6 month growing season in mid latitudes.

Water Use Efficiency

Changes in the water use efficiency (WUE, ratio of crop yield to total crop evapotranspiration) under climate change with the beneficial direct effects of CO_2. In general, the WUE declined at all locations, with highest reductions in the central regions (main production areas) and lowest in the coldest region (Zamora) (Fig. 13-3). Nevertheless, this is not a realistic future situation since farmers and agricultural institutions will adapt to changing climate conditions. An important adaptation strategy is the early sowing under a warmer climate, allowing the crop to develop during a period of the year with cooler temperatures and partially offseting crop yield reductions. A combination of adaptation strategies (such as changes in hybrids and sowing dates), maintains or improves the water use efficiency in most regions under the climate change scenarios (Fig. 13-3). The WUE increase in the Zamora site is due to the choice of a hybrid with a longer crop growth duration accompanied by an earlier sowing date.

Irrigation

Changes in irrigation requirements under the climate change scenarios are presented in Figure 13-4. Because of the assumptions made for their calculation, only relative changes are considered. Irrigation requirements not only reflected changes in crop evapotranspiration but also the predicted changes in precipitation during spring mainly. In general irrigation demand showed great variability, and the magnitudes of the changes varied with sites and GCMs predictions (Fig. 13-4). The decreases in irrigation water requirements in some sites have to be carefully interpreted, since they may only reflect the decrease in crop evapotranspiration due to the significant shortening of the crop growing season which is accompanied with significant yield reductions. Therefore we analysed relative changes in irrigation use efficiency (IUE=crop yield/total irrigation water) (Fig. 13-5). There is a consistent decrease in IUE at all sites under all scenarios considered, compared with base.

If the adaptation strategies are considered, the simulated IUE under all scenarios does not change significantly from base (range of 10%) in some regions and IUE of main production regions (Albacete, C. Real and Badajoz) remained 10% lower. Therefore adaptation strategies may be able to offset the negative impact on IUE of climate change.

13-6. WATER USE OPTIMIZATION WITH ALTERNATIVE MANAGEMENT

In the main production central area (i.e. Albacete site), new management strategies are being evaluated in order to assess alternatives that would lower water consumption.

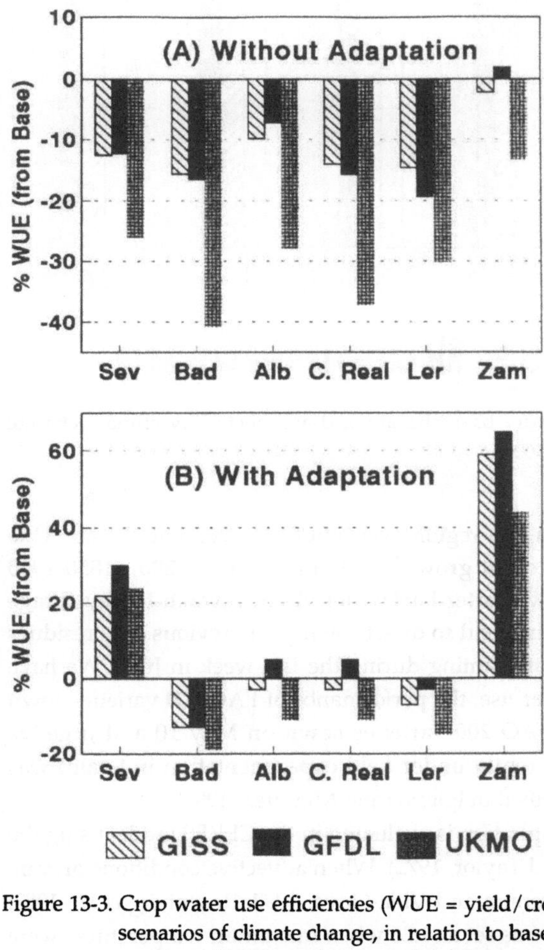

Figure 13-3. Crop water use efficiencies (WUE = yield/crop ET) for the six sites, under the three scenarios of climate change, in relation to base climate. (A) With current management practices. (B) When adopting adaptation strategies

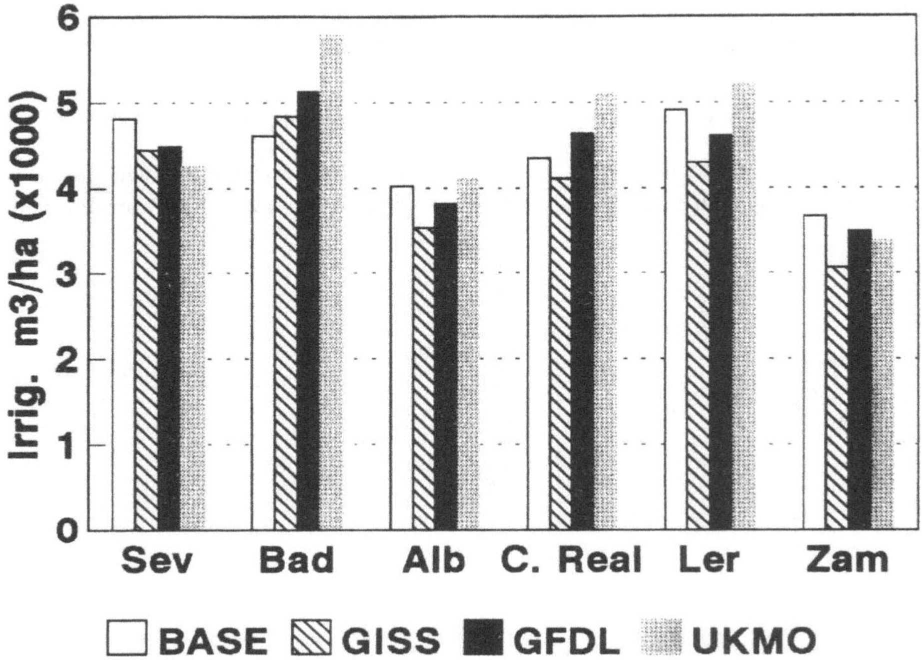

Figure 13-4. Net irrigation demand under base climate and as affected by climate change scenarios generated by the GCMs

Water availability is often limited in this region even under the present climate. One strategy considers the inclusion of short growth cycle maize (FAO 200) (1850-1950 thermal units) as a second crop sown after barley, lentils or a vetch-barley forage mixture. A delay in sowing date lowers yield so direct sowing on previous crop residues is being implemented. This allows for planting during the last week in June. We have compared, in terms of yield and water use, the performance of FAO 700 varieties sown May 10 (current strategy) versus FAO 200 varieties sown on May 10 and June 25. Calibration for FAO 200 hybrid (currently under field experimentation in Spain) was done using the same procedure described in Iglesias and Mínguez (1994).

As stated above, crop evapotranspiration is evaluated in the CERES model using the Priestley-Taylor relation (Priestley and Taylor, 1972). When advective conditions prevail, α increases (Kanemasu *et al.*, 1976; Pereira and Villa-Nova, 1992; Rosenberg *et al.*, 1983; Shouse *et al.*, 1980). The α coefficient was set as 1.26 when maximum temperatures were under 35°C and 1.45 over this temperature (calibration data not shown ere). In this central region, maximum temperatures over 35°C correspond to the summer months of July and August where rainfed crops that are cultivated around irrigated areas, are matured or harvested. When short cycle maize hybrids are sown as second crop in late June or July, the crop is subject to high temperatures before reaching full ground cover. This implies that the advective and micro-advective (between rows) processes occur increasing crop ET.

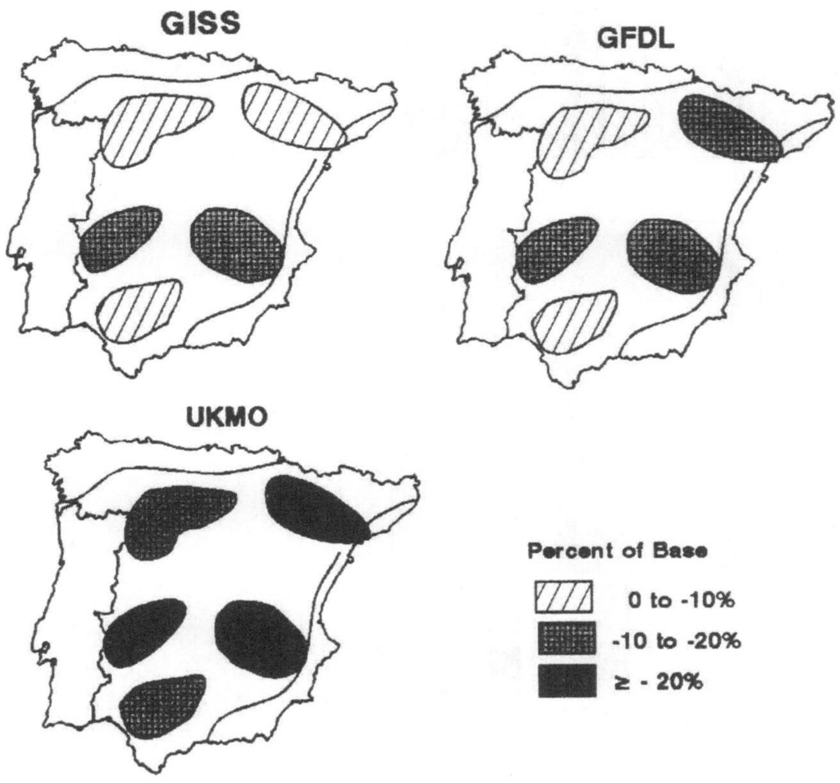

Figure 13-5. Changes in the irrigation use efficiencies (IUE= yield/net irrigation demand) in relation to base climate under the GCMs scenarios with current management strategies

Figure 13-6 shows that a short-season hybrid clearly reduced irrigation demand when sown in the traditional spring season (200 1st in the Figure) and it further reduced irrigation demand if sown later in the year (200 2nd in the Figure). In the case of maize 200 as 2nd crop, maximum water requirements coincided with both lower temperature

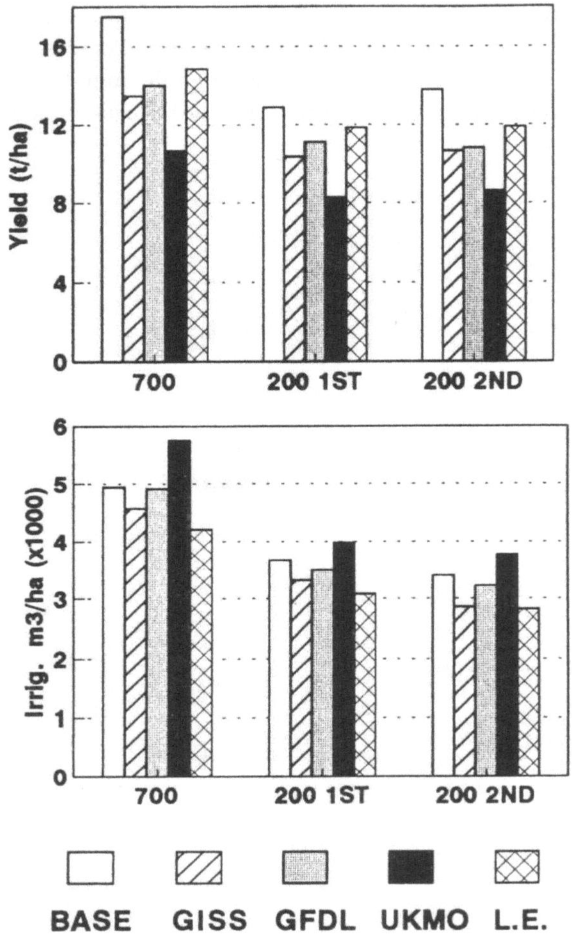

Figure 13-6. Comparisons among net irrigation demand of long-growing cycle variety (700) sown on May 10 (current strategy), short-growing cycle sown on May 10 (200 and in June 25 (200 2nd) under base climate, GCMs and low end (L.E.) scenarios for the Albacete site

and higher precipitation period (end of summer, begining of autumn). But a shorter-season variety entailed yield reductions when compared with the traditional varieties (700 in the Figure). We have evaluated the changes in yield, net irrigation and irrigation use efficiency resulting from this crop management alternative, both under current climate and climate change scenarios (the direct beneficial effects of CO_2 were

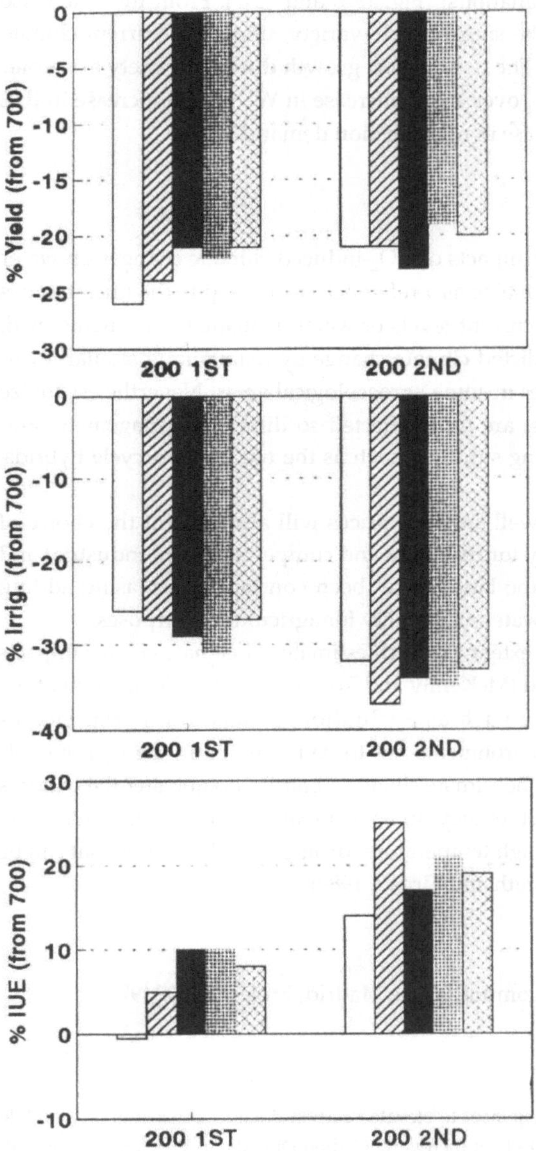

Figure 13-7. Changes in yield, irrigation and irrigation use efficiency (IUE) under base climate, GCMs and low end scenarios when implementing the new strategy based on the choice of a short-growing cycle variety taking the traditional long-growing cycle variety as a reference. Same legend for Scenarios as Figure 13-6

considered in all climate change simulations) (Fig. 13-6 and 13-7). From the water use perspective, there was a benefit of the shorter-cycle variety, under the current climate and all climate scenarios considered. The use of short growth duration variety as second crop resulted in a 20% yield decrease, over a 20% increase in WUE, 15% increase in IUE and aproximately 2,000 m^3 ha^{-1} decrease in net irrigation demand.

13-7. CONCLUSIONS

This study assessed the potential impacts of CO_2-induced climatic changes on water use in irrigated areas in Spain, using maize as a reference crop. Despite the uncertainties associated with the use of GCMs, if current levels of water availability are maintained, there is scope for adaptation for predicted climate change by maintaining similar levels of water and irrigation use efficiencies in some agroecological areas. Nevertheless maize main production areas (central areas) are more affected so different management strategies that imply less water use during summer, such as the use of short cycle hybrids, should be considered.

Water and production costs as well as grain prices will also delimit the choice of management strategies. Water supply for irrigation and competition from industrial and urban uses are hard to predict but the latter have been constantly increasing adding another uncertainty factor to future water availability for agricultural purposes.

Work is needed for revealing the extent to which estimates of climate change impacts may depend on the methodology used (McKenny and Rosenberg, 1993). Simulation of the processes of the hydrological cycle such as evaluation of maximum crop evapotranspiration in arid and semiarid environments has to be improved in the crop models (Rosenberg *et al.*, 1983). Furthermore, the climate change scenarios do not alter the patterns of events in the base climate and therefore they do not simulate changes in the underlying variability (e.g., extended periods of high temperature, drought incidence, etc) that can be crucial for crops (Pittock *et al.*, 1991; Smith and Tirpak, 1989).

Aknowledgements

This work was financed by The Comunidad de Madrid, Project n§057/92.

REFERENCES

Acock B. and Allen Jr L.H. (1985). Crop responses to elevated carbon dioxide concentrations. *In*: B.R. Strain and J.D. Cure (Eds.). Direct Effects of Increasing Carbon Dioxide on Vegetation. DOE/ER-0238. U.S. Department of Energy. Washington, D.C. pp. 53-97.

Adams R.M. and Rosenzweig C., Peart R.M., Ritchie J.T., McCarl B.A., Glyer J.D., Curry R.B., Jones J.W., Boote K.J., and Allen Jr L.H. (1990). Global climate change and US agriculture. Nature, 345(6272): 219-224.

Allen L.H. Jr, Jones P., and Jones J.W. (1985). Rising atmospheric CO_2 and evapotranspiration. *In*:

Advances in Evapotranspiration, pp. 13-27, American Society of Agricultural Engineers, St. Joseph, Mich.

Allen R.G, F.N. Gichuki, and C. Rosenzweig. (1991). CO_2-induced climatic changes and irrigation-water requirements. J. Water Res. Plan. Manag., 117: 157-178.

Bignon J. (1990). Agrometeorology and the physiology of maize. Publication: EUR 13041 EN. Office for Official Publications of the EC, Series: An Agricultural Information System for the EC. Luxembourg.

Carter T.R., Parry M.L., and Porte J.Rr. (1991a). Climatic change and future agroclimatic potential in Europe. Int. J. Climatol., 11: 251-269.

Carter T.R., Porter J.R., and Parry M.L. (1991b). Climatic warming and crop potential in Europe: Prospects and uncertainties. Global Environ. Change 1: 291-312.

Cure J.D. (1985). Carbon dioxide doubling responses: A crop survey. *In* B.R. Strain and J.D. Cure (Eds.), Direct Effects of Increasing Carbon Dioxide on Vegetation. U.S. Department of Energy. DOE/ER-0238. Washington, D.C. pp. 33-97.

Font-Tullot I. (1983). Climatologia de Espana. Instituto Nacional de Meteorologia. M.O.P.T. Espana.

Goudriaan J. and Unsworth M.H. (1990). Implications of increasing carbon dioxide and climate change for agricultural productivity and water resources. *In:* Kimball, B.A., N.J Rosenberg, L.H. Allen Jr. (Eds.), Impact of Carbon Dioxide, Trace Gases, and Climate Change on Global Agriculture. ASA Special Publication Number 53. American Society of Agronomy, Inc., pp. 71-82.

Hansen J., Russell G., Rind D., Stone P., Lacis A., Lebedeff S., Ruedy R., and Travis L. (1983). Efficient three-dimensional global models for climate studies: Models I and II. April Mon. Weather Rev.,111: 609-662.

Hansen J., Fung I., Lascis A., Rind D., Lebedeff S., Ruedy R., and Russell G., (1989). Global climate changes as forecasted by Goddard Institute for space studies three-dimensional model. J. Geophys. Res., 93: 9341-9364.

Iglesias A. and Mínguez M.I. (1995). Perspectives of maize production in Spain under climate change. *In*: Harper, L., Hollinger, S., Jones, J. and Rosenzweig, C. (Eds). American Society of Agronomy Special Publication, Madison, in press.

Imerson A., Dumont H., and Sekliziotis S. (1987). Impact Analysis of Climatic Change in the Mediterranean Region. Volume F: European Workshop on Interrelated Bioclimatic and Land Use Changes. Noordwijkerhout, The Netherlands.

IPCC (1990) Climate Change. The IPCC Impacts Assessment. W.J. McG. Tegart, G.W. Sheldon and D.C. Griffiths (Eds.), Intergovernmental Panel on Climate Change. World Meteorological Organization and United Nations Environmental Program. Bracknell, UK.

Jones C.A. and Kiniry J.R. (Eds.). (1986). CERES-Maize: A Simulation Model of Maize Growth and Development. Texas A&M University Press. College Station, TX., pp. 194.

Kanemasu E.T., Stone L.R., and Powers W.L. (1976). Evapotranspiration model tested for soybean and sorghum. Agron. J., 68:569-573.

Kellogg W.W. and Zhao Z.C. (1988). Sensitivity of soil moisture to doubling of CO_2 in climate model experiments. J. Clim. Change, 1: 348-366.

Kenny G.J. and Harrisson P.A. (1992). Thermal and moisture limits to grain maize in Europe: model testing and sensitivity to climate change. Clim. Res., 2: 113- 129.

Kimball B.A. (1983). Carbon dioxide and agricultural yield. An assemblage and analysis of 430 prior observations. Agron. J., 75: 779-788.

Körner C. (1990). CO_2 fertilization: the great uncertainty in future vegetation development. *In*: A.

Salomon and D. Reidel (Eds.), Global Vegetation Change. Hingham, Mass.

Le Houerou H.N. (1990). Global change: vegetation, ecosystems and land use in the Mediterranean basin by the mid twenty-first century. Israel J. Bot., 39: 481- 508.

Manabe S. and Wetherald R. (1987). Large-scale changes of soil wetness induced by an increase in atmospheric carbon dioxide. J. Atmos. Sci., 44: 1601-1613.

McKenny M.S. and Rosenberg N.J. (1993). Sensitivity of some potential evapotranspiration estimation methods to climate change. Agric. For. Meteorol., 64:81-110.

MOPT. (1993a). Plan Hidrolõgico Nacional. Memoria. Ministerio de Obras Pùblicas y Transportes. Secretaria de Estado para las Politicas del Agua y el Medio Ambiente.

MOPT. (1993b). Anteproyecto de Ley del Plan Hidrolõgico Nacional. Ministerio de Obras Publicas y Transportes. Secretaria de Estado para las Politicas del Agua y el Medio Ambiente.

Overpeck J. and Bartlein P.J. (1990). Assessing the response of vegetation to future climate change: Ecological response surfaces, and paleoecological model validation. *In:* The Potential Effects of Global Climate Change on the United States, Appendix D, Forests, J.B. Smith and D.A. Tirpak (Eds.), US Environmental Protection Agency, Washington, D.C., pp. 1- 32.

Parry M.L., Carter T.R., and Konijn N.T. (Eds.). (1988). The Impact of Climatic Variations on Agriculture. Vol 1, Assessments in Cool Temperate and Cold Regions. Vol. 2, Assessments in Semi-arid Region. Kluwer, Dordecht, Netherlands. pp. 876 and 764.

Parry M.L., Carter T.R., Porter J.R., Kenny G.J., and Harrison P.A. (1992). Climate Change and Agricultural Suitability in Europe. Environmental Change Unit. University of Oxford. Report No. 1. Oxford, UK.

Peart R.M., Jones J.W., Curry R.B., Boote K., and Allen L.H. Jr. (1989). Impact of climate change on crop yield in the southeastern U.S.A. In Smith, J.B. and D. Tirpak (Eds.). The Potential Effects of Global Climate Change on the United States. Report to Congress. EPA-230-05-89-050. U.S. Environmental Protection Agency. Washington, DC. Appendix C-1, pp. 2-1 to 2-54.

Pereira A.R. and Villa-Nova N.A. (1992). Analysis of the Priestley-Taylor parameter. Agric. For. Meteorol., 61:1-9.

Pittock A.B., Fowler A.M., and Whetton P.H. (1991). Probable changes in rainfall regimes due to the enhanced greenhouse effect. International Hydrology & Water Resources Symposium. Perth. pp. 182-186.

Priestley C.H.B. and Taylor R.J. (1972). On the assessment of surface heat flux and evaporation using large scale parameters. Mon. Weather Rev., 100: 81-92.

Revelle R. and Waggoner P.E. (1983). Effects of a carbon dioxide induced climatic changes on water supplies in the western US. *In*: Changing Climate 419-432. National Research Council, National Accademy Press, Washington, D.C.

Rind D., Rosenzweig C., and Goldberg R. (1992). Modelling the hydrological cycle in assessments of climate change. Nature, 358: 119-122.

Rogers H.H., Bingham G.E., Cure J.D., Smith J.M., and Surano K.A. (1983). Responses of selected plant species to elevated carbon dioxide in the field. J. Environ. Qual., 12: 569-574.

Rose E. (1989). Direct (physiological) effects of increasing CO_2 on crop plants and their interactions with indirect (climatic) effects. *In*: The Potential Effects of Global Climate Change in the US: Appendix C- Agriculture Volume 2. EPA 230-05-89-053. J.B. Smith and D.A. Tirpak (Eds.), US Environmental Protection Agency, Washington, DC.

Rosenberg N.J. (1981). The increasing CO_2 concentration in the atmosphere and its implication on agricultural productivity. *In*: Effects on Photosynthesis, Transpiration and Water use Efficiency. Clim. Change, 3: 265-279.

Rosenberg N.J., Blad B.L., and Verma S.B. (1983). Microclimate, The Biological Environment, 495 pp.

John Wiley, New York.

Rosenberg N.J., Kimball B., Martin P., and Cooper C. (1989). Climate change, CO_2 enrichment and evapotranspiration. Climate and Water. *In*: P.E.Waggoner (Ed.), Climate Change, Climatic Variability, and the Planning and Management of US Water Resources. John Willey and Sons, New York, pp. 131-147.

Rosenzweig C. (1990). Crop response to climate change in the Southern Great Plains: A simulation study. Professional Geographer, 42: 20-37.

Rosenzweig C., Parry M.L., Baer B., Baethgen W., Brklacich M., Chou T.-Y., Curry R.B., Dellecolle R., Downing T.E., Eid H.M., Escano C.R., Iglesias A., Jones J., Karim Z., Koval L., Liverman D., Menzhulin G.V., Meyer M.S., Muchena P., Qureshi A., Rao G., Ritchie J.T., Sala O., Seino H., de Siqueira O.J.F., Tongyai M.L.C., and Zhiqing J. (1992). Implications of climate change for international agriculture: Global food trade and vulnerable regions. Crop modeling study. U.S. Environmental Protection Agency and U.S. Agency for International Development. Washington, DC.

Rowntree P.R. (1990). Estimates of future climatic change over Britain. Part 2: Results. Weather, 45: 79-89.

Santer B. (1985). The use of General Circulation Models in climate impact analysis-a preliminary study of the impacts of a CO_2- induced climatic change on Western European Agriculture. Clim. Change, 7: 71-93.

Shaw R.H. (1977). Water use requirements of maize- a review. *In*: Agrometeorology of the Maize (corn) Crop. Proceedings of Symposium, July 1976. Ames, Iowa, USA. WMO-N¨ 481, WMO, Geneva, Switzerland, pp. 119-134.

Shouse P., Jury W.A., and Stolzy L.H. (1980). Use of deterministic and empirical models to predict potential evapotranspiration in an advective environment. Agron. J., 72: 994-998.

Smith J.B. and Tirpak D. (Eds.). (1989). The Potential Effects of Global Climate Change on the United States. Report to Congress. EPA-230-05-89-050. U.S. Environmental Protection Agency. Washington, DC, pp. 423.

Tegart W.L. McG, Sheldon G.W., and Griffiths D.C. (Eds.). (1990). Intergovernmental Panel on Climate Change. Impacts Assessment Report. Working Group II.

Wilson C.A. and Mitchell J.F.B. (1987). A doubled CO_2 Climate Sensitivity Experiment with a Global Model Including a Simple Ocean. J. Geophys. Res., 92: 13315-13343.

CHAPTER 14

COASTAL OCEAN WATER RESOURCES:
LINKAGES WITH TERRESTRIAL FRESHWATER RESOURCES, ANTHROPOGENIC INFLUENCES AND CLIMATE CHANGE

TERRY A. NELSEN

NOAA-AOML-OCD,
4301 Rickenbacker Causeway, Miami,
Florida 33149, USA

14-1. INTRODUCTION

"The population of the world at mid-century was 2.5 billion; some time in 1987 it passed five billion. The increase in the past 40 years has equaled the total increase over the millions of years from when the human species emerged until 1950" (Keyfitz, 1989). This fact places in perspective the urgency for mankind's attention to the changes occurring around us and, either directly or indirectly, effecting our collective lives. The magnitude of this problem and changes due to it will tend to overprint problems of a local to regional nature. Added to this is the potential for both natural and anthropogenic-induced climate change. Although a multitude of social, economic and environmental problems can, and are, resulting from our exploding population, and vary depending on whether we inhabit a developed or a developing nation, one irreplaceable need is common to us all: freshwater.

Freshwater is essential to all life on land and provides for domestic and industrial needs but, most importantly, for the production of crops to feed this ever increasing world population. Global water resources traditionally include terrestrial-associated rivers, lakes and subterranean aquifers. In a less traditional, though significant manner, the coastal ocean also serves as a water resource for mankind. Most importantly it serves as a fisheries habitat and hence a source of vital protein for many nations which may be considered analogous to land-based agriculture as a source for protein.

On a global scale, water is stored in eight major reservoirs that include the world's oceans (97.41%), glaciers (1.98%), groundwater (0.59%), lakes (0.007%), soil moisture (0.005%), atmosphere (0.001%), rivers (0.0001%) and biota (0.0001%) (Speidel and Agnew, 1988). A dynamic interchange, at variable rates, exists between these reservoirs. One such important interchange from the oceans to the land is in the form of oceanic evaporation, atmospheric transport and continental precipitation. Coupled with

NATO ASI Series, Vol. I 36
Diachronic Climatic Impacts on Water Resources
Edited by Andreas N. Angelakis and Arie S. Issar

precipitation from evapo-transpiration of terrestrial origin, approximately 35% of the continental precipitation returns back to the coastal oceans (Shiklomanov, 1993) carrying with it both natural and anthropogenic components of the continents. On average, freshwater runoff from the world's continents to the coastal oceans is approximately 45,000 km^3/year (Shiklomanov, 1993). Globally, there is a significant inequality as to where and when these discharges occur (Table 14-1).

Table 14-1. Characteristics of global river discharge to the coastal oceans

Region	Asia	South America	North America	Africa	Europe	Australia and Oceania
Runoff volume[a]	10,485	10,377	6,945	3,808	2,321	2,011
Runoff maximum[b]	August	June	July	January & March	May	March

a = volume in cubic kilometers; source: World Resources 1990, p. 330
b = Shiklomanov, 1993

Just as earth's processes are dynamic, so are mankind's. Although subject to debate, one of society's most dynamic aspects is the current population increase, with consequences ranging from regional to global scales. In more tangible terms, the worldwide population increase was approximately 77% from 1960 (3.0 billion people) to 1990 (5.3 billion people), with a projected increase by an additional 60% by the year 2025 (8.5 billion people). Regionally, countries surrounding the Mediterranean basin show similar trends, as shown in Figure 14-1. Globally, and regionally, population increases of this magnitude continue to place stresses on natural resources which, in turn, induce change. These changes take the form of land use for agriculture and deforesting that modify freshwater runoff, sediment transport and the flux of natural and anthropogenic nutrients and pollutants coastward. In concert with this, dam building and sewage disposal further modify traditional runoff patterns to the coastal ocean.

As the interface and buffer between continents and deep oceans, the coastal zone is a transition region in which large amounts of dissolved and particulate material are received, transformed, transported, deposited and exported. This zone is commonly recognized as extending from coastal wetlands and estuaries past the surf zone and seaward to the continental shelfbreak (approximately 200 m water depth) and is an area of high biological productivity. As such, the coastal ocean supplies man with both fisheries and recreational resources and therefore what we deliver to, and deprive from, the coastal oceans ultimately will affect man's interests.

Mankind's dependence on food from the sea is less than on land-based resources but, none-the-less, constitutes a tangible contribution to our dietary needs. Worldwide,

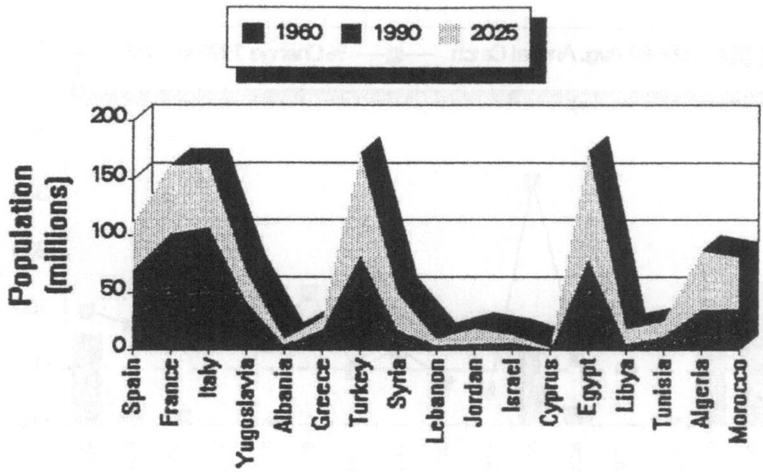

Figure 14-1. Past, present and projected populations for countries at the rim of the Mediterranean Sea

16% of total animal protein comes from fish while Africa, Western Europe, the Near East and North America draw 21.1%, 9.7%, 7.8% and 6.6% respectively (FAO Circular No. 853, 1992). Regionally, the coastal ocean plays a significant role as a food source for countries rimming the Mediterranean Basin (Fig. 14-2). Although the coastal ocean constitutes only <8% of the surface of the globe, it accounts for as much as 25% of the global biological production and more than 99% of the global landings of marine fish take place within 320 km of the coastline (Clark, 1981). Thus, loss or diminution of this resource will place additional burdens on terrestrial food resources which are already becoming stressed regionally if not globally (Crosson and Rosenberg, 1989; Postel, 1992, 1993).

Moreover, approximately 50% of world population today is believed to currently live within 60 km of the coastline. This is driven in part by traditional habitation patterns and partially by expanding global populations. With a growing proportion of people living juxtaposed to the sea, anthropogenic impacts of pollution, and environmental alteration of the coastal zone, have accelerated over the last decades. Within the Mediterranean Basin, the total coastline is approximately 48,400 km compared to that of the United States which is approximately 19,900 km. Moreover, nearly 54% of urban population in countries rimming the Mediterranean, and approximately 41% of the United States' urban populations, live in large coastal cities (World Resources, 1990, p. 336). Given this, mankind has ample opportunity to influence the coastal ocean in both

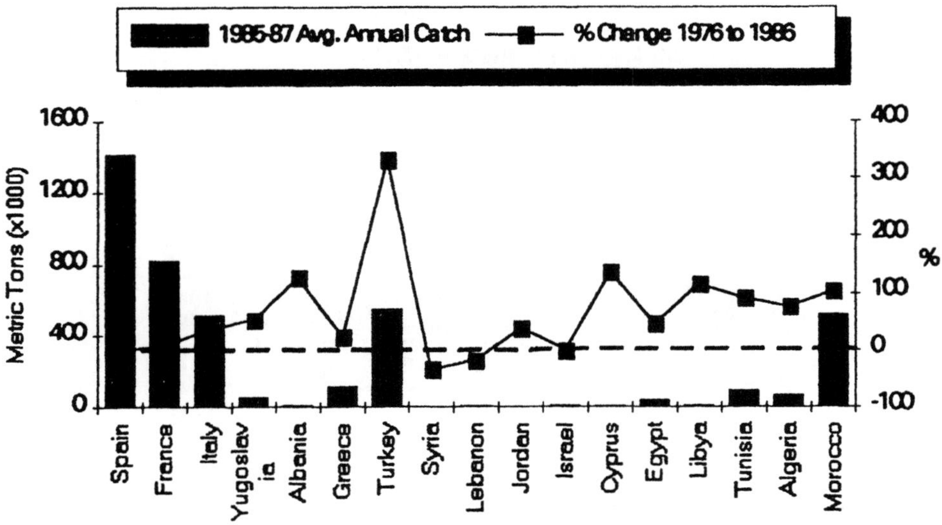

Figure 14-2. The average annual fisheries catch for nations surroundings the Mediterrean Basin as well as the percent change in fisheries landings over the decade 1976-1986

positive and negative ways.

It will therefore be the objective of this chapter to place the coastal ocean within the context of water resources and show how changes induced by both man and climate can and are influencing the coastal ocean. Because of its importance, the coastal ocean will be examined for how it is influenced by freshwater input, what changes have taken place, and how changing patterns of freshwater input may affect this economically important zone. Examples will be drawn from the Mediterranean Sea and the United States' Gulf of Mexico coastal zone.

14-2. CONTINENTAL INFLUENCES

Global

Now, as throughout geologic time, earth's precipitation has fallen on continents where it has been accumulated in drainage basins, each with its own characteristics of natural and, more recently, anthropogenic components. Ultimately, these waters transition to the coastal oceans via rivers. Through this passage to the sea, the delivery of each river's natural and anthropogenic burdens occurs. If we examine man's use of terrestrial freshwater over two decades from the 1980s to a projected use in the year 2000

(Table 14-2), it is obvious that no major percentage shift between the three primary consumptive uses (agriculture, industry and domestic) is projected.

This creates the illusion that a static situation may be assumed for the coastal ocean, which in reality is invalid for two reasons. First, studies have shown that river-borne pollution, such as nitrogen is proportional to population size within a given river's drainage basin (Peierls *et al.*, 1991). Secondly, most pollution sources are on the continents with most contaminants being water-borne in rivers, outfall pipes and storm runoff. The primary use of global freshwater is agriculture (Table 14-2) and it is there that profound changes in the anthropogenic impact are experienced. Agricultural useage augments the industrial and domestic (sewage) sources that eventually are transported via rivers to the coastal ocean. As such, the agricultural source will be the focus of ensuing discussions, although similar arguments could be made for the other sources as well.

Table 14-2. Consumptive use of fresh water for the 1980s versus projected use in the year 2000

	1980s			2000		
	Agriculture	Industry	Domestic	Agriculture	Industry	Domestic
Europe	110	193	48	125	250	56
Africa	120	6.5	10	160	33	30
USSR	260	117	23	300	145	35
N. America	330	294	66	390	365	90
S. America	70	30	24	90	105	40
Asia	1300	118	88	1500	330	200
Oceania	16	1.4	4.1	20	3.25	5.5
total =	2206	759.9	263.1	2585	1231.25	456.5
	68%	24%	8%	60%	29%	11%

Source = World Resources 1990, p. 167 (data in cubic km)

The operational definition of a "point" source is typically piped flows such as sewage and industrial outfalls, while "non-point" sources are the more diffuse sources such as river catchment basins which eventually coalesce into channelized river flows. Data indicate that the primary (64%) non-point source pollution affecting rivers and lakes in the United States, originates from agricultural sources (Fig. 14-3A) which are rich in nutrients and that nutrients form one quarter (approximately 24%) of all non-sediment pollution in these water bodies (Fig. 14-3B). These nutrients primarily consist of nitrogen (N) and phosphorus (P) with estimates for the annual global discharge to the coastal oceans between 7-35 x 10^6 and 0.6-3.5 x 10^6 metric tons respectively (World Resources, 1992, p. 176). Nitrate pollution, primarily from fertilizers and animal manure, constitute a serious problem in both North America and Europe. The following

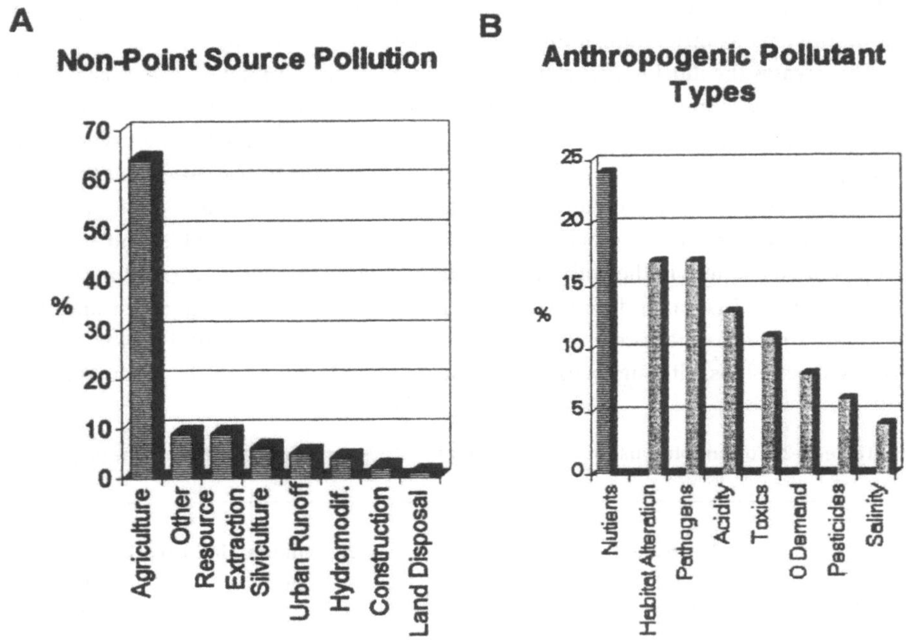

Figure 14-3. A. Major non-point source pollution, and B. Leading anthropogenic pollutant types in affected United States rivers and lakes

discussion will therefore focus on nitrate.

Simple percentages, as seen in Figure 14-3, tell only part of the story in that they give no clues as to the rate of change for these parameters. The rate of change in introducing fertilizers into the environment can best be seen, for individual continents and the world scale, in Figure 14-4. These data show an increased use of fertilizers by approximately 200% over the two decades spanning from the mid-1960s to the mid-1980s. Because nitrate, a major portion of such fertilizers, has high solubility in water and thus low tendency to bind to soil particles, it is easily transported via runoff to rivers and lakes. As a consequence, river-borne nitrate burdens have risen dramatically over the last few decades (Table 14-3), as shown for the Mississippi, Po, Ebro and Rhône Rivers. This rise in application (Fig. 14-4), transported by rivers to the coastal oceans, has contributed significantly to changes in the water quality and ecosystem health of the vital coastal zone.

In the two sections to follow, examples will be given which elucidate trends observed in the coastal zones adjacent to the rivers cited in Table 14-3, both as a function of their freshwater transport and associated anthropogenic burdens. In doing so the operational definitions of "point" and "non-point" sources will be extended beyond their traditional convention of outflow pipes and basin-wide inputs, respectively. In a

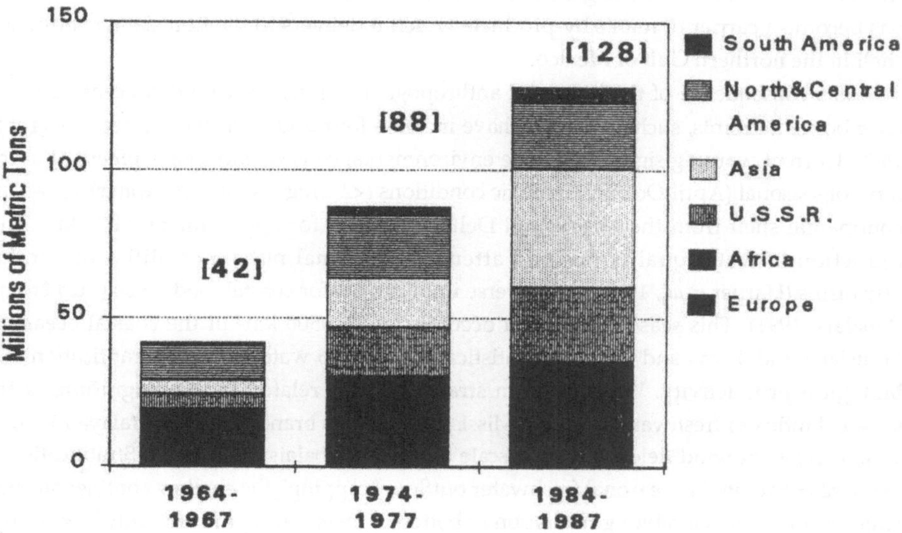

Figure 14-4. The globally increasing use of artificial fertilizer, by continent, for the period 1964 through 1987

manner analogous to the conventional definitions given above, rivers entering a water body such as the coastal ocean act as, and will be considered, a "point" source, while general distributive coastal runoff is diffuse and will be viewed as a "non-point" source.

Table 14-3. The averaged dissolved nitrate burden of one United States and three Mediterranean Basin rivers discussed in text (Source: World Resources, 1988, p. 320)

Nitrate (mg/l)	1970	1975	1980	1985	1988
Mississippi	0.36	1.04	1.3	1.23	1.47
Po	0.95	1.35	1.63	3.28	2.55
Ebro			5.3	10.16	8.92
Rhône			0.52	0.54	0.46

The Mississippi River and Adjacent Shelf

Point sources. The Mississippi River has a drainage basin of approximately 3.2×10^6 km^2 which integrates the natural and anthropogenic runoff of approximately 41% of the surface area of the continental United States and constitutes the largest single source of fresh water into the U.S. coastal zone. Since the beginning of the Industrial Revolution, as well as the onset of farming and the increased commercial use of fertilizers, the river has become a carrier of man's by-products which it delivers to the Louisiana continental shelf in the northern Gulf of Mexico.

As a consequence of the increased anthropogenic inputs, measured concentrations of river-borne nutrients, such as nitrates, have increase four-fold over the last decades (Table 14-3). Current warning signs of adverse environmental consequences have emerged in the form of seasonal (April-October) hypoxic conditions (<2.0 mg O_2/liter sea water) across the continental shelf from the Mississippi Delta westward for approximately 275 km. This reduction in water quality has had attendant seasonal mass mortalities of benthic organisms (Harper *et al.*, 1981) and adverse implications for coastal food webs (Turner and Rabalais, 1991). This seasonal hypoxia occupies up to 9500 km^2 of the coastal ocean off Louisiana and Texas and has been statistically linked to water-column stratification and biological productivity. Water-column stratification is related to the magnitude of the seasonal influx of freshwater from the Mississippi, and its branch, the Atchafalaya River, as well as regional wind fields and shelf-scale currents (Rabalais *et al.*, 1991). Stratification is primarily attributed to seasonal freshwater outflow "capping" the shallow continental shelf water column, thus inhibiting reaeration of bottom waters during the normally low-energy summer season. Within areas affected by a stratified water column, hypoxia is believed to be a direct consequence of high rates of near-surface primary production which subsequently sinks, increases bottom-water and benthic respiration rates, which in turn deplete dissolved oxygen down to hypoxic levels (Rabalais *et al.*, 1991).

The occurrence of seasonal hypoxia has been systematically monitored for less than a decade on the Louisiana continental shelf and it is therefore reasonable to ask if hypoxia has been prevalent there since earlier times, or is it linked to anthropogenic causes? The latter has been suggested by reports of enhanced river-borne nutrient fluxes (Bratkovich and Dinnel, 1992) and high present-day rates of primary productivity (Fahnenstiel *et al.*, 1992; Lohrenz *et al.*, 1992). Ongoing measurements and spotty historical records of primary productivity and shelf hypoxia were inadequate to provide historical perspective. However, continental shelf sedimentary sequences can serve as historical repositories of natural and anthropogenic inputs. Moreover, delta-associated continental-shelf sediments typically accumulate at a rate of two to four orders of magnitude faster than slope and abyssal deposits and therefore provide ideal very-high resolution sediment records. As such, retrospective analysis of Mississippi Delta and associated shelf sediments was undertaken under the auspices of the United States' National Oceanic and Atmospheric Administration's Coastal Ocean Program. The objectives of this study were to test hypotheses that: a) by-products of hypoxia/anoxia events have left characteristic markers which produce a time history of such events in the sediment record; and b) observed riverine nutrient enhancement has promoted a shift in the foraminiferal community structure of the Mississippi/Louisiana Shelf and this shift is preserved in the sediment record. The results of this study (Nelsen *et al.*,

1994) will be summarized below to elucidate the coupling between terrestrial freshwater outflow and its attendant and rising nutrient burden in the coastal ocean.

Sediments recovered west of the Mississippi Delta, both within and outside areas of known hypoxia, contained continuous sequences radiochemically dated back to approximately the turn of the century (approximately 1900). They were thus ideal for evaluating anthropogenic influences believed to have commenced in the last five to six decades. Results indicated that the solid phase glauconite occurred above detection limits only in a sediment core from the hypoxia area and with a major increase in abundance coincident with the onset (approximately 1940) of increased use of commercial fertilizers in the United States. In the same hypoxic-area core, benthic foraminifera species-diversity decreased upcore from around the turn of the century to present in a manner concurrent with fertilizer application in the drainage basin, and with glauconite abundance (not shown) increases in the core (Fig. 14-5). Dominant benthic foraminifera species, known for their opportunistic nature in stressed environments (i.e. hypoxic), increased upcore from approximately 31%, near the core base, to approximately 60% of the total population at the top of the core. Benthic foraminifera population and diversity change and glauconite were not apparent throughout a "control" core outside the area of documented hypoxia. From these, and other data, (Nelsen *et al.*, 1994) concluded that a

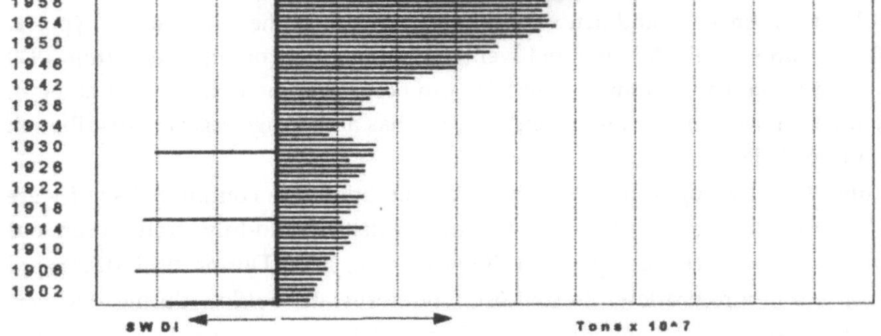

Figure 14-5. Benthic foraminiferal diversity data, in the form of the Shannon-Weiner Diversity Index (SWDI), and the tonnage of commercial fertilizer applied within the United States versus time (years 1900-1990). The foraminifera data was obtained from a core recovered within an area of known seasonal hypoxic conditions (See text for full discussion)

strong argument exists for linking hypoxia-related upcore benthic foraminifera species shifts and implied glauconite authigenesis to anthropogenic activities such as fertilizer usage in the continental United States. Moreover, data also implied that U.S. fertilizer usage, and its attendant rise in river-borne nutrient transport, has enhanced coastal primary productivity with an inception date of approximately 1940 for anthropogenically-induced hypoxic conditions in this segment of the continental shelf. It should be noted here that enhanced primary productivity can be both detrimental and beneficial. In later sections, dealing with the Mediterranean, enhanced primary productivity appears to be detrimental within the northern Adriatic, while current rises off Egypt's Nile Delta appear to be revitalizing coastal fisheries.

Studies such as Rabalais *et al.* (1991) have indicated a linkage between discharged freshwater quantity and hypoxia, while those of Nelsen *et al.* (1994) have established linkages of water quality, based on proxy fertilizer-application data, with hypoxia and species shifts. However, an additional and seldom recognized factor, i.e. the timing of freshwater introduction, may also play a significant role in the creation of hypoxic events. The following example will illustrate this potential linkage.

In river-dominated segments of coastal oceans, such as adjacent to the Mississippi River, stratification is primarily a function of freshwater outflow (halocline) and secondarily thermal warming (thermocline). Maximum outflow during the winter through the mid-spring months will tend to form weak or no stratification due to water-column mixing induced by winter storms and the lack of reinforcement from a seasonal thermocline. In contrast, maximum freshwater outflow during late-spring to summer months will be subject to less storm-induced water-column mixing, reinforced by the presence of a thermocline, and thus yield more stable and persistent stratification conducive to hypoxia formation. As such, the timing of freshwater delivery to a coastal ocean segment can potentially have significant influence on the occurrence and extent of water-column stratification and hence conditions favorable to the formation of hypoxia. Water flow records for the Mississippi River (U. S. Army Corps of Engineers), from 1930 to present, indicate that maximum river flow can occur anytime from the beginning of January through late June for any given year, but has an average seasonal distribution peaking in April (Fig. 14-6A).

In June, 1988 hydrographic conditions on the Louisiana continental shelf were similar to those observed in previous years except that by mid-July stratification and seasonal hypoxia were essentially absent (Rabalais *et al.*, 1991). During 1988 Mississippi River flow was at a record low. In contrast, total freshwater outflow during 1987 was only approximately 3% greater than 1988 (Table 14-4), yet hypoxia developed on a seemingly normal basis. This suggested that flow characteristics other than total water volume, introduced into the coastal ocean, can control hypoxia development.

To test this hypothesis, Mississippi River flow records were examined for all years from 1930 to 1993 for both normal yearly-flow characteristics and yearly flows equal to

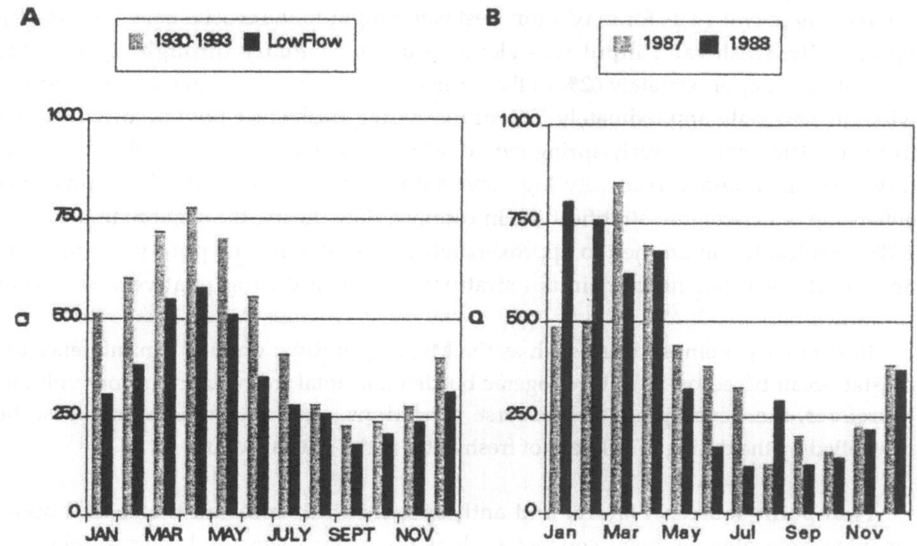

Figure 14-6. A. Comparison of normal seasonal flow for the Mississippi River (labeled 1930-1993) with selected low-flow years between 1930-1993, B. Comparison of monthly flow records for a year with strong seasonal Louisiana shelf hypoxia (1987) and for a year of essentially no shelf hypoxia (1988). (Q = monthly mean flows in thousands of cubic feet per second)

or less than that of 1988. The objective was to establish what additional characteristics, if any, may distinguish low-flow from high-flow years. Eighteen low-flow years were found to meet the criterion and were compared with normal flow from the same time interval. The results (Fig. 14-6A) suggest that there is no apparent difference between low-flow and normal years other than freshwater volume.

Table 14-4. Mississippi River flow (sum of monthly mean flow in thousands of cubic feet per second) during hypoxia monitoring years

Year	1985	1986	1987	1988	1989	1990	1991	1992
Flow	7089	6167	4805	4675	6939	7402	7774	5675

Frequently summary data, such as in Figure 14-6A, obscure details crucial to understanding complex natural systems. Therefore, a comparison of monthly flow data was done for the cited hypoxic (1987) and non-hypoxic (1988) years. Even though their total yearly flows were essentially identical, results indicated a significant temporal shift between these two years for maximum freshwater input to the coastal zone (Fig. 14-6B). Specifically, freshwater input was elevated during January through April, 1988, accounting for approximately 62% of that year's freshwater outflow to the coastal ocean, while in 1987 only approximately 53% of that year's freshwater flow occurred during that same late-winter to early-spring period. More importantly, however, flow for May-July, 1987, maintained relatively high levels during periods of vernal warming thus enhancing water-column stratification. In contrast, flow during the same time period in 1988 significantly diminished to approximately 50% of that for the prior year and were apparently insufficient to maintain stratification, even during relatively quiescent summer months.

In summary, point sources, such as the Mississippi River outflow can influence the coastal ocean based on its anthropogenic burden and total freshwater outflow volume. Moreover, the development of adverse conditions, such as hypoxia, may also be controlled by the timing of delivery of freshwater to the coastal ocean.

Non-point sources. Natural and anthropogenic influences on the coastal ocean emanate from obvious point sources, such as the major rivers, and more subtle non-point sources. At and near coastal boundaries, precipitation, nutrients, and pollution frequently enter the coastal ocean as numerous small, localized streams normally through wetland and estuarine gateways and by atmospheric wet and dry deposition. These sources typically lack effective regional organization and as such receive and transport natural and anthropogenic materials into the coastal zone as a coastwide series of effectively diffuse, non-point sources.

The coastal zone, and most often estuaries, provide a valuable linkage between non-point-source (as defined above) influx of freshwater and nutrients, and coastal and/or open-ocean fisheries. Within the southern Gulf of Mexico, Deegan *et al.* (1982) showed that fishery harvest per unit of open water was approximately proportional to the annual inflow of freshwater to the Gulf of Mexico. The linkage is provided by estuaries which serve as nursery grounds for numerous fisheries species, including offshore species not normally associated with estuarine influences. Browder and Moore, (1981) note that freshwater inputs to estuaries may enhance the production of marine organisms based on the highest marine standing stocks along shorelines adjacent to or in estuaries. For the Gulf of Mexico, Rounsefell (1975) estimated that approximately 95% of commercial fisheries production, as well as a significant proportion of recreational fisheries, are estuary dependent. Fisheries production, however, normally is a result of complex interactions of many factors including supply of food to larval fish, predation, nutrient supplies as well as changes in freshwater input. However, freshwater input, and variations thereof, can have profound effects on coastal fisheries and is a considerable factor influencing mankind in dietary and economic terms. An example from the Gulf of Mexico illustrates this.

In the northern Gulf of Mexico the shrimp family *Penaeidae* constitutes the most valuable shrimp product in the United States. The brown (*Penaeus aztecus*) and white (*Penaeus fluviatilis*) shrimp are the dominant species of this family landed along the Texas and Louisiana coasts. Studies by Gunter and Edwards (1969) address the relationships of the annual catches of *Penaeus aztecus* and *Penaeus fluviatilis* with rainfall and runoff over the state of Texas. Using a 38 year long data base (1927-1964) they determined statistical relationships for rainfall versus shrimp catches. For the annual catch of *Penaeus fluviatilis*, off Texas, correlations range from a modest positive value ($r = 0.30$) for rainfall and catch in the same year, to their strongest positive value ($r = 0.67$) for catch with rainfall in the two previous years. Similar calculations for *Penaeus aztecus* showed no significant correlations.

To link these relationships, Gunter and Edwards (1969) point out that shrimp spawn at sea, but the young develop in bays where they are restricted to the low salinity estuarine areas. Each species prefers different salinity optima; specifically *Penaeus fluviatilis* shows a preference for lower salinities than *Penaeus aztecus*. Thus, they have concluded that, on the Texas coast, increased rainfall and consequent dilution of the coastal marine waters occurs creating a more favorable habitat so that elevated populations and landings for *Penaeus fluviatilis* occur. They speculate that improved correlation with a temporal lag may be due in part to a lag in the rise and fall of coastal salinities as a function of runoff lag. Moreover, it should be realized that increase runoff to the coastal zone may affect coastal ocean temperatures, though this issue was not addressed in their study.

The Mediterranean Basin

Basin-wide trends. Data for a wide variety of social and scientific parameters need to be viewed in perspective to provide a unified overview of regional and global trends and their potential impact on mankind. Past, present and projected trends of population growth, both globally and for countries around the Mediterranean Basin (Fig. 14-1), provide unambiguous evidence that there will be an ever growing demand on regional natural resources to provide water, food and living space for these expanding populations. To meet the challenge of feeding our inhabitants, the world has turned increasingly to artificial fertilizers to enhance food production (Fig. 14-4). In doing so we have also caused unforeseen problems with water quality (Table 14-3) that have been shown to propagate to distal coastal environments. Coastal waters are increasingly effected by algal blooms (toxic and non-toxic) which pose a potential threat to other marine life. Considerable scientific evidence suggests that an increase in the frequency of blooms is related to nutrient enrichment of coastal oceans and inland seas on a global scale (Gosselin *et al.*, 1989). Eutrophication - the over enrichment of waters by nutrients, primarily nitrogen and phosphorus compounds - provides the "food" for explosive algal blooms and can be of two forms, natural and anthropogenic euthrophication. The former results from natural erosion of continents, the time scale of which allows ecological adaptations, while the latter typically occurs in the decadal to sub-decadal time scale that produces stressed, rapidly evolving ecosystems.

Such trends are also related with current problems within the Mediterranean Basin and act as harbingers for developing and unrealized future maladies. The rate of ecological change in the coastal zones of the Mediterranean Basin has accelerated over the last decade, due to both progressively increasing anthropogenic runoff and intensity of fishing, at rates that should concern Mediterranean countries. Two examples of such trends, eutrophication and fisheries decline, provide evidence of the magnitude and urgency implied by available data.

As previously defined, eutrophication is driven by excessive nutrients delivered, in the Mediterranean's case, to the coastal zone. Significant sources for Mediterranean coastal-ocean nutrient enrichment comes from two anthropogenic sources: human waste and fertilizers. Data for 700 Mediterranean coastal cities and towns show that 70% of all sewage is untreated, and in addition to being a source of excess nutrients it is also a source for bacteria, parasites, and viruses which threaten human health (World Resources, 1992, p. 168). An estimated 85% of all effluents (sewage and other) from large Mediterranean cities are discharged directly into the sea untreated (World Resources, 1990, p. 183). This situation is in turn compounded by the increasing use of artificial fertilizers.

The application of artificial fertilizers is not only increasing on a global scale but this is reflected in trends around the entire Mediterranean Basin as well as is clearly indicated in Figure 14-7. As with the examples given for the Mississippi River, nitrate burdens for major rivers entering the Mediterranean have shown both exceptionally high concentrations and increasing levels over the last several decades (Table 14-3). Unlike the Mississippi River, which is still experiencing increasing concentrations, the northern Mediterranean rivers (Po, Ebro, and Rhône) show a decreasing nitrate burden since 1985 which may reflect, and be a consequence of, decreased use of fertilizer for that same time interval (Fig. 14-7, Northern Mediterranean). Such interpretations however must be made cautiously because such downturns may simply reflect temporary or temporally-modified short-term trends and not true long-term trends.

Ecological consequences for the Mediterranean coastal ocean, from rising nutrient burdens, is observed basin-wide by documented occurrences of eutrophication, to some extent, everywhere along the Mediterranean coast where river and urban outflows occur (Cruzado, 1987). However, examples of larger-scale and more persistent features include the Gulf of Lion (Bellan-Santini and Leveau, 1987), Saronikos Bay (Friligos, 1987), NW. Adriatic (Marchetti *et al.*, 1987) and the Emilia Romagna coastal region of Italy (Todini and Bizzarri, 1987). For the latter, eutrophication has progressed to such an extent that during mid-1989, large areas of the coastal ocean were essentially devoid of oxygen

(World Resources 1990, p. 197). It needs to be emphasized here that eutrophication, in itself, does not normally cause hypoxia and/or anoxia (no oxygen), but rather is a significant contributing factor to a set of prerequisites including water column stratification and low-energetic atmospheric conditions. Perhaps most importantly, consequences of eutrophication and hypoxia extend beyond these phenomena with direct implications to coastal food webs and fisheries yields.

Because the coastal ocean and associated estuarine regions serve as fisheries habitats, the fate of these zones is closely linked with our fisheries industries and hence affect a tangible portion of our food productions. Figure 14-8 shows the commercial fisheries landings over nearly two decades for all the recognized fisheries subdivisions of the Mediterranean Sea. These data reveal a dual trend of increased (316-459 tons ≈ 45% increase) landings from 1970-1985 and a subsequent decrease (459-430 tons ≈ 9%) after

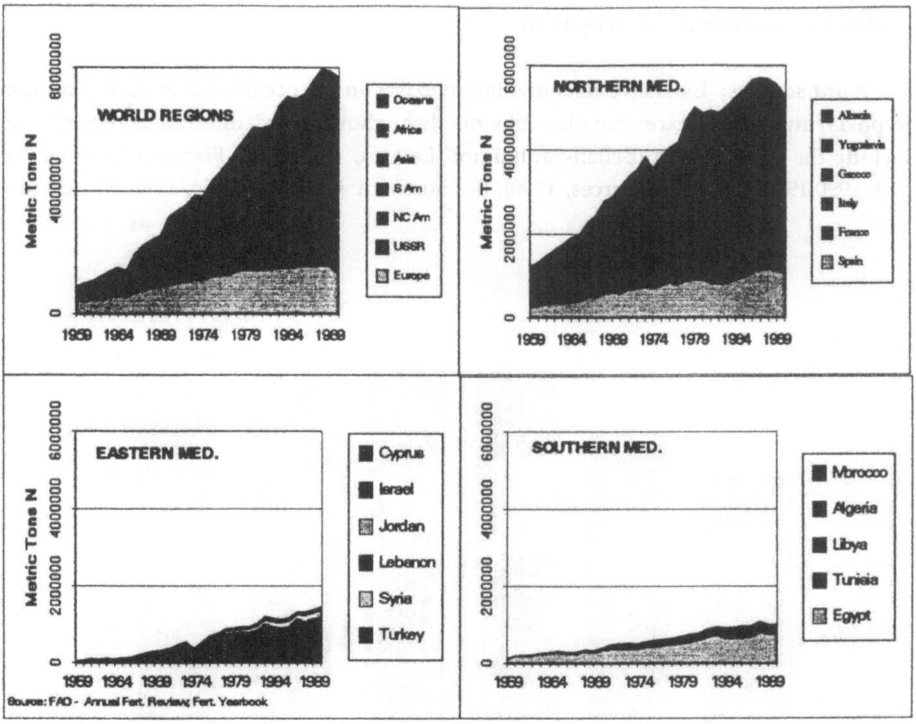

Figure 14-7. Fertilizer use, expressed as N, over three decades for both the globe and countries surrounding the Mediterranean Basin

1985. Both the initial increase and subsequent decrease have dual origins. In brief, the increase has been attributed to both enhanced effort and improved fishing methodology, which together have provided one of the two reasons attributed to the overall decline: one is overfishing, also known as exceeding maximum sustainable yield (FAO, Fisheries Circular No. 710:1987, p. 13; 1992, p. 35); the latter, fishing methodology, is beyond the scope of this chapter but will be briefly addressed in a following section. The second reason for decline, and most germane to this chapter is coastal eutropocation and hypoxia. In the sections to follow, mankind-induced water quality changes (eutrophication, hypoxia), within the Mediterranean Basin, will be placed in context with changes for point and non-point sources with linkage of these to fisheries trends. As noted above (Table 14-3 and Fig. 14-7) for the nitrate trends, decreases should be viewed with caution in that they may be part of a short-term, as opposed to a long-term, phenomenon that only a longer time series can evaluate. However, current fisheries statistics lists the entire Mediterranean Sea as one of three global fisheries regions that exceeds the maximum sustainable yield (World Resources, 1992, p. 179) thus adding credence to the cited trends (Fig. 14-8).

Point sources: Extensive documentation exists on the occurrence of eutrophication, hypoxia/anoxia and excessive algal blooms throughout the Mediterranean Basin. They include the Gulf of Lion (Bellan-Santini and Leveau, 1987; FAO Fisheries Circular No. 710, 1990,1992; World Resources, 1990), the northern Adriatic (Justic, 1991; Marchetti *et*

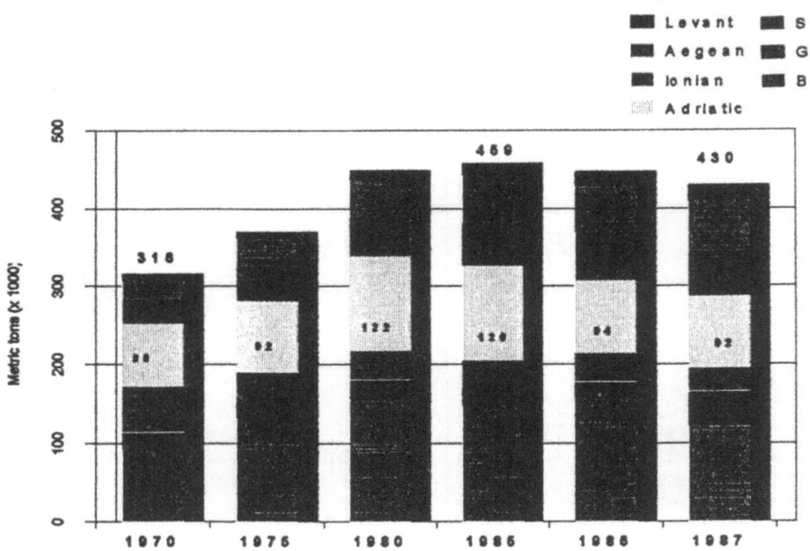

Figure 14-8. The total landings of coastal pelagic fisheries for the major fishing regions of the Mediterrannean Sea. (Plot based on data from the FAO Fisheries Circular 710, 1990)

al., 1987; FAO, Fisheries Circular No. 710, 1990; World Resources, 1990), the Emilia Ronagna region of Italy (Todini and Bizzarri ,1987), the northern Aegean (FAO, Fisheries Circular No. 710:1987;1992) and the Tunisian Gulf of Gabes (FAO, Fisheries Circular No. 710, 1990). In the works cited above, some areas are within the direct influence of major rivers while others are not. It must be kept in mind that clear divisions of influence between point sources (rivers), as used in this section, and the more distributive, non-point sources is difficult at best. However, because of their spatial and temporal persistence, point sources must be considered for their contributions to the Mediterranean Basin system.

As with the northern Gulf of Mexico example given earlier, the Mediterranean Basin has four major point sources, the Ebro, Rhône, Po and Nile Rivers. Unlike the Mediterranean, however, the Gulf of Mexico has a more open exchange with adjacent water bodies which is potentially enhanced by water transport of the Gulf of Mexico Loop Current (Maul and Vulkovich, 1993). Data in Table 14-3 indicate that the nitrate burden transported to the coastal ocean, for the first three rivers, has increased through 1985, after which the concentrations for all have declined. Similar data for the Nile is not available but fertilizer applications within the Nile drainage basin (Fig. 14-9) will provide a basis for implied trends in its river burden. Correlation of fertilizer application (FAO, 1979-1986; data not shown here), for the drainage basins of these rivers, with their river-transported nitrate burden could provide insight as to the relationship between each trend. Unfortunately, only the Po River has a marginally adequate data series (5 points)

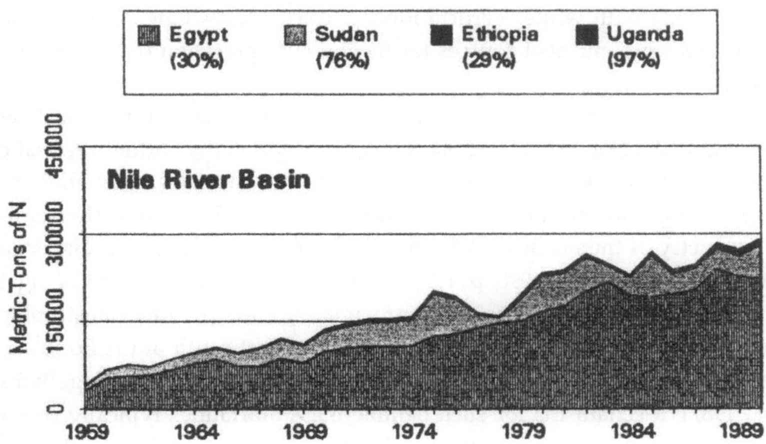

Figure 14-9. Application of commercial fertilizers to the four primary countries within the Nile River Basin. Percentages represent proportion of that country's area drained by the Nile River (Source: FAO 1979-1986 Fertilizer Yearbooks)

from which a strong positive correlation (r = 0.81) was obtained between fertilizer application and river nutrient burdens. Extending this result to the Ebro and Rhône implies a coupling between respective drainage-basin fertilizer application and the data-limited trends in Table 14-3. Moreover, similar reasoning implies continued elevation of Nile River nitrate burden based on increasing fertilizer application within its drainage basin (Fig. 14-9).

For the Ebro River and adjacent coast, clearcut point source eutrophication and/or hypoxia have not been reported as a major regional problem despite highly elevated nitrate concentrations (Table 14-3). However, it is significant to note that between 1935 and 1951 a 17% drop (average discharge drop from 600-500 m^3/sec.) in river discharge occurred as a result of increased damming and withdrawal for irrigation (Maldonado, 1975). The importance of this lies with changes in coastal salinity balances and fisheries habitat as noted above for the United States' Louisiana coastal region. At present, this author is unaware of any widely available data and/or reports addressing either water quality or fisheries related problems associated with the Ebro River.

The Rhône River, with an average discharge of approximately 1700 m^3/sec. (at Beaucaire), provides approximately 93% of the very eutrophic freshwaters flowing into the Gulf of Lion (Bellan-Santini and Leveau, 1987). It has been progressively modified by man through channelization and damming with the addition of contaminants accompanied by a concomitant reduction in biological diversity (Fruget, 1992). Moreover, the Rhône has been recognized as the major nutrient source for the Gulf of Lion (FAO, Fisheries Circular No. 710, 1992). With an annual input of approximately 150 x 10^3 and 7 x 10^3 tons of nitrogen and phosphorus respectively, it is generally considered the dominant of four (along with winter vertical mixing, coastal upwellings, and coastal town and industry effluents) nutrient sources for fueling eutrophication (Bellan-Santini and Leveau, 1987).

Within the Mediterranean Sea, eutrophication, hypoxia/anoxia and large-scale algal blooms are most highly developed and regionally extensive within the northern Adriatic and along the contiguous Emilia Ronagna coast of Italy. Two factors are generally to blame for this problem: nutrient and waste burdens of the Po River, and the semi-enclosed basin geometry of the northern Adriatic which inhibits wide-spread dispersal of contaminants (World Resources, 1990, p. 197). Although the historical record of algal blooms in the northern Adriatic goes back to approximately 1870, the first catastrophic dieoff of benthic organisms happened a century later (1975), just south of the Po River, because of massive algal blooms consisting mainly of diatoms and dinoflagellates (Marchetti *et al.*, 1987). Mechanisms for such benthic mass mortalities typically result from dying algal masses which sink into the benthic zone during stratified conditions, consume oxygen during decomposition, and thus create hypoxic to anoxic environments for essentially non-motile benthic organisms. In the Adriatic, river-borne inputs of nitrogen and phosphorus are estimated at 250 x 10^3 and 79 x 10^3 tons/year respectively

(Marchetti *et al.*, 1987) along with an average freshwater discharge of 2900 m^3/sec, with 60% of these inputs coming from the Po River (Justic, 1991). The seasonal cycles of coastal primary productivity are coincident with Po discharge cycles. Moreover, strong linkages between increases in coastal primary productivity (planktonic algal blooms), rises in surface oxygen and contemporaneous decreases in benthic oxygen levels, and the anthropogenic utilization of nitrogen and phosphorus compounds used in the Po drainage basin have been established (Justic, 1991). As in the Louisiana coastal zone, hypoxia and benthic mortalities are coincident with a stratified water column and low-velocity (reduced vertical mixing) wind conditions (Justic, 1991).

One of the most important features of regional circulation and nutrient input to the southeast Mediterranean, before the 1964 closure of the Aswan High Dam, was the annual Nile flood. Since 1964, essentially no sediment and drastic reductions in freshwater discharge have resulted in changes in coastal salinity patterns (Sharaf El Din, 1977) and the onset of extensive coastal erosion (Stanley and Warne, 1993). Coastal-ocean waters off the Nile River are essentially oligotrophic (nutrient deficient), especially since the emplacement of the Aswan High Dam. This effectively eliminated discharge of nutrients to the coastal ocean resulting in a sharp decrease in coastal primary productivity (Wahby and Bishara, 1981) with attendant catastrophic drops in coastal fisheries (discussed below). Subsequent to the mid-1960s, re-elevations of nutrients in coastal waters have been attributed to human activities, specifically drainage water from agriculture and domestic sewage from urban populations (FAO, Fisheries Circular No. 710, 1992; FAO, Studies and Reviews, 1990).

Non-point sources. In the Mediterranean, estimates for anthropogenic inputs from distributive non-point sources are spotty. For sectors of the Spanish Balearic Islands, the French-Italian coast to the islands of Tuscany and Corsica limited data indicate urban and industrial coastal effluents may constitute only approximately 10% of the river-borne nutrient inputs for that region (Bellan-Santini and Leveau, 1987).

As an essentially enclosed basin, the Mediterranean Sea has limited exchange of water masses with adjacent oceans and seas, a narrow continental shelf, and a great mean depth with limited vertical exchange between surface and nutrient-depleted deep waters. These factors make the non-coastal surface waters of the Mediterranean Sea a largely oligotrophic (nutrient depleted: i.e. 0.03 and 0.1 μmol/dm^3 P-PO_4 and N-NO_3 respectively) environment compared with coastal eutrophic (i.e. 0.15-0.3 and 0.5-≥ μmol/dm^3 P-PO_4 and N-NO_3) environments (Unesco, 1987). Normally, the first trophic level to respond to increased nutrients in the coastal ocean is the primary producers (phytoplankton) and this is expressed as accelerated growth rates and elevated, to highly-elevated, abundances.

As a consequence of nutrient data paucity, remotely sensed images (Tragana *et al.*, 1983) and proxy indicators such as phytoplankton distributions, can provide valuable environmental insight. Evidence from phytoplankton distributions, within the Mediterranean, suggest basin-wide abundances of nutrient supplies within the coastal zones (Fig. 14-10). From this figure it is clear that phytoplankton abundance, and

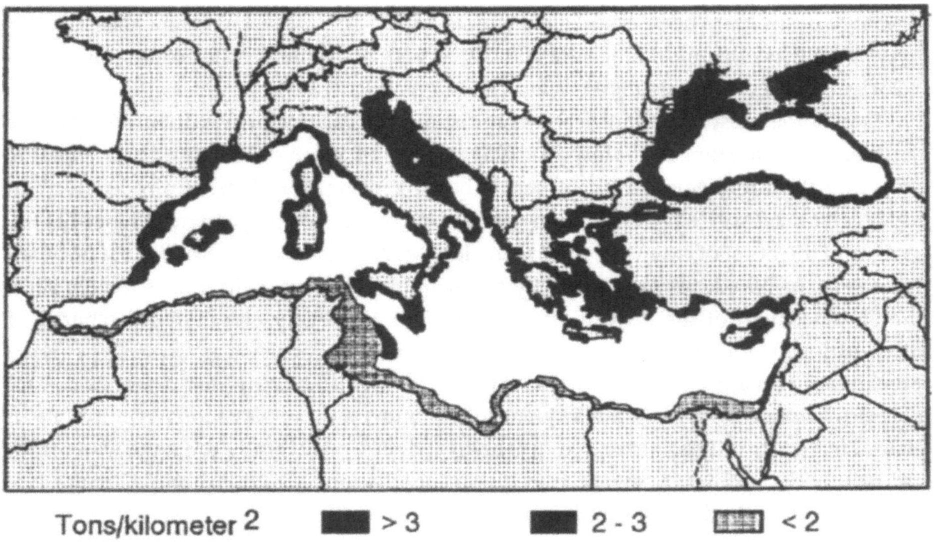

Figure 14-10. Distribution of phytoplankton (in relative units) in the Mediterranean and Black Seas based on 30 NSF/NASA images during May, 1980. (Source: FAO Fisheries Circular, No. 710, 1992)

implicitly nutrient availability, has the greatest aerial extent in the Spanish-French, the northern Adriatic and northern Aegean coastal zones with generally narrower, but nonethe-less present, phytoplankton production zones rimming essentially the entire Mediterranean Sea. It is important to point out that nutrients in these non-point-source regions reflect supplies from domestic (sewage), agricultural, nutrient recycling, and coastal upwelling; with the former being the highly dominant supply. In the Mediterranean, the problem of treatment and disposal of municipal liquid wastes is considered urgent. During the tourist season, Mediterranean coastal regions have a population of approximately 200 million people, 50% of whom are not served by sewerage systems (World Resources, 1990).

In situ recycling has been shown to be an important nutrient source of phytoplankton production but requires an initial terrestrial source. Nutrient recycling can also come from coastal upwelling and has been shown to be an important nutrient-supply process in many major fisheries areas (Codispoti, 1981). Currently it is poorly understood in the Mediterranean, even in limited areas such as the Gulf of Lion (Bellan-Santini and Leveau, 1987) and essentially unquantified for the balance of the area shown in Figure 14-10. As such, data based on remotely-sensed images (Fig. 14-10) provide us with basin-wide distributions of primary productivity and hence implications to nutrient availability which, to date, remain largely unquantified for the majority of the Mediterranean coastal ocean.

Coastal fisheries. The dependence, and growing importance, of coastal fisheries for the nations rimming the Mediterranean has been shown in Figure 14-2. Since approximately 99% of the global marine fish landings occur within 320 km of a coastline (Clark, 1981), the link between the health of the coastal oceans and fisheries resources is of outmost importance for mankind. Approximately 24% of the Mediterranean's surface area lies over the continental shelf (≤ 200 m depth) and they are, by volume, the most fertile waters (approximately 40 times greater than deep water) in the Mediterranean Sea (FAO, Studies and Reviews, 1990). Fish production is ultimately limited by primary production (phytoplankton) which in turn is normally limited by nutrient and light availability. Increasing fisheries production over the Mediterranean continental shelves, from 1972-1986 (approximately 0.96-3.2 tons/km^2) (FAO, Studies and Reviews, 1990), suggests a coupling with rising nutrients within this same zone as implied by the geographic coincidence of highest fisheries-catch areas (Fig. 14-11), areas of documented high primary production (Fig. 14-10) and eutrophication (i.e. northern Adriatic). However, post-1986 data (Fig. 14-8) indicates an overall decline in total landings of coastal pelagic fishes for both the Mediterranean Sea as a whole and major fishing regions there. Existing data suggest that the post-1986 trend results from both overfishing and eutrophication-based habitat modification and/or destruction. In brief, fisheries data indicate that overfishing can, in part, account for declining fish landings in the Adriatic Sea, Gulf of Lion, and around Greece, Cyprus and Egypt (FAO, Fisheries Circular No. 710, 1987,1990,1992). Examples from the Adriatic Sea and Egypt's Nile coastal zone will focus next on anthropogenic influences on coastal fisheries.

Nutrient enrichment, introduced into oligotrophic waters, can benefit fisheries production as discussed above, and implied from Figure 14-8. However, extension of this process to highly eutrophic conditions can have the opposite effect and may actually reduce fisheries stock. This can be accomplished when primary production rises to "bloom" proportions, water turbidity tends to increase and settling blooms tend to foul the bottom water and sediment substrate with both biomass and eutrophication-based oxygen consumption. This leads to hypoxic/anoxic bottom-water conditions, as documented for the northern Adriatic (Marchetti *et al.*, 1987). Subsequent benthic mortalities result in great losses of bottom-dwelling biomass (Harper *et al.*, 1981), direct loss of commercially valuable species, reduced productivity and disrupted food chains

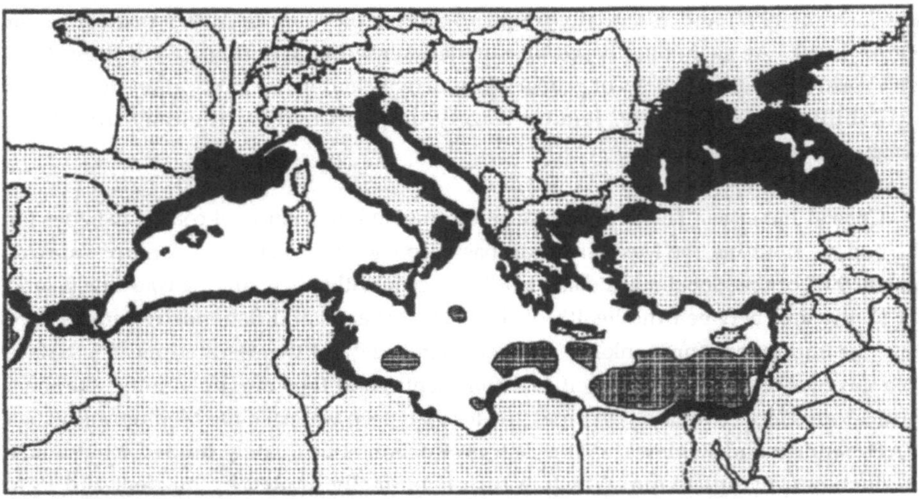

Figure 14-11. Total shelf (≤ 200 m contour) production of pelagic and demersal fishes in 1989. (Source: FAO, Fisheries Circuclar No. 710, 1992)

which lead to reduced food sources for both demersal and pelagic fish (Stachowitsch and Avcin, 1987). When events such as this recur with increasing frequency, ecosystem diversity is eventually lost, yielding ecologically impoverished communities (Stachowitsch and Avcin, 1987). Moreover, extreme plankton blooms, sometimes called "red tides", can not only poison marine animals, resulting in mass mortalities, but cause human illness from ingestion of infected benthic organisms or by exposure to aerosols from these waters (Aubert and Aubert, 1987).

Linkage between the above sequences of anthropogenically-induced ecological changes accompanied by declining fisheries stocks, has been well documented for the Black Sea and the Sea of Azov (FAO, Fisheries Circular No. 710, 1990, 1992; FAO Studies and Reviews, 1990) and provides a sobering parallel to the evolving ecological and fisheries conditions in the Adriatic Sea as presented above and shown in Fig. 14-8. Estimates of increasing biomass, for the Black Sea, suggest continued benefits from excessive eutrophication but actually reflect replacement of fisheries stocks with medusae (jellyfish) and ctenophores of no commercial value. Fisheries data for the Adriatic (Fig. 14-8) show that catches in this area are second in volume only to the Balearic, and indicate a pattern of increasing and subsequent declining catches that mimic the overall pattern for the Mediterranean Sea. Because of the severity of coastal eutrophication and hypoxia in the Adriatic region, linkage of these phenomenon, especially in light of a similar evolution of the Black Sea, should be viewed with alarm.

Arguments to present have focused on deleterious water quality, ecological and fisheries changes instigated by progressive increases in anthropogenically-derived nutrients in the coastal ocean. The antithesis of this, catastrophic decreases in nutrients, can also be deleterious to the coastal ocean's ecosystems and fisheries. Within the Mediterranean, Egypt's Nile coastal zone serves as the best example. With the final closure of the Aswan High Dam in 1964, yearly transport of freshwater, nutrients and sediment during the Nile's yearly flood runoff, were catastrophically reduced (Sharaf and Din, 1977; Wahby and Bishara, 1979). Prior to closure, freshwater discharge at the mouths of the Nile Delta exhibited a two-layer esturarine flow with outflowing surface freshwaters, transported eastward along the coast and extending up to 80 km offshore, balanced by bottom-inflowing marine waters (Sharaf and Din, 1977). Post-closure modified this to a one-layer marine inflow at all depths with a very thin seaward surface discharge of domestic wastes and sewage (Sharaf and Din, 1977). With decreased flow came decreased advection of nutrient-rich waters into this section of the coastal ocean and a concomitant sharp decline in primary productivity (Wahby and Bishara, 1979). As a consequence, the coastal ocean off Egypt experienced a rapid (approximately 62%) reduction in overall fisheries landings (1962: approximately 38,000 tons vs. 1968: 14,000 tons) and a total collapse (approximately 97% reduction) of the Egyptian sardine fisheries (1962: approximately 18,166 tons vs. 1968: 464 tons) (Wahby and Bishara, 1979). Subsequently, and as an example of beneficial eutrophication, nutrient-rich agricultural and urban domestic sewage runoff have reintroduced adequate nutrients into the coastal zone to revive coastal fisheries to total annual catches of approximately 9000 tons of which approximately 78% is sardines (FAO, Studies and Reviews, 1990). Recent data, however, indicate that these anthropogenic runoffs are laden with organic and inorganic pollutants (Stanley and Warne, 1993) which potentially present new problems to all trophic levels from plankton upward to mankind.

14-3. SUMMARY AND FUTURE CONSIDERATIONS

Assessing the Present

"Population growth, ignorance and poverty, along with poor agricultural practices, have endangered water resources" (Maurits la Rivière, 1989). This quote summarizes a group of practices that should be of concern to all of mankind. Although this reference was primarily to terrestrial water sources, it has been the aim of this chapter to extend this concept to the coastal ocean while presenting linkages between population growth, the need to understand potential seaward transfers from the continents as well as how this, linked with some human practices such as agriculture, can impact our coastal ocean water resource.

Data for both global and regional (Fig. 14-1) population increases indicate a growing need to feed more people with finite resources. This, in part, can be and is, fulfilled by the sea (Fig. 14-2). Conceptually, we understand and accept the process of continental rain and snowfall transiting through various drainage basins to the ocean. To this however, we must add an understanding that there can be, and is, a co-transport of both natural and anthropogenic pollutant types (Fig. 14-3). It is worth considering the idea of "pollutant". The addition of anthropogenic components to natural systems, in themselves, are not necessarily pollutants, but should be considered so when the elevated concentrations of these anthropogenic inputs reach toxic or otherwise harmful levels.

Although pollutants are numerous in nature, this chapter has focused on nitrogen, one of the major pollutants associated with the globally explosive increased use of artificial fertilizers (Fig. 14-4) and concomitant rise in river nitrogen levels (Table 14-3). In one example, drawn from the United States' Mississippi River outflow region, a temporally-coherent teleconnection was established between the application of fertilizers and decline of species diversity (Fig. 14-5) as well as the apparent onset of continental-shelf hypoxic conditions. Drawn from several studies, data suggests linkage, or potential linkage, between freshwater inflow into the coastal ocean and hypoxic bottom-water conditions resulting from water quality, quantity and timing of water arrival (Fig. 14-6).

In the Mediterranean Basin, trends in fertilizer usage (Fig. 14-7) strongly reflect global increases. In conjunction with other sources of pollutants, such as domestic sewage, eutrophication and declining freshwater quality have been linked. Increasing landings for Mediterranean fisheries have, within less than a decade, declined basinwide (Fig. 14-8) as a function of declining coastal water quality, attendant degradation of coastal-fisheries habitat and overfishing (exceeding maximum sustainable yield). Within the highly oligotrophic waters of the Mediterranean Sea coastal primary production, as represented by phytoplankton production, outlines essentially the entire coastal zone (Fig. 14-10). Nutrient sources for this production, such as rivers, are more easily quantified than more distributive sources, but together suggest spatial correlation with fisheries landings (Fig. 14-11), phytoplankton relative abundance (Fig. 14-10) and northern Mediterranean coastal nutrient inputs. Downturning trends in basinwide fertilizer usage, in the northern Mediterranean (Fig. 14-7) are not reflected in similar patterns within the Nile River Basin.

These examples, from the Gulf of Mexico and the Mediterranean Sea, provide an illustration of conditions that exist or are developing, at a global scale, within our coastal oceans. Given this as an environmental foundation, the quality of mankinds' future requires recognition of this decline, in conjunction with careful planning to control continued degradation as well as accommodating future climate change.

Assessing the Future

Unlike other natural resources, such as oil and minerals, water has no substitute and the magnitude of the global water crisis has been presented thoroughly in many sources

(Gleick, 1993; Postel, 1992; Speidel *et al.*, 1988; Waggoner, 1990). Similarly, the global coastal oceans must be considered a valued natural resource for both recreation and commercial fisheries, but it too is undergoing crisis-scale modifications as documented in areas like the northern Adriatic. At present, knowledge of the function of the coastal ocean, on a global scale is limited, and the feedbacks to the global system, resulting from external forcings is imperfectly understood (IGBP, 1993). Global climate change will most likely add new stresses to those already imposed by man.

Climate change and water. Measured changes in greenhouse gases over the last several decades have spurred interest in climate change and its impact on mankind. In an attempt to quantify these changes scientists have attempted to predict the future through both numerical global climate models (GCMs) and scenario models. At present, however, both are at an imperfect state of development with generally higher degrees of success for global than for regional predictions. Given this, it has been estimated that changes in temperature, based on continued anthropogenic increases of these gases, would be +2 to +5°C globally and -3 to +10°C regionally (Schneider *et al.*, 1990). Moreover, it is generally recognized that these higher temperatures will increase evaporation and subsequently precipitation, the latter estimated at +7 to +15% globally and -20 to +20% regionally (Schneider *et al.*, 1990). As a consequence of elevated temperatures and evaporation, elevated precipitation would include rainfall and snowfall, along with modifications in snowmelt and runoff patterns (Gleick, 1993). In areas with significant snowfall and snowmelt, temperature increases would have three effects (Gleick, 1993):

a) increase the ratio of rain to snow in cold months
b) decrease the overall length of snow season
c) increase the intensity and rate of snowmelt.

As a result of these factors, average winter runoff and peak runoff would increase and occur earlier with a concomitant increased risk of flooding. With changing runoff patterns there will be direct effects on water supply, hydroelectric energy generation, human health, commercial and industrial development, and environmental conditions (Gleick, 1993). Several aspects of the latter will be consider next.

Linkage - connecting the ends. "We now have quantitative evidence that, on a global scale, human population within a river's watershed is strongly related to the concentrations of nitrate in rivers that discharge to coastal ecosystems... Projected global population growth will potentially result, therefore, in increased nitrate export from rivers, with often undesirable consequences for coastal ecosystems" (Peierls *et al.*, 1991). To date, widespread failure to connect upstream with coastal pollution results from watersheds typically being fragmented into several political jurisdictions and this is compounded by scientific study with a propensity toward mono- as opposed to multi-disciplinary investigations, the latter tending to promote a broader, more system-oriented approach.

Scientifically, we need to understand and approach environmental studies on the regional and system-level and consider, in a unified manner, the effects from the watershed to the coastal ocean. For example, if climate-change predictions become reality, increased and/or earlier runoff, in conjunction with increased transport of pollutants, could influence the development and/or intensity of continental shelf hypoxic conditions. Traditional strategies for river control typically include channelization and dam construction. Continuation of this *modus operandi*, to control runoff increases, will exacerbate existing problems for upstream migration of spawning fish, prevent coastward transport of beneficial nutrients, as in the Nile fisheries example, and alter the salinity balance of coastal estuaries and the habitat for coastal fisheries ecosystems. Moreover, changing patterns of rainfall may alter inter-basin precipitation and change freshwater and nutrient/pollutant inflow to segments of the coastal zone with attendant shifts in coastal primary production and fisheries habitat. Because the total amount of global freshwater is essentially fixed, exploding populations will reduce the per capita supply. To adapt to this, and for populations currently in arid climates, desalinization is becoming increasingly popular (World Resources, 1992, p. 164). However, brine and salt residues potentially present disposal problems. Land based disposal will need to consider leaching into, and salinization of, existing groundwater and soils while disposal into the coastal zone can potentially modify estuarine salt balances and hence fisheries habitats and coastal ecosystems.

From these few examples it is clear that we must stand back and be scientifically and politically more cognizant of the entire hydrological cycle and all the anthropogenic "passengers" that board and disembark during this cycle. Moreover, we need to consider how local actions and decisions (dam construction, fertilizer applications, domestic waste disposal) potentially modify seemingly distant coastal oceans. Finally, we need to examine the influence of climate change to ensure an ocean environment that sustains our fisheries and recreational needs for the predictable future.

Acknowledgements

Drs. P. Blackwelder, T. Hood, B. McKee and Mr. C. Zarikian contributed data to Figure 14-5. Lt(jg) Jonathan Klay provided valuable assistance by generating most figures and tables herein. Drs. D. Atwood and G. Maul reviewed and made valuable comments on a draft of this text. An anonymous reviewer made suggestions that improved the text. I thank the above for their contributions.

REFERENCES

Aubert M. and Aubert J. (1987). Effets pathologie ques de l'eutrophisation marine. *In*: Eutrophication in the Mediterranean Sea: receiving capacity and monitoring of long-term effects, Report and Proceedings of a Scientific Workshop, Bologna, Italy. Unesco Rep. Mar. Sci. 49: 81.

Bratkovich A. and Dinnel S. (1991). Lower Mississippi River historical influx nitrate flux and Mississippi River outflow. *In*: Nutrient Enhanced Coastal Ocean Productivity, Proceedings of a Workshop at LUMCON. Texas A & M Sea Grant Pub 92-109, pp. 37.

Bellan-Santini D. and Leveau M. (1987). Eutrophication in the Gulfe du Lion. *In*: Eutrophication in the Mediterranean Sea: receiving capacity and monitoring of long-term effects, Report and Proceedings of a Scientific Workshop, Bologna, Italy. Unesco Rep. Mar. Sci. 49:107-115.

Browder J. and Moore D. (1981). A new approach to determining the quantitative relationship between fishery production and the flow of fresh water to estuaries. *In*: Cross R., Williams D. (Eds), Proceedings of the National Symposium on freshwater inflow to estuaries. U.S. Fish and Wildlife Service, Office of Biological Services, 1:403-430.

Clark C. (1981). Bioeconomics of the ocean. BioSci , 31:231-238.

Codispoti L. (1981). Some chemical consequences of upwelling. *In*: Richards F. (Ed.) Coastal upwelling. Amer. Geophy. U., p. 209.

Crosson P. and Rosenberg N. (1989). Strategies for agriculture. Sci. Amer., 261:128-135.

Cruzado A. (1987). Eutrophication in the pelagic environment and its assessment. In: Eutrophication in the Mediterranean Sea: receiving capacity and monitoring of long-term effects, Report and Proceedings of a Scientific Workshop, Bologna, Italy. Unesco Rep. Mar. Sci. 49: 57-63.

Deegan L., Day J., Gosselink J., Yáñez-Arancibia A., Chávez S., and Sánchez-Gil P. (1986). Relationships among physical characteristics, vegetation distribution and fisheries yield in Gulf of Mexico estuaries. *In*: Wolfe D. (Ed), Estuarine Variability. Academic Press Inc. p. 83.

Fahnenstiel G., Marcovitz M., McCormick M., Redalje D., Lohrenz S., Carrick H., and Dagg M. (1992). High growth and microzooplankton-grazing loss rates for phytoplankton populations from the Mississippi River plume region. *In*: Nutrient Enhanced Coastal Ocean Productivity, Proceedings of a Workshop at LUMCON. Texas A and M Sea Grant Pub. 92-109, pp. 111.

FAO. (1979-1986). Fertilizer Yearbooks. FAO Statistics Series. Food and Agricultural Organizatin of the United Nations, pp. 29-36.

FAO Circular No 710. (1987). Review of the State of World Fishery Resources. Food and Agriculture Organization of the United Nations, Revision 5.

FAO Circular No 710. (1990). Review of the State of World Fishery Resources, Part 1 - Marine Resources. Food and Agriculture Organization of the United Nations, Revision 7.

FAO Studies and Reviews. (1990). Recent Trends in the Fisheries and Environment in the GFCM Area, No. 63. Food and Agriculture Organization of the United Nations.

FAO Circular No 710. (1992). Review of the State of World Fishery Resources, Part 1 - Marine Resources. Food and Agriculture Organization of the United Nations, Revision 8.

FAO Circular No 853. (1992). Marine Fisheries and the Law of the Sea: A Decade of Change. *In*: Special chapter (revised) of The State of Food and Agriculture 1992, Food and Agriculture Organization of the United Nations, pp. 35.

Friligos N. (1987). Eutrophication of the Saronikos Bay. *In*: Eutrophication in the Mediterranean Sea: receiving capacity and monitoring of long-term effects, Report and Proceedings of a Scientific Workshop, Bologna, Italy. Unesco Rep. Mar. Sci. 49: 123-132.

Fruget J. (1992). Ecology of the lower Rhône after 200 years of human influence: a review. Regulated Rivers: Res. and Manag., 7:233-246.

Gleick P. (1993). Water in the 21st century. *In*: Gleick P. (Ed.), Water in Crisis: A guide to the world's fresh water resources. Oxford University Press, pp. 105.

Gosselin S., Fortier L., and Gagné. (1989). Vulnerability of marine fish larvae to the toxic dinoflagellate Protogonyaulax tamarensis. Mar. Ecol. Prog. Ser., 57:1-10.

Gunter G. and Edwards J. (1969) .The relation of rainfall and freshwater drainage to the production of the Penaeid shrimps (*Penaeus fluviatilis* SAY and *Penaeus aztecus* IVES) in Texas and Louisiana Waters. *In*: Proceedings of the world scientific conference on the biology and culture of shrimps and prawns. FAO Fisheries Rep No 57, 3:875-891.

Harper D., McKinney L., Nance J., and Salzer R. (1991). Recovery responses of two benthic assemblages following an acute hypoxic event on the Texas continental shelf, northwestern Gulf of Mexico. *In*: Tyson R., Pearson T. (Eds.) Modern and Ancient Continental Shelf Anoxia. The Geol. Soc. London, pp. 49.

IGBP. (1993). Global change: Land-ocean interactions in the coastal zone, Science Plan, Report No 25, pp. 50.

Lohrenz S., Redalje D., Fahnenstiel G., and Lang G. (1992). Regulation and distribution of primary production in the northern Gulf of Mexico. *In*: Nutrient Enhanced Coastal Ocean Productivity, Proceedings of a Workshop at LUMCON. Texas A & M Sea Grant Pub 92-109, pp. 95.

Justic D. (1991). Hypoxic conditions in the northern Adriatic Sea: historical development and ecological significance. *In*: Tyson R, Pearson T (Eds.), Modern and Ancient Continental Shelf Anoxia. The Geol. Soc. London, pp. 95.

Keyfitz N. (1989). The growing human population. Sci. Amer., 261:119-126.

Marchetti R., Gaggino G., and Provini A. (1987). Red tides in the northwest Adriatic. *In*: Eutrophication in the Mediterranean Sea: receiving capacity and monitoring of long-term effects, Report and Proceedings of a Scientific Workshop, Bologna, Italy. Unesco Rep. Mar. Sci., 49: 133-139.

Maldonado A. (1975). Sedimentation, stratigraphy, and development of the Ebro Delta, Spain. *In*: Broussard M. (Ed.) Deltas, models for exploration. Huston Geol. Soc., pp. 311.

Maul G. and Vukovich F. (1993). The relationship between variations in the Gulf of Mexico loop current and Straits of Florida volume transport. Phy. Ocean., 23:785-796.

Maurits la Rivière J. (1989). Threats to the world's water. Sci. Amer., 261:80-94.

Nelsen T., Blackwelder P., Hood T., McKee B., Romer N., Zarikian C., and Metz S. (1994) Time-based correlations of biogenic, lithogenic, and authigenic sediment components with anthropogenic inputs to the Gulf of Mexico NECOP study area. Estuaries, 17:813-885.

Peierls B., Caraco N., Pace M., and Cole J. (1991). Human influence on river nitrogen. Nature, 350:386-387.

Postel S. (1992). Last oasis: facing water shortage. W. W. Norton and Co, pp. 239.

Postel S. (1993). Water and agriculture. *In*: Gleick P (Ed.) Water in Crisis: A guide to the world's fresh water resources. Oxford Univ. Press, pp. 56.

Rabalais N., Turner R., Wiseman W., and Boesch D. (1991). A brief summary of hypoxia on the northern Gulf of Mexico continental shelf: 1985-1988. *In*: Tyson R., Pearson T. (Eds.), Modern and Ancient Continental Shelf Anoxia. The Geol. Soc. London, pp. 35.

Rounsefell G. (1975). Ecology, utilization, and management of marine fisheries. C. Mosby Co.

Schneider S., Gleick P., and Mearns L. (1990). Prospects for climate change. *In*: Waggoner P. (1990) Climate Change and U.S. Water Resources. John Wiley and Sons, pp. 41.

Sharaf E.l. and Din S. (1977). Effect of the Aswan high dam on the Nile flood and on the estuarine and coastal circulation pattern along the Mediterranean Egyptian coast. Limn. Ocean., 22:194-207.

Shiklomanov I. (1993). World fresh water resources. *In*: Gleick P. (Ed.), Water in Crisis: A guide to the world's fresh water resources. Oxford Univ. Press, pp. 13.

Speidel D. and Agnew F. (1988). The world water budget. *In*: Speidel D., Ruedisili L., Agnew A. (Eds.), Perspectives on Water Uses and Abuses. Oxford Univ. Press, pp. 27.

Speidel D., Ruedisili L., and Agnew A. (Eds). (1988). Perspectives on Water Uses and Abuses. Oxford Univ Press, pp. 388.

Stachowitsch M. and Avcin A. (1987). Eutrophication-induced modifications of benthic communities. *In*: Eutrophication in the Mediterranean Sea: receiving capacity and monitoring of long-term effects, Report and proceedings of a scientific workshop, Bologna, Italy. Unesco Rep Mar Sci 49, pp. 67.

Stanley D. and Warne A. (1993). Nile Delta: Recent geological evolution and human impact. Sci., 260:628-634.

Todini E. and Bizzarri A. (1987). Eutrophication in the coastal area of the regione Emilia Romagna. *In*: Eutrophication in the Mediterranean Sea: receiving capacity and monitoring of long-term effects, Report and Proceedings of a Scientific Workshop, Bologna, Italy. Unesco Rep. Mar. Sci., 49: 143.

Traganza E., Silva V., Austin D., Hanson W., and Bronsink S. (1983). Nutrient mapping and recurrence of coastal upwelling center by satellite remote sensing: Its implication to primary production and the sediment record. *In*: Suess E., Thiede J. (Eds.), Coastal Upwelling, its Sediment Record. Plenum Press, pp. 61.

Turner R. and Rabalais N. (1991). Changes in Mississippi River water quality this century - implications for coastal food webs. BioSci., 41:140-147.

Unesco. (1987). Guidelines for monitoring, assessment and control of eutrophication in the Mediterranean Sea. *In*: Eutrophication in the Mediterranean Sea: receiving capacity and monitoring of long-term effects, Report and Proceedings of a Scientific Workshop, Bologna, Italy. Unesco Rep. Mar. Sci., 49:31.

Waggoner P. (1990). Climate change and U.S. water resources. John Wiley and Sons, pp. 496.

Wahby S. and Bishara N. (1981). The effect of the river Nile on Mediterranean water, before and after the construction of the high dam at Aswan. *In*: Martin J., Burton J., Eisma D. (Eds.), River Inputs to Ocean Systems, Proceedings of a review workshop, FAO headquarters, Rome, pp. 311.

World Resources. (1988). A report by The World Resources Institute in collaboration with The United Nations Environment Programme and The United Nations Development Program. Oxford Univ Press, pp. 372.

World Resources. (1990). A report by The World Resources Institute in collaboration with The United Nations Environment Programme and The United Nations Development Program. Oxford Univ Press, pp. 383.

World Resources. (1992). A report by The World Resources Institute in collaboration with The United Nations Environment Programme and The United Nations Development Program. Oxford Univ Press, pp. 385.

CHAPTER 15

IMPACT OF CLIMATIC CHANGES ON COASTAL WATER QUALITY

STYLIANOS RAFAILIDIS[1], JACQUES GANOULIS[1], ISTVAN BOGARDI[2], ISTVAN MATYASOVSZKY[3] AND LUCIEN DUCKSTEIN[4]

[1]*Dept. of Civil Eng., Division of Hydraulics and Environmental Engineering, Aristotle Univ. of Thessaloniki, GR-540 06 Thessaloniki, Makedonia, Greece.*
[2]*Dept. of Meteorology, Eotvos Lonard Univ,. Budapest, Hungary.*
[3]*Dept. of Civil Engin., Univ. of Nebraska, Lincoln, NE 68588-0531, USA.*
[4]*Systems and Industrial Engin., Univ. of Arizona, Tucson, AZ 85721,USA.*

15-1. INTRODUCTION

Of considerable concern is the increase in atmospheric CO_2 in recent years due to its influence on global climate. The studies so far have concentrated primarily on precipitation and air temperature changes from an assumed doubling in CO_2 (the $2xCO_2$ scenario). For water resources, more specifically, climatic changes affect both quantity and quality. The effects on water quantity are presented in detail elsewhere in this volume and are not discussed here. The same applies also to other indirect substantial effects like changes in coastline, wetlands, etc., caused by sea level changes, flooding or erosion from coastal runoffs (Stakhiv *et al.*, 1991).

In this Chapter the emphasis is placed mainly on coastal water quality. As coastal waters we usually consider that part of the sea which is influenced significantly by runoffs and human activity based on the shore. In the Mediterranean context, in view of the relatively shallow continental shelf, the coastal band accounts for a significant proportion of the overall sea body.

The importance of preserving the quality of coastal waters for both economic activities and ecological reasons is firstly discussed. Then likely scenarios of climate change in the region are presented along with the impacts on precipitation and ambient temperature. Subsequently, a detailed discussion is followed on the mechanisms by which climatic change may influence water quality in the coastal region. Finally, the case study of Thermaikos Gulf in Makedonia, Greece, is presented.

15-2. IMPORTANCE OF PRESERVING COASTAL WATER QUALITY

In the Mediterranean basin, human settlements have been concentrated traditionally

NATO ASI Series, Vol. I 36
Diachronic Climatic Impacts on Water Resources
Edited by Andreas N. Angelakis and Arie S. Issar

along the coast. The almost 50,000 km of Mediterranean coastline is some 2.5 times longer than that of the U.S. Furthermore, some 54% of the population of the region live in the coastal countries (Nelsen, 1994). In Greece, more specifically, along the 16,000 km coastline concentrate 60% of the population, 70% of industrial activity, 40% of agricultural and 90% of tourist activity (Golfi *et al.*, 1993). Clearly, therefore, any deterioration in coastal waters would influence large numbers of people.

Preserving coastal water quality safeguards human well-being and health, whilst maintaining diversity in the biota. Fishing and other aqua cultures are traditional and very important economic activities in the region, employing and feeding large populations, especially in islands. Tourism is a basic element of the economy of many Mediterranean countries directly related to the quality of marine resources. The importance of these aspects is discussed below.

Human Well-Being and Health

Though there is little direct impact of sea water quality on the actual health of coastal populations, indirect influences exist through the food chain and are extremely important. Fish catches and oyster farming in polluted areas introduce bacterial or toxic metal contamination into the food chain of humans, causing epidemiological occurrences. Even in cases where contamination remains still tolerable, the presence of pollutants may cause abnormal growth of certain algae, causing oxygen depletion (eutrophication). Fish feeding on these algae may suffer adverse changes in taste or odour, becoming unsuitable for human consumption. On the other hand, decaying algae produce H_2S and other odorous substances which may affect the well-being of the population living along the coast. The important interplay between water quality and human settlement on the coast may be exemplified by the total absence of permanent habitation around the Dead Sea. The quality of waters there is so poor that it not only does not attract humans, but actually turns them away.

Ecological Impacts - Biodiversity

A rich variety of marine organisms inhabit the world's coastal and oceanic waters (Fig. 15-1). Generally, these may be divided into producers (e.g., phytoplanktonic diatoms, flagellates, etc.) and consumers of organic matter (e.g., zooplankton, nekton, benthos, etc.). In addition, different types of bacteria also exist, in concentrations ranging from one per litre to more than 10^8 per millilitre. Generally, bacteria do not contribute significantly to nutrient recycling in the water column but mainly in the sediments (Odum, 1971).

Coastal zones play an important role in solar energy trapping and biological matter transformations (Table 15-1). In fact, the relatively shallow coastal waters and up-welling zone have productivities which are orders of magnitude larger than the deep ocean or

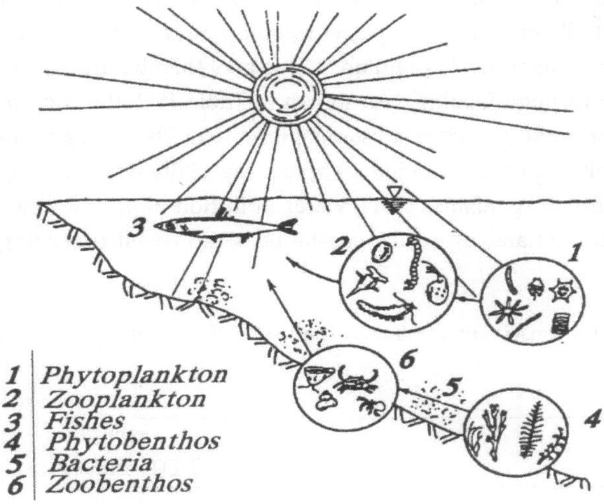

Figure 15-1. Main biological organisms in coastal ecosystems (Ganoulis, 1988a)

the terrestrial biosphere. They thus contribute most to an ecosystem which may remain sustainable in energy and organic matter.

Species diversity in the water column is related directly to water quality. Studies by Copeland and Bechtel (1971) have shown a biodiversity paucity close to effluent outfalls, with the effect diminishing with distance. Also, water toxicity was found to be correlated inversely proportional to species diversity in the water body.

Table 15-1. Estimated annual gross primary production in the biosphere (Odum, 1971)

Ecosystem	Area [10^6 km^2]	Gross Annual Primary Productivity [kcal/m^2]	Total Annual Gross Production [10^{16} kcal]
Marine			
Open ocean	326.0	1,000	32.6
Coastal zones	34.0	2,000	6.8
Up-welling zones	0.4	6,000	0.2
Estuaries and reefs	2.0	20,000	4.0
Subtotal	362.4		43.6
Terrestrial			
Subtotal	135.0		57.4
Total for biosphere (excl. ice caps)	500.0		100.0

Copeland (1966) has reported that in polluted waters the metabolic rates of various industrial wastes in fish increase, through the effluent has not reached toxic levels yet. The reduced dissolved oxygen concentration, due to the simultaneous discharge of biological matter around the outfall, forces marine species to pump more water through their gills and thus absorb greater quantities of pollutants. This may then have a knock-on effect to the rest of the biota through feeding. As shown in Table 15-2, the concentration of a polluting constituent rapidly reaches substantial (and toxic) levels higher in the food chain. This is particularly important for higher species, especially when feeding is based predominantly on marine organisms. Thus, water pollution along the coast might have deleterious effects to humans, especially coastal or island dwellers feeding predominantly on fish.

Table 15-2. Bioaccumulation of a conservative pollutant (pesticide DDT) in the food chain (Woodwell *et al.*, 1967)

	ppm DDT residues[a]
Water	0.00005
Plankton	0.04
Silverside Minnow	0.2
Sheephead Minnow	0.9
Pickerel (predatory fish)	1.3
Needlefish (predatory fish)	2.1
Heron (feeds on small animals)	3.6
Tern (feeds on small animals)	3.9
Herring Gull (scavenger)	6.0
Fish Hawk (Osprey) egg	13.8
Merganser (fish-eating duck)	22.8
Cormorant (feeds on larger fish)	26.4

[a] Parts per million of total residues (DDT + DDD + DDE) on a wet weight, whole organism basis

Impacts on Economic Water Uses

Fishing and oyster farming. Sea water quality is very important for fishing and the aqua culture industry, especially shellfish growing. Marine organisms are well known to accumulate in their flesh pollutants from the surrounding water and pass them into the food chain. This is particularly so for mussels, oysters and other stationary marine animals, growing in polluted waters. Because of this, for some time now legislation stipulates the allowable quality of oyster farming waters (Table 15-3).

Table 15-3. Water quality criteria for oyster farming waters (EEC Directive 30/10/79) (Ganoulis *et al.*, 1993b).

Quality Criteria	Desirable Limit	Higher Allowable Limit
Dissolved Oxygen [% of saturation]	≥ 80	70-100
Faecal Coliforms [MPN per 100 *ml*]	< 70	
(in the flesh)	≤ 300	

Effects on Recreational Uses

Aesthetics and heritage. The Mediterranean basin is a region with rich history, since a number of the most important human civilizations nurtured there during the last six millennia. For some of these people interaction with the coast and sea was of profound importance. For others (i.e., island civilizations) lack of good quality coastal resources, and the advantages they offer to human settlements, would have made their historical existence problematic, if not impossible.

Furthermore, the aesthetic characteristics of the coast had, beyond doubt, a profound impact on the achievements of these civilizations. For example, the characteristics of the Aegean sea coastline with its multitude of inundations and islands has contributed to make Greek aesthetics different from that of the Pharaohs. These differences are clearly reflected in the respective architectural and intellectual achievements of each one of these civilizations.

The importance of coastal waters for the heritage of Mediterranean countries may be exemplified by Greece, where 60% of the human settlements which have been declared traditional are island or coastal communities. No less than 35% of them are concentrated in the Aegean sea (Cyclades and Dodecanese) whereas most of the remaining ones are also dispersed on islands (Golfi *et al.*, 1993).

Tourism. The Mediterranean basin benefits from a unique combination of benign climate, sunshine, hospitable people and rich cultural heritage. Since these are the main criteria determining the movement of tourists to different destinations, the Mediterranean countries enjoy substantial tourist influxes. It is estimated that as much as 1/3 of the world tourist traffic concentrates there (Golfi *et al.*, 1993). The coastal strip has become a major attraction for tourist recreation, in the form of bathing, sport fishing and water sports.

As a result, in areas lacking other significant natural resources (e.g., island communities), tourism has become a vital contributor to local economy. This advantage will be definitely jeopardised by any deterioration in coastal water quality. An example of the importance of good water quality for water-based tourism was, in the late 1980s, the damage to the tourist industry on the North Adriatic sea-coast due to the occurrence of severe seasonal algal blooms.

15.3. MODELLING CLIMATE CHANGE

Overview

To determine the likely effect of an increase in CO_2 on ambient temperature and precipitation, a stochastic model has been developed. The methodology (Bogardi *et al*, 1993a) consists of:

a) Classification and stochastic modelling of historical data of daily macro-scale circulation patterns (CP), based on global circulation models (GCM).
b) Space-time modelling of historical daily precipitation and temperatures, conditioned on the various CP types.
c) Consideration of the spatial average pressure height of each chosen CP type.
d) Stochastic simulation of spatial daily precipitation, temperature, pressure height, etc., using the linkage between daily CP types under the classification defined in step 1. Details of these steps are discussed below.

Atmospheric Circulation Patterns

Spatial meteorological variables are often classified on the basis of CPs. The main meteorological characteristics used are spatial distributions of pressure at sea level, or at some low or middle tropospheric pressure heights. There are several possibilities for classifying daily CPs and considerable experience has been accumulated using them. The earliest approach was subjective classification, based on meteorological experience. Such classification schemes are subjective because they depend strongly on the attitude of the meteorologist, but are also objective in the sense that they are based on observations of the physical behaviour of the atmosphere.

Early CP classifications were developed using "synoptic experience." For instance, Baur *et al.* (1944) have defined 29 types for Europe; a list of these and their corresponding meteorological descriptions may be found in Hess and Brezowsky (1969). Another classification scheme for the extra-tropical latitudes of the Northern Hemisphere was developed by Dzerdzeevskii (1968 and 1970). Lamb (1972) introduced a circulation pattern scheme for the British Islands, a relatively small area. For the United States, on the other hand, Krick (1943) and Elliott (1943; 1949) have prepared catalogues of weather types.

The availability of computers and the appearance of modern mathematical clustering algorithms have made possible to use of the so-called "objective" methods (Anderberg, 1973, and Hartigan, 1975). Objectivity means here that a given algorithm makes data processing automatic, although the selection of the algorithm itself introduces subjectivity. Sequential clustering or hierarchical algorithms are relatively simple, but need huge computer storage even for moderate sample sizes. The most common hierarchical techniques are based on correlation coefficients or sum of squares differences of maps (Lund, 1963 and Kirchofer, 1973). A mathematically well-developed and computerized non-hierarchical technique is the k-means method (MacQueen, 1967).

In recent years, principal component analysis (PCA) (Craddock and Flood, 1969 and Kutzbach, 1970) and PCA coupled with clustering (Key and Crane, 1986) have been applied intensively. Wilson *et al.* (1992) compared four methods of PCA classification: k-means clustering, fuzzy clustering, principal components, and principal components coupled with k-means. An information measure was developed to quantify the extent to which each of the classification schemes discriminates the meteorological events. Evaluation of the four algorithms using this measure shows that all methods perform approximately equally well. In the present work the principal component technique coupled with k-means is used, because this combination provides the most separable system of even densely concentrated clusters (Bartholy, 1992 and Gong and Richman, 1992).

We have used the U.S. National Meteorological Centre (NMC) grid point analyses. The summer and winter periods from April to September and from October to March were examined separately. Values at each grid point were standardized to exclude the annual cycle inherent in meteorological phenomena. Standardization (subtraction of the expectation and division by the standard deviation) was done by fitting sine and cosine waves with periods of one year and one half year. Time series, to be fitted for calculating the mean, consisted of the grid values whereas for the variance, the squared differences of grid values and their estimated means were used.

A PCA was performed and then the k-means method was used by defining a Euclidean distance. The aim was not to extract characteristic spatial patterns from the data and reduce the dimensionality, but to obtain new uncorrelated variables for the classification: here, 20 principal components were maintained to explain 97-98% of the total variance. The concept resembles the use of Mahalanobis distance (Anderberg, 1973), but the PCA also delivers a dimension reduction.

The k-means clustering algorithm starts from a system of initial cluster centres and the first grid is examined by calculating the distances between this grid and each cluster centre. The grid is classified into the cluster which is the closest. This cluster centre is recalculated to account for the new cluster member. The other grids are considered in the same manner. After classifying all grids, the cluster centres are considered initial centres and the procedure is reiterated. These steps are repeated until cluster centres stabilize.

The number of CP types was chosen as a compromise between increasing the number of clusters and decreasing the sum of inner distances within each cluster. It is well-known that standard deviations of pressure heights appear to be greater in winter than summer, but the decrease in inner distances is more rapid in winter. This means that winter circulation is easier to characterize. Nevertheless, defining only 9 types for both winter and summer half-years seems to suffice for the whole year.

Effects on Precipitation

To determine climate change effects, three types of daily CP data have been used:

a) Historical data, represented by the NMC grid point analyses of the height of 500 hPa pressure fields. The analysis is based on daily values at 12:00 GMT, on a diamond grid covering the sector of interest, for the period January 1948-June 1989.
b) A 10-year long data series for the same pressure level has been obtained from the Canadian Climate Centre GCM (Boer *et al.*, 1984), to correspond to the 1xCO_2 scenario;
c) An analogous time series has been obtained using the 2xCO_2 scenario.

For precipitation, the methodology developed by Bardossy and Plate (1992) for Germany has been used, modified for the climatic conditions in the area of interest. In the original model the catalogue of daily CP types corresponds to the "subjective" classification of Hess and Brezowsky (1969). In the present application, CPs are classified using "objective" classification. In contrast to Bardossy and Plate (1992), the daily CP types can be modelled with Markov chains instead of more complicated semi-Markov chains. A simple power transformation between a normally distributed random variable and daily precipitation event is an approach which works under the specific climatic conditions of the original model. In this work, however, a general transformation has been used instead, which is applicable independently of local climatic conditions.

An important question is whether or not CP types obtained from historical data may be used in 1xCO_2 and 2xCO_2 scenarios. For this reason, two kinds of CP-type catalogues of GCM-produced pressure fields were derived. First, cluster centres obtained from historical data (case 1) were used; then, cluster centres were also obtained from GCM-produced pressure fields (case 2), using the same clustering algorithm. The mean of the differences between the cluster centres and cluster elements for each CP type under either scenario were finally computed. For example, by applying the technique to the Great Plains of the U.S., Matyasovszky *et al.* (1993a) show that the mean of squares corresponding to case 2 is smaller than for case 1, but the difference is not substantial. For instance, in the 1xCO_2 scenario in summer, the mean of squares in case 2 is 7% less than case 1. In winter, the relative difference is even smaller, not exceeding 1.2%. The 2xCO_2 case exhibits negligible further differences, of the order of 0.2% in summer and 0.3% in winter (Fig. 15-2).

These facts suggest that the same CP types may be used for the observational data and for the two GCM outputs and such an approach has been taken in this work. The relative frequencies of the nine selected CP types in both seasons and for the three data sets used were quite uniform (Fig. 15-3). Contrary to initial expectations, the difference between relative frequencies corresponding to the 1xCO_2 and 2xCO_2 cases is not greater than that between historical data and the 1xCO_2 scenario.

Given the set of CPs, the next step is to develop a stochastic model to describe the sequence of these patterns. The properties of persistency, i.e., sequencing of CP types, are described by a Markov model. Markov chains are characterized by their transitional

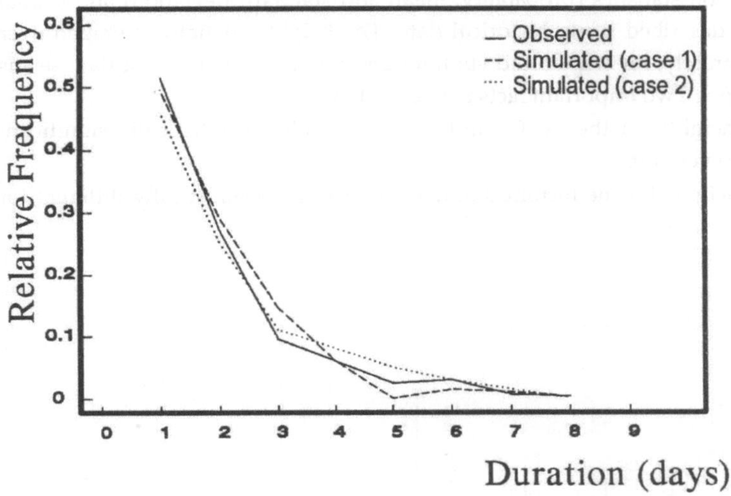

Figure 15-2. Duration of, for example, CP type #3 in the winter half-year in the U.S. Great Plains. It has been calculated from the observed and simulated data series using transition probability matrices corresponding to the historical (case 1) and GCM-produced data under the $2xCO_2$ scenario (case 2) (Matyasovszky *et al.*, 1993a)

probabilities, that is, the probability of transition from one type of circulation pattern to another. An example of how such a Markov model can describe statistically the duration of CP types is shown in Figure 15-2. A good fit is seen between the observed and GCM-produced CP durations in the $2xCO_2$ case. Also, the $1xCO_2$ case appears to yield very similar relative frequencies of CP type duration.

Originally, this approach was intended to simulate space-time series of local climatic variables reflecting climate change. However, the difference between the frequencies of the $1xCO_2$ and $2xCO_2$ cases is relatively small, compared with the difference between the frequencies of the $1xCO_2$ case and historical data (Fig. 15-3). On the other hand, some authors argue that, within a CP type, the characteristic which influences precipitation may change. This means that knowledge of CP type alone is not enough but additional parameters have to be introduced to describe CPs under climate change. Since CPs in the $2xCO_2$ case are characterized, in general, by larger pressure level heights, it is reasonable to link meteorological variables to the type and height of CPs. The height is a physically meaningful indicator of the atmospheric pressure (high or low) and temperature (warm or cold).

The assumption is that the annual cycle of pressure height is analogous to the difference between the present and the warmer $2xCO_2$ climates. The relationship between precipitation statistics (probability, mean and standard deviation) and spatial average height is described using historical data. The heights spatially averaged over each CP type separately, for winter and summer and for the three cases of data sets is shown in Figure 15-4. Two important facts may be noted:

a) The average heights for the $1xCO_2$ and $2xCO_2$ cases have statistically significant differences for every type,
b) The average heights for the historical and $1xCO_2$ cases are statistically different for

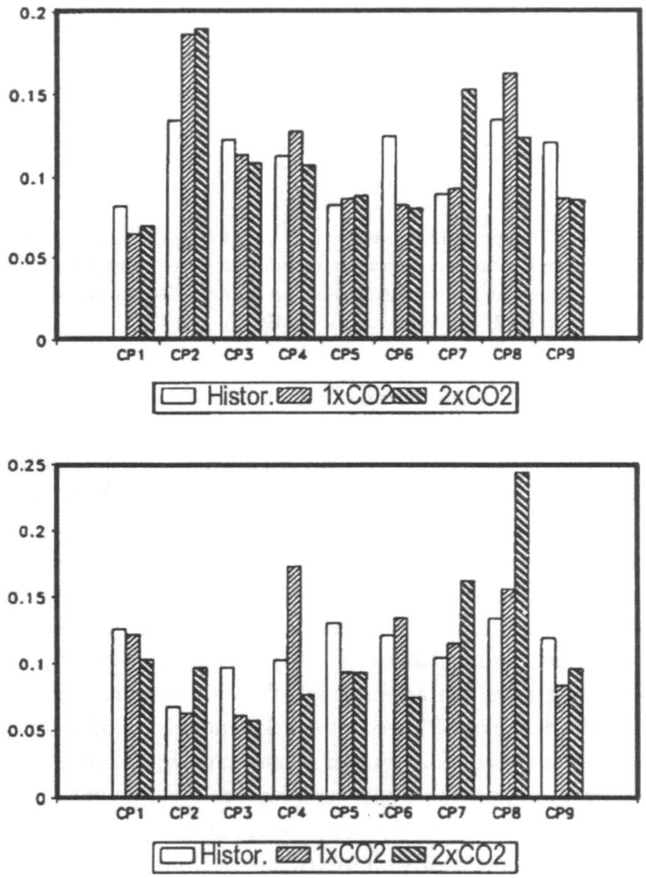

Figure 15-3. Relative frequency of the 9 chosen daily CP types for the winter (top) and summer (bottom) half years for Nebraska (Matyasovszky *et al.*, 1993a)

several CP types. The direction of differences in heights suggests that the GCM used in this paper systematically underestimates pressure height.

Because of these observations only the differences between the heights corresponding to the two CO_2 scenarios and not the heights themselves were used, under the hypothesis that the difference between daily precipitation amounts, corresponding to the "true" $2xCO_2$ pressure heights and the present climate heights, is equal to the difference between precipitation amounts corresponding to the GCM-produced $2xCO_2$ and $1xCO_2$ heights. Consequently, local precipitation statistics reflecting the $2xCO_2$ scenario have been estimated in two steps.

a) The regression curve of these statistics on the spatially averaged pressure height was estimated using historical data and a non-parametric regression technique.

b) Statistics taken from these curves, at heights corresponding to the $1xCO_2$ and $2xCO_2$ cases, were used to adjust precipitation statistics obtained from the whole historical data set, in order to arrive at statistics reflecting the "true" $2xCO_2$ scenario.

For further details on the analysis outlined above one may refer to Bogardi *et al.* (1993a) and Matyasovszky *et al.* (1993a).

Effects on Ambient Temperature

The same approach may be used to model other local climatic variables such as temperature, wind and evaporation (Bardossy, 1991). Presumably, it may also be extended to stream flows and other related hydrological events, such as floods (Duckstein *et al.*, 1993) or lake level changes. Prediction of regional/local effects of climate change scenarios involves the following steps:-

a) The probability distribution function of the climatic variable of interest (e.g., daily precipitation, temperature, etc.) is estimated from observation data, conditioned to the CP types.

b) A Markov model of CP types, obtained from a GCM, is constructed.

c) The stochastic linkage model is used to simulate space-time series of the climatic variable reflecting climate change.

The use of results from the Canadian Climate Centre GCM to obtain 10-year series of daily CP appears to give the most accurate output from all GCMs (Crane and Hewitson, 1991). Disaggregation of GCM output to regional and local level is a topic of great current interest. The methodology used here may be a useful operational tool to achieve such disaggregation. To this end, the present classification method should be modified in order to measure the difference between CPs reflecting present conditions and climate change scenarios. One possibility is to return to the use of "subjective" classification (Hess and Brezowsky, 1969). In fact, Bardossy and Caspary (1990) using this scheme have been able to detect changes in the time series of CP types. Additional information from synoptic experience may also be used for better results.

Other Effects (e.g. surface runoff, sediment yield, etc.)

These may be determined, again, with the following general methodology:-

a) Select a model using hydrometeorological input to estimate sediment and nutrient yields.

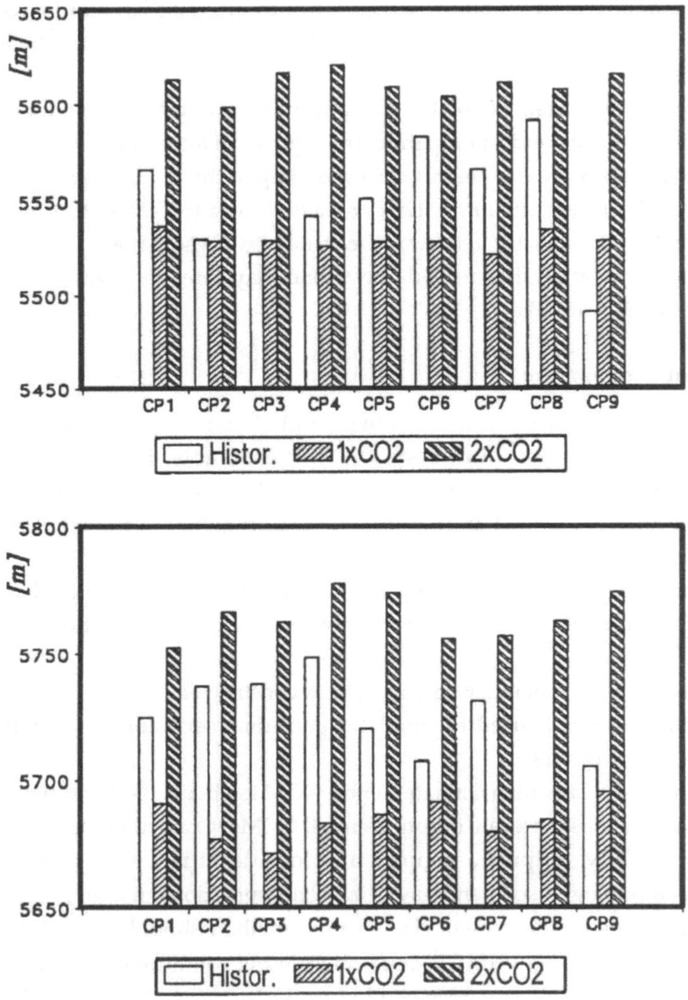

Figure 15-4. Spatially averaged height of the 500 hP· pressure surface in Nebraska for winter (top) and summer (bottom) half years, for different CP types (Matyasovszky *et al.*, 1993a)

b) Calculate the hydrometeorological input corresponding to global climate change.
c) Use model (1) with hydrometeorological inputs (historical and $2xCO_2$ case) to assess the effect of climate change.

There are four basic system components: precipitation, runoff, sediment, and phosphorus (Bogardi and Duckstein, 1978; Duckstein *et al.*, 1978). The last three all depend on precipitation which varies highly in space, amount, duration and time between events. As explained before methods have been developed to predict precipitation time series under different climate change scenarios, thus the remaining three components may be estimated.

The direct runoff Q_r from a watershed is (SCS, 1972)

$$Qr = \frac{(P_r - 0.2S)^2}{(P_r + 0.8S)^2} \tag{15-1}$$

where

P_r is the amount of precipitation and S the potential precipitate retention, depending upon the soil characteristics, land use, etc. The amount of sediment lost from a watershed during each precipitation event (e.g., storm) is (Williams, 1975)

$$Z = 95\,(Q_v\,Q_p)^{0.56}\ LS{\cdot}K{\cdot}C{\cdot}PF \tag{15-2}$$

where

Z	= sediment lost, [tons],
Q_v	= volume of runoff, [acre-feet],
Q_n	= peak discharge, a function of precipitation, time of concentration, watershed discharge characteristics, etc., [acre-feet],
A	= watershed area, [feet^2],
LS	= watershed slope and steepness factor, [-],
K	= soil erodibility factor, [-],
C	= land use and management factor, [-],
PF	= erosion control practice factor, [-].

Phosphorous yield, due to precipitation drainage, results from two general transport mechanisms. One is the movement associated with runoff from *in-situ* soils to streams, whereas the other is advective transport of P by the stream to the waterbody of interest (lake or coastal area). Details of these mechanisms are given in Bogardi and Bardossy (1984) and Higgins (1990).

Estimation of Climate Change in the North Aegean Sea

The approaches for climate change prediction described above have been used to determine the likely changes at a coastal area in the North Aegean Sea, Eastern Mediterranean (Fig. 15-5). Using the above methodologies Bogardi et al. (1993b) and Matyasovszky *et al.* (1993b) show that global climate change will not lead to significant rainfall changes in the region of interest. This is shown in Figure 16-6, for both the winter and summer periods.

The corresponding impact on temperatures is shown in Figure 15-7. The general tendency is for variable amounts of temperature increase within the year. The highest temperature rise is about +4°C in January, April and September, while there is little change in February, March, June, July, August, October and November.

15-4. INTERACTION BETWEEN CLIMATE CHANGE AND COASTAL WATER QUALITY

Overview

Even without the substantial amounts of man-made pollution discharged into water bodies in modern times, nature itself has already provided for continuous recycling of biological matter in coastal waters. A simplified diagram of the process is shown in Figure 15-8.

In addition to the inherent biological loading from natural recycling of carbonaceous matter, further inputs from the shore may arise in the form of followings:

a) large amounts of sediments, resulting from increased soil erosion due to substantial deforestation, especially in Mediterranean countries.
b) inorganic and organic pollutants, mainly as nitric or phosphoric fertilisers, pesticides or herbicides used in farming. These result in a substantial contribution estimated to account for most of the overall coastal water pollution (USEPA, 1984; ASIWPCA, 1985.
c) organic, microbial or toxic man-made pollutants such as heavy metals or greases discharged from sewers.

Of these loads, heavy metals and toxic constituents tend to be chemically inactive and are removed mainly by mechanical or physical processes (e.g., sedimentation, adsorption on solid particles or surfaces immersed in the water, etc.). On the contrary, organic and other inorganic substances decay by numerous and very complicated chemical and biological processes.

All pollution loads, either natural or man-made, are subject to the influence of water circulation currents. This results in advection and turbulent dispersion in the water body, following the laws of conservation of mass for each constituent i in the system

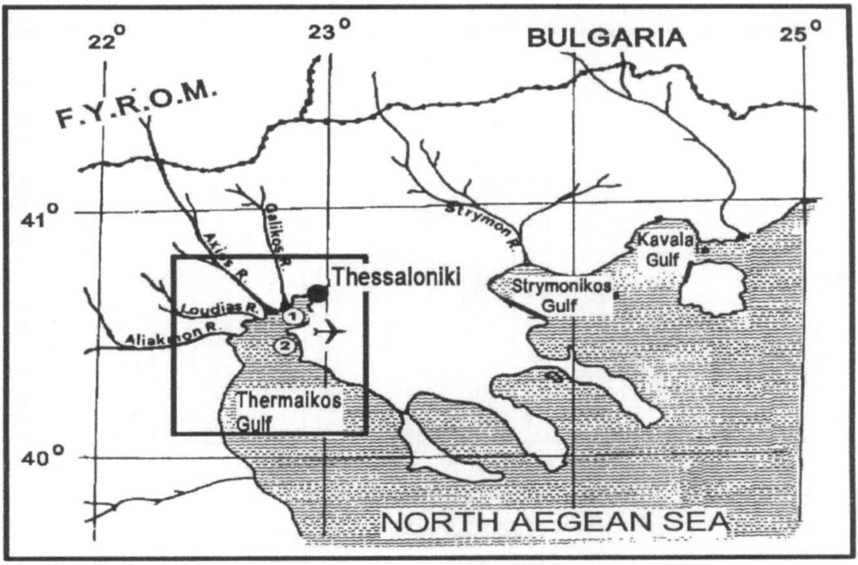

Figure 15-5. Geography of North-West Aegean Sea showing Thermaikos Gulf and Thessaloniki airport (Mikra); (1) Thessaloniki Bay, (2) outer gulf

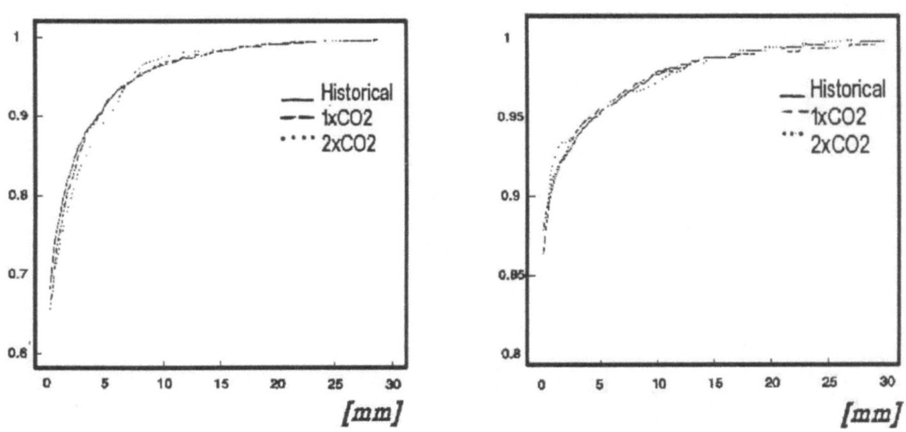

Figure 15-6. Probability distribution functions of daily precipitation at Thessaloniki airport (cf. Fig. 15-5) for the periods of January-March (left) and July-September (right) (Bogardi *et al.*, 1993b)

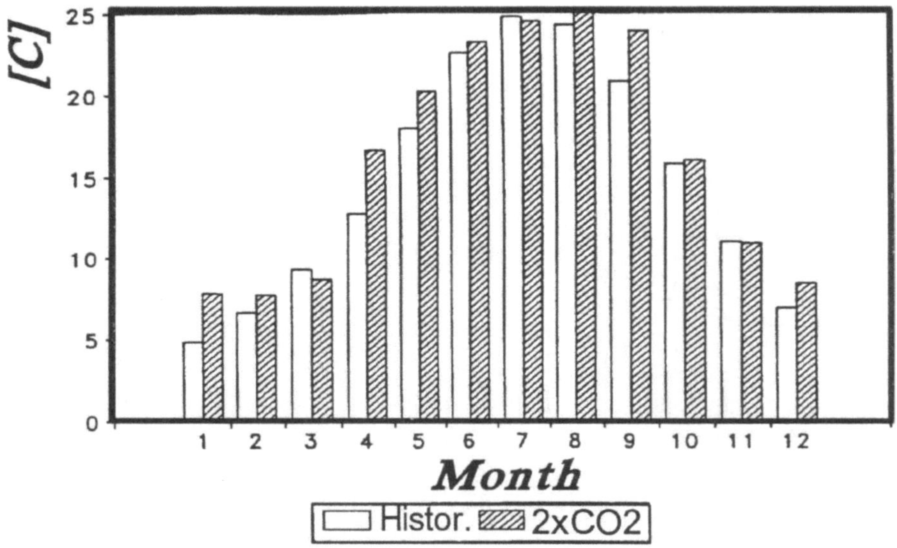

Figure 15-7. Monthly daily mean temperatures at Thessaloniki airport under the historical trends and a $2xCO_2$ climate change scenario (Ganoulis *et al.*, 1993a and Matyasovszky *et al.*, 1993b)

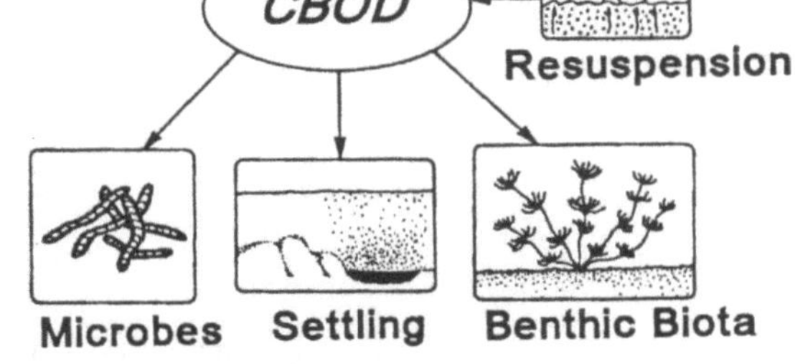

Figure 15-8. Overview of the organic matter cycle in a water reservoir (Bowie *et al.*, 1985)

$$\frac{\partial C_i}{\partial t}+\frac{u\partial C_i}{\partial x}+\frac{v\partial C_i}{\partial y}+\frac{w\partial C_i}{\partial z}-\frac{\partial}{\partial x}\left(K_x\frac{\partial C_i}{\partial x}\right)-\frac{\partial}{\partial y}\left(K_y\frac{\partial C_i}{\partial y}\right)-\frac{\partial}{\partial z}\left(K_z\frac{\partial C_i}{\partial z}\right)=\sum S_i \tag{15-3}$$

where

C_i = mean concentration of constituent i, mass/volume,
u, v, w = mean velocities in -x, -y, -z directions, length/time,
K_x, K_y, K_z = eddy diffusion coefficients, length^2/time,
ΣS_i = sum of source/sink rates of constituent, mass/(volume-time)
t = time.

Advection is by turbulent mass transport within the water, while additional diffusion and turbulent dispersion of pollutants takes place. In addition, the pollutants are subject to different types of decay, such as:

- Chemical, from the oxidizing effect of oxygen dissolved in the water, and from mutual neutraliszing between acidic and alkaline pollutants, and
- biological, arising from metabolism by microbes, phyto- or zooplankton.

Overall, all these processes are extremely complicated and not well known in great detail, with the exception of water advection and circulation. Therefore, much of the following discussion is based predominantly on empirical findings from experiments. First, the effect of a possible climate change on water circulation will be discussed together with the most important meteorological factors affecting it. Then, the current knowledge on the interplay between climatic conditions and water quality will be outlined. The interaction between these two aspects will be illustrated by an example of a typical case of a Mediterranean coastal water body, the Thermaikos Gulf in Greece.

Climate Effects on Water Circulation and Dispersion of Pollutants

Both naturally occurring inherent "pollutants" (e.g., recycled dead marine life) or man-made discharges are transported and dispersed into the water body by sea currents which are, in general, three dimensional. In the relatively shallow coastal Mediterranean waters vertical dispersive transport is much less significant than the horizontal one and circulation may be considered two-dimensional, as a first approximation. Little mass exchange occurs between vertical layers with the exception, of course, the settling of solid sediments, phytoplankton or decaying matter.

The water circulation patterns arise from the interaction between the Earth's rotation and the meso- and micro-scale circulation developed in the water body. The latter is caused by differentials in surface elevation, due to wind-driven water accumulation in different parts of the domain, enhanced by temperature-driven density differences or precipitation-driven salinity gradients. Sea tides also play a role, in the Mediterranean, however, they are not significant. Precipitation changes, caused by climatic changes, are reflected in the amount of shore runoffs entering into the water body. This affects the salinity and/or temperature locally at the outlet of rivers, irrigation channels, etc., resulting in changes of, again, the circulation pattern.

Changes in climate affect all these mechanisms. The most important interactions are:

a) Adverse winds, which change the surface shear on the water, "feeding" down to lower layers and altering the overall three-dimensional water circulation.
b) Ambient temperature gradients, influencing the temperature of the water and generating localized density gradients.
c) Local barometric pressures, affecting the evaporative heat transfer at the surface and altering the eventual water temperature and density.
d) Droughts on land, which alter the shore runoffs, the pollutant inflows and their concentration.
e) Increased rainfall, which increases the freshwater input and reduces salinity.

Pollutant dispersion and diffusion are also influenced by meteorology. The interaction between the wind microclimate and the vertical geomorphology of the shoreline may increase air turbulence above the coastal zone; this, in turn, "feeds down" into the water body, increasing turbulent eddy diffusion and dispersion.

Climate Effects on the Decay of Dispersed Pollutants in the Ecosystem

Overview. Every pollutant entering the water body will dissolve, disperse and diffuse, and will be subject to physical, chemical and biological transformations under the influence of sunlight, chemical neutralisation, oxidation, or metabolism by marine microbes, fauna and flora. In general, each constituent i will fulfil the mass balance

$$\begin{aligned}\frac{\partial C_i}{\partial t} = &-\frac{\partial}{\partial x}(uC_i) - \frac{\partial}{\partial y}(vC_i) - \frac{\partial}{\partial z}(wC_i) + \\ &+\frac{\partial}{\partial x}(E_x\frac{\partial C_i}{\partial x}) + \frac{\partial}{\partial y}(E_y\frac{\partial C_i}{\partial y}) + \frac{\partial}{\partial z}(E_z\frac{\partial C_i}{\partial z}) + \\ &+ S_L + S_B + S_K\end{aligned} \tag{15-4}$$

where

E_x, E_y, E_z = dispersion coefficients, area/time,
S_L = direct and diffuse loading rate, mass/volume,
S_R = boundary loading rate (upstream, downstream, benthic, atmospheric, etc.), mass/volume,
S_K = total kinetic transformation rate, mass/volume.

The three groups of terms in Eq. 15-4 account for transport, diffusion and loadings, respectively. In the practice of water resources engineering of particular interest are the concentrations of:

- The Carbonaceous Biochemical Oxygen Demand (CBOD). This is an indicator of overall "loading" of the aquatic system due to the oxidation needs of organic pollutants. It also includes the respiration demand of marine microbes which metabolise organic and fix inorganic matter (e.g., nitrates, inorganic phosphorus, etc.) (*cf.* Fig. 15-8).
- The Dissolved Oxygen content (DO). This parameter is more critical because it shows whether there is sufficient oxygen in the water for marine life to survive. The actual DO content reflects the equilibrium between reaeration at the surface added to photosynthetic oxygen generation by chlorophyll in the water body, minus the biological and any chemical oxygen demand. These very important mechanisms will be discussed in detail later.Generally, most marine fauna will swim away from waters in which DO has fallen to less than about 5 mg/l. Nevertheless, some types of worms have been still found surviving in virtually anoxic sediments in river deltas or heavily polluted areas around effluent outfalls.
- The concentrations of nutrients (ammonia, nitrates, phosphates, inorganic nitrogen or phosphorus) are linked directly to non-point source runoff from agricultural watersheds, as a consequence of soil fertilisation, insecticide or pesticide spraying, etc. Nutrients are metabolised by marine micro-organisms and the inorganic elements are fixed to more complex compounds. Algae play a very important role in these processes, enhancing water denitrification (release of N_2 to the atmosphere) or nitrification (capture of N_2 from air).Ammoniacal compounds are antagonistic to nitrates, both competing for algal uptake. On the other hand, the simultaneous presence of phosphorus enhances algal growth, leading to eutrophication, i.e., abnormal growth of algae and marine flora. This is particularly troublesome in enclosed waters (e.g., lakes and lagoons) but also occurs in coastal areas suffering from large pollution inflows and suppressed natural circulation and flushing.
- The coliform bacteria concentration. Although these micro-organisms are not pathogenic and exist naturally in the intestines of humans, their presence indicates pollution due to urban sewage effluents. However, doubts have been voiced about the suitability of this indicator to show pathogenic potential in coastal waters (Sobsey and Olson, 1983). This is because pathogenic viruses have lower decay rates than coliforms, and also smaller doses of them are required to cause infection. Furthermore, whereas coliforms are human-sourced, some opportunistic pathogens (e.g., *Pseudomonas aeruginosa, Legionella pneumophila*) often have non-faecal sources too and can grow in waters naturally (Bowie *et al.*, 1985). Upon discharge into the water body environmental conditions such as temperature and sunlight determine the eventual fate of coliform bacteria through a multitude of processes (e.g., photo-

oxidation, sedimentation, pH, predators, algae, bacteriophages, etc.). These mechanisms will be discussed in greater detail.

Apart from the above pollutants, sediments in the water column may also cause environmental problems as they bury benthic flora, or choke the gills of marine invertebrates. In fact, coastal areas at the deltas of large rivers typically suffer from anoxic conditions (Nelsen, 1994) due to oxygen demand caused by the large sediment and biological matter transport. This is additional to the polluting effects of any other organic or inorganic nutrients carried by the river.

In summary, then, CBOD indicates the overall organic pollution of the water, DO shows whether marine life may be sustained there, whereas nutrient concentration gives the potential for eutrophication. Coliform counts indicate the danger of disease for humans using the water for bathing or recreation. It is beyond the scope of this discussion to present the effect of climate change on all the numerous and complicated parameters which eventually determine water quality. In the following, we concentrate only on the most important mechanisms. Some of these aspects have been discussed in more detail by Ganoulis *et al.* (1993a).

Climate Effects on Reaeration. The capture of atmospheric oxygen at the free surface is one of the main oxygen replenishing mechanisms. Reaeration safeguards that the water remains habitable by living organisms. In a well mixed water column of depth H the oxygen influx F is

$$F = \frac{k_L}{H}(C_{sat} - C) \tag{15-5}$$

i.e., it depends upon the surface reaeration coefficient k_L and the difference between the actual and the saturation concentrations of dissolved oxygen. Because DO levels in natural waters are usually below saturation, oxygen is transferred from the atmosphere to the sea water column. However, in equatorial waters with strong photosynthesis during daylight, the net transfer may be reversed and the sea then becomes a significant oxygen source.

The reaeration rate coefficient $k_2=k_L/H$ in Equation 15-5 is a function of wind speed, water shear rate below the surface and the kinematic viscosity of water. Depending on the wind speed, the surface profile also changes, with "choppy" waves or foaming developing. The resulting increase in reaeration is depicted in Figure 15-9 and is additional to the influence of any surfactants, suspended particles, hydraulic structures, etc., which alter reaeration further. For a more detailed discussion of wind effects on reaeration one is referred to Gulliver and Stefan (1981) and Frexes *et al.* (1984).

On the other hand, APHA (1985) suggests that the saturation concentration of

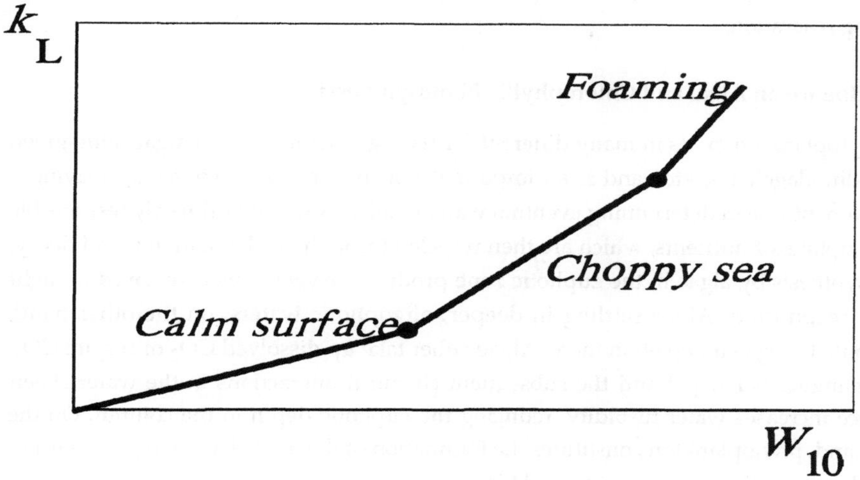

Figure 15-9. Qualitative representation of the relationship between wind velocity at 10 m above sea surface W10 and the surface reaeration coefficient k_L

oxygen in water Csat at different temperatures and salinities, at equilibrium with water-saturated air, is

$$ln(C_{sat}) = -139.344 + (1.576x10^5 / T) - (6.642x10^7 / T^2) + \\ + (1.244x10^{10} / T^3) - (8.622x10^{11} / T^4) - \\ - CHL\{(3.193x10^{-2}) - (19.43 / T) + (3.867x10^3 / T^2) \qquad (15\text{-}6)$$

where chlorinity is defined as

$$CHL = \frac{Salinity}{1.80655} \qquad (15\text{-}7)$$

Although salinity includes both chloride and bromide salts, it is sufficient to consider that all salinity may be expressed in terms of chloride only (APHA, 1985).

Both salinity and temperature are strongly climate dependent. For instance, ambient temperature and barometric pressure influence water temperature. This is also affected by solar insolation which, in turn, depends on the clarity of the atmosphere, another meteorological parameter. On the other hand, salinity is reduced by rainfall on the water body and drainage of precipitation-caused floodwaters and shore runoffs, thus finally affecting reaeration.

Influence on Algal and Chlorophyllic Photosynthesis

Phytoplankton exists in many different forms (e.g., diatoms, green algae, blue-green algae, dinoflagellates, etc.) and is an important part of the water ecosystem, playing a predominant role in determining eventual water quality. Algae are primarily responsible for the uptake of nutrients, which are then recycled through algal respiration and decay. Photosynthesis by algae in the euphotic zone produces oxygen; this is reversed at night due to respiration. Algae settling in deeper, oligophotic waters, on the other hand, contribute to oxygen depletion there. Algae either take up dissolved CO_2 or respire CO_2, thus changing water pH and the subsequent chemical interactions in the water. Their presence increases water turbidity, reducing the euphotic depth in the column. On the other hand, phytoplankton constitutes the foundation of the food chain of higher species, virtually supporting the marine animal biota.

Because of the variety of different algae, it is customary to consider algal concentrations in terms of chlorophyll-a concentrations. Then, phytoplankton dynamics (Fig. 15- 10) follows the mass balance equation

$$\frac{dA}{dt} = (\mu - r_A - e_x - s - m_A)\, A - G_A \tag{15-8}$$

where

A = phytoplankton concentration, mass/volume,
i = phytoplankton gross growth rate, 1/time,
r_A = respiration rate, 1/time,
e_x = excretion rate, 1/time,
s = settling rate, 1/time,
m_A = non-predatory mortality, 1/time,
G_A = loss rate due to grazing, mass/volume-time.

Algal growth is a function of temperature, solar insolation and nutrients (phosphorus, nitrogen, carbon, and silicon for diatoms). Other essential nutrients such as iron, manganese, sulphur, zinc, copper, cobalt, molybdenum and vitamin B_{12} may also be important, especially in oligotrophic waters. The general algal growth is

$$\mu = \mu_{max}(T_{ref})\ f(T)\ f(L, P, N, C, Si) \tag{15-9}$$

where

$i_{max}(T_{ref})$ = maximum phytoplankton gross growth rate under optimal conditions, at a certain reference temperature Tref, 1/time,

f(T) = temperature adjustment function of growth rate,

f(L,P,N,C,Si) = growth limiting function for light (L) and nutrients (Phosphorus, Nitrogen, Carbon and Silicon).

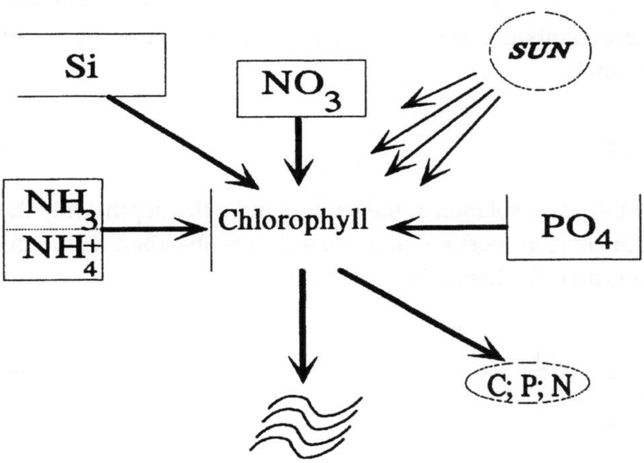

Figure 15-10. Phytoplankton kinetics

The phytoplankton growth rate increases with temperature, above a reference temperature Tref = 20°C. The major types of temperature response of algal growth are shown schematically in Figure 15-11. The temperature adjustment function f(T) follows the Arrhenius equation

$$f(T) = \theta^{(T-Tref)} \tag{15-10}$$

with

$$\theta = \frac{k_{T_2}}{k_{T_1}}^{\left(\frac{1}{T_2-T_1}\right)};$$

k is the reaction rate at the corresponding temperature.

As a rule of thumb, reaction rates of phytoplankton double with every 10°C rise in temperature. The algal growth remains relatively modest at low temperatures, increasing rapidly above 20-30°C. At higher temperatures growth virtually ceases and algae start decaying by protein denaturation.

In addition to temperature effects, algal growth is affected by solar insolation and nutrient availability. The exact mechanisms of influence are very complicated, especially when the presence of various nutrients is taken also into account. For a full discussion one is referred to Bowie *et al.* (1985) and De Groot (1983).

The water temperature influences each nutrient uptake mechanism according to Equation 15-10. Thus, any long-term climate warming will have a significant effect on chlorophyllic photosynthesis and algal nutrient uptake. On the other hand, climate also affects the light limitation mechanism through solar insolation. Generally, light penetration into the water column is

$$I(z) = I_o e^{-\gamma z} \tag{15-11}$$

with *ã* the extinction coefficient, I_o the insolation at the surface and z the depth. Only the visible spectrum is usually of interest, as most radiation outside it is absorbed within the first meters of depth. The light growth limitation is given by Steele (1965)

$$f(L) = \frac{I}{I_{sat}} e^{\left(1-\frac{I}{I_{sat}}\right)} \tag{15-12}$$

with I_{sat} being the optimum (saturating) light intensity. Care must be taken in that, because at low light intensities encountered below the euphotic depth, algae become more sensitive to light by changing the proportion of chlorophyll in their cells. This has to be taken into account in deeper waters also.

Returning to Equation 15-8, phytoplankton respiration and excretion combined take the form

$$r = r|_{T_{ref}} f_r(T) \tag{15-13}$$

where the temperature function generally follows Equation 15-10. Settling rates depend on the density, shape, size and physiological state of the cells, the viscosity and density of the water, and the turbulence and velocities of the coastal flow field. Usually, to determine settling velocities the Stoke's law is used, modified to account for non-spherical sizes (Scavia, 1980).

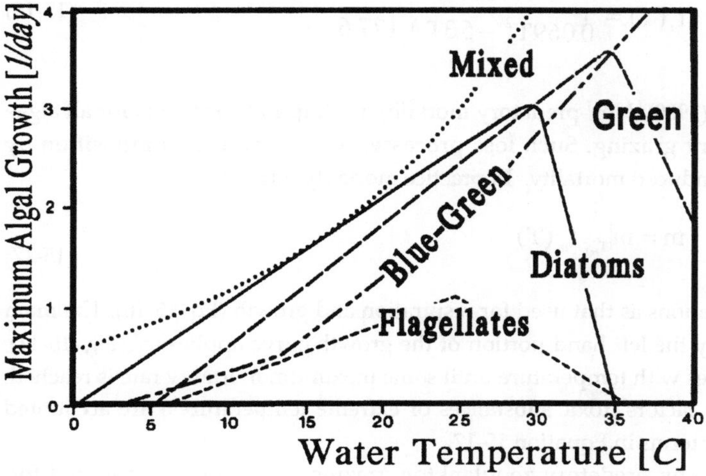

Figure 15-11. Temperature response of the growth rate of various algal types (Canale and Vogel, 1974)

$$W_s = \frac{2}{9} \frac{g R^2 (\rho_p - \rho_w)}{\nu F_s} \qquad (15\text{-}14)$$

where

W_s = settling velocity, length/time,
g = gravitational acceleration, length/time^2,
ϱ_p, ϱ_w = densities of cell and water, mass/volume,
ν = kinematic viscosity, mass/(length-time),
F_s = shape factor, [-].

The settling rate is then found from the settling velocity W_s and the column depth H

$$s = \frac{W_s/_{T_{ref}}}{H} f_s(T) \qquad (15\text{-}15)$$

A typical example of an empirical temperature adjustment function $f_s(T)$ (Eq. 15-15) is

$$f_s(T) = \frac{157.5}{0.069T^2 - 5.3T + 177.6} \tag{15-16}$$

given by Tetra Tech (1980). Non-predatory mortality m (Eq. 15-8) accounts for all algae lost not by predatory grazing. Such loss processes are bacterial decomposition by parasites and stress-induced mortality. In practice mortality follows

$$m = m|_{T_{ref}} f_m(T) \tag{15-17}$$

using similar formulations as that used for respiration and growth (Eq. 15-10). The main difference is that only the left hand portion of the growth curve applies (*cf.*, Fig. 15-11), i.e., mortality increases with temperature until some maximum mortality rate is reached. Additional mortality factors (toxic substances or extreme temperatures) are accounted for by adding further terms in Equation 15-17.

Algae is lost also by predatory zooplankton grazing. This can be accounted for, either by adding further terms in Equation 15-17 or, explicitly, assuming a loss rate proportional to algal density (Raytheon, 1974)

$$G = e_z|_{T_{ref}} \; f_g(T) \; A \tag{15-18}$$

where

e_Z = zooplankton grazing rate coefficient, 1/time,
$f_g(T)$ = grazing rate temperature function, 1/time,
A = algal biomass density, mass/volume.

The temperature functions for grazing are usually similar to those used for growth, respiration or mortality. Climate changes influence algal kinetics either directly, through temperature interactions and changes in salinity due to precipitation variations, or indirectly through atmospheric clarity which alters the amount of solar insolation available for photosynthesis. Any variations in prevailing winds also alter water turbidity, affecting light penetration, algal growth and decay, in the euphotic depth at least.

Climate change and zooplankton growth. Zooplankton is part of the same biomass pool as phytoplankton. Zooplankton dynamics depends on growth, reproduction, respiration, excretion and non-predatory mortality. Similar considerations as those in provious Section also apply to zooplankton. The main difference is that zooplanktonic organisms are mobile, so that settling does not occur. Furthermore, zooplankton migrates vertically following the diurnal cycle, adding another complication.

In similarity to Equation 15-8 zooplankton equilibrium takes the form

$$\frac{dZ}{dt} = (g_z - r_z - m_z)\, Z - G_z \tag{15-19}$$

where

Z = zooplankton concentration, mass/volume

g_Z = gross zooplankton growth rate, 1/time

r_Z = respiration and excretion rate, 1/time

m_Z = non-predatory mortality rate, 1/time

G_Z = predatory loss rate, mass/(volume-time).

Growth, consumption, respiration and non-predatory mortality are direct functions of temperature. Predation is also related to the consumption rates by higher predators. In general, the trends shown in Figure 15.11 for phytoplankton may be used here too, with linear, exponential or optimum temperature growth. Zooplanktonic animals are typically filter feeders. Therefore, zooplankton growth may be simply considered proportional to the available food concentration

$$g_Z = C_f F_T E \tag{15-20}$$

where

C_f = zooplankton filtration rate, water volume/(mass zooplankton-time),

F_T = total available food concentration, mass/volume,

E = assimilation efficiency, [%].

The assimilation efficiency acknowledges that zooplankton has different preferences or ingestion efficiencies for different food types (Leidy and Ploskey, 1980). The detailed analysis of this aspect is further complicated by factors such as zooplankton species, sex, size, age or nutritional stage. Respiration, excretion and non-predatory mortality follow the same lines as for phytoplankton (Eqs. 15-13 and 15-17). Non-predatory mortality is much more difficult to assess, however, because of the spatial movement of higher predators, carnivorous zooplankton or fish. Usually, overall average total predatory mortality rates are used, to account for losses from all zooplankton consumers. Because of the intimacy, through the food chain, between phytoplankton and zooplankton dynamics, any climate effects on the former will also have an impact on zooplankton populations. This, in turn, will cascade to higher predators, eventually affecting humans too.

Influence on chemical decay rates. In general, the complex chemical reactions (i.e., oxidation, reduction, pH neutralisation, etc.) which occur in the water depend upon water temperature T according to an Arrhenius-type relationship

$$k = k_{20°C}\, T^{\,(T-20°C)} \tag{15-21}$$

and show similar temperature dependencies as all other chemical decay mechanisms (e.g., denitrification, phosphorus uptake, etc.). A simplified overview of the interactions between CBOD, dissolved oxygen and nutrients is depicted in Figure 15-12.

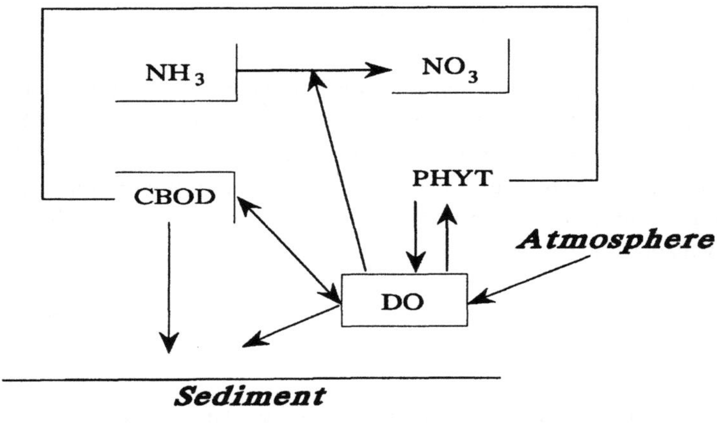

Figure 15-12. The cycle of dissolved oxygen and carbonaceous biochemical oxygen demand

The equilibrium between CBOD and DO is

$$\begin{aligned}\frac{\partial C}{\partial t}+u\frac{\partial C}{\partial x}+v\frac{\partial C}{\partial y}+w\frac{\partial C}{\partial z}&=D_x\frac{\partial^2 C}{\partial x^2}+D_y\frac{\partial^2 C}{\partial y^2}+D_z\frac{\partial^2 C}{\partial z^2}-k_1C-k_3C\\ \frac{\partial D}{\partial t}+u\frac{\partial D}{\partial x}+v\frac{\partial D}{\partial y}+w\frac{\partial D}{\partial z}&=D_x\frac{\partial^2 D}{\partial x^2}+D_y\frac{\partial^2 D}{\partial y^2}+D_z\frac{\partial^2 D}{\partial z^2}+k_1D-k_2D\end{aligned} \tag{15-22}$$

where

C	= the CBOD concentration, mass/volume,
D	= oxygen deficiency below saturation=DOsat-DO, mass/volume,
u,v,w	= advective velocities, length/time,

D_x, D_y, D_z = dispersion coefficients, length^2/time,
k_1 = coefficient of decay, 1/time,
k_2 = reaeration coefficient, 1/time,
k_3 = settling coefficient, 1/time.

Like all biochemical processes, the CBOD decay rate increases with temperature, following Equation 15-21.

Influence on bacteriological decay. Coliform growth and decay depends mainly upon environmental conditions, through a variety of mechanisms. These include physical (e.g., photo-oxidation, coagulation, sedimentation), physico-chemical (e.g., pH, osmotic effects) and biochemical-biological (e.g., nutrient levels, predators, algae, bacteriophages) factors. Table 15-4 gives a summary of some of the possible interactions; all are influenced, to a greater or lesser extent, by climatic factors.

The interplay between these factors is poorly understood, especially quantitatively. Because of this limitation, coliform growth and decay have been assessed, traditionally, on the basis of a first-order approach (e.g., Baca and Arnett, 1976, Chen *et al.*, 1975, Anderson *et al.*, 1976)

$$C_t = C_o e^{-kt} \qquad (15\text{-}23)$$

where
C_t = coliform concentration at time t, count/volume,
C_o = initial coliform concentration, count/volume,
k = disappearance rate constant, 1/time,
t = time, time.

This typically results in the T-90 measured values, i.e., the exposure required to ensure 90% mortality. Care must be taken, however, in that bacterial decay in the dark is only approximately half of that at midday (Bowie *et al.*, 1985). Apart from this sensitivity to light, measurements have also detected a sensitivity to water salinity.

Temperature, like in all other physico-bio-chemical processes, plays an important role here too (Ganoulis, 1992).

Table 15-4. Factors affecting coliform disappearance rates (Bowie *et al.*, 1985)

Factor	Effects
Sedimentation	Important with regard to water column coliform levels, particularly where untreated or primary sewage effluent or storm water is involved, and under low vertical mixing conditions. May adversely affect shellfish beds by depositing coliforms and faecal matter into benthos.
Temperature	Probably the most influential factor modifying all other factors.
Adsorption, Coagulation, Flocculation	Inconclusive.
Solar Radiation	Important; high levels may cause more than ten-fold increases in disappearance rates over the corresponding rate in the dark. Rates are increased further in freshwater environments.
Nutrient Deficiencies	Appear to accelerate disappearance. Numerous studies have indicated that increasing nutrient levels in sea water decrease disappearance rates.
Predation	Several species of organisms (bacteria, amoebae) have been shown to attack and destroy E. Coli. Importance of predation depends strongly on the concentration of predators.
Bacteriophages	Apparently not important.
Algae	Bactericidal substances are known to be produced by planktonic algae. Substances may be photoactivators, mediating the influence of light on coliform disappearance. Another hypothesis is that algal predators with blooms concomitant with algal blooms may produce substances toxic to *E. Coli* or may prey on them.
Bacterial Toxins	Antibiotic substances produced by indigenous bacteria are not believed to be important in coliform disappearance.
Physico-chemical Factors	Apparently pH, heavy metals, and the presence of organic chelating substances mediate coliform disappearance rates. The importance of each, however, is poorly understood at present. Salinity strongly enhances the effect of solar radiation.

Summary

From the discussion above it has become clear that, apart from its effects on pollutant dispersion (cf. Section 4-2), climate may also have significant effects on pollutant decay. The most important meteorological parameters affecting the pollution levels are:

a) Local barometric pressures, because they affect the evaporative heat transfer at the surface and thus affect the water temperature.
b) Higher than normal air temperatures, which influence water temperature causing changes in the biological and chemical equilibrium constants, as well as altering water circulation.
c) Atmospheric diffusivity and solar absorption, which change the intensity of irradiation incident on the water; this influences directly chlorophyllic photosynthesis.
d) Shore droughts, which alter the quantity of shore runoffs and increase the concentration/toxicity of the pollutant loads flowing into the water body.
e) Increased rainfall, which increases the freshwater input and reduces salinity at the mouth of rivers or irrigation channels. This affects the eventual water circulation patterns and has a slow-down effect on the decay of individual pollutants. In extreme cases it may even affect the suitability of the water for certain marine species to survive (e.g., shellfish).

Meteorological and local climatic information is, therefore, essential in analyzing the long-term quality characteristics of coastal waters. More specifically, with respect to any global warming, it is useful to see the likely effect of the speculated climate change scenarios on water quality. This may be studied by simulation, as presented below for a typical case in Thermaikos gulf, Aegean sea, Mediterranean, Greece.

15-5. A STUDY CASE - THE LIKELY EFFECT OF CLIMATE CHANGE ON THERMAIKOS

The question is: What would the consequence be on the water quality of Thermaikos from a doubling in the carbon dioxide content of the atmosphere (2xCO_2 scenario)? In the following, only the direct influence of temperature changes on water quality will be considered. Indirect effects, caused by variations in the amount of runoff or rainfall precipitation entering the water body, have not been included.

Description of Thermaikos Gulf

Thermaikos gulf is located at the north-west corner of the Aegean sea with a width of 15 km at its maximum opening between *Aherada* Peninsula, on the west, and *Epanomi* on the east (*cf.* Fig. 15-5). The maximum "height" of the gulf, from north to south, is 45 km and its total surface 473 km^2; Figure 15-13 shows its bathymetry. Thermaikos is open only on the south side. It constitutes the discharge basin for one major (*Axios*) and three minor (*Aliakmon, Loudias, Galikos*) rivers (cf. Fig. 15-5). All three carry water year-round,

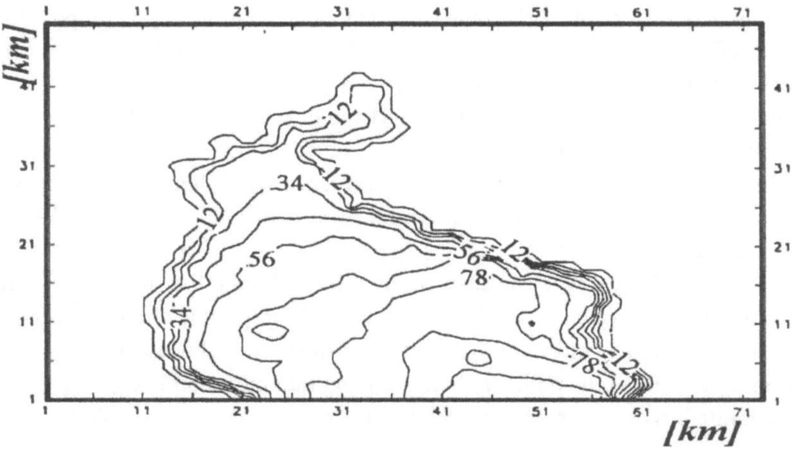

Figure 15-13. Bathymetry of Thermaikos Gulf (in [m])

with flow rates varying between 10 m^3/s and 400 m^3/s from summer to winter. The flow rates vary greatly also due to irregular drainage for agricultural irrigation. In addition, the sewage from the town of Thessaloniki (1,000,000 inhabitants) is also discharged into the gulf (Ganoulis, 1991).

Table 15-5. Meteorological characteristics in Thermaikos Gulf: temperature during period 1930-75 (Ganoulis, 1988a)

Temp [°C]	J	F	M	A	M	J	J	A	S	O	N	D	Year
Min	3.0	2.9	6.3	12.1	17.5	22.6	25.4	25.4	20.2	14.2	9.5	5.2	15.3
Max	10.5	11.3	13.7	17.4	22.3	25.4	28.3	28.4	25.4	21.5	14.5	11.5	17.5
Ave	6.0	7.3	10.0	14.8	19.6	24.0	26.8	26.5	22.4	17.2	12.4	8.0	16.2

Table 15-6. Meteorological characteristics in Thermaikos Gulf: precipitation during period 1930-75 (Ganoulis, 1988a)

Month	J	F	M	A	M	J	J	A	S	O	N	D	Year
Precip [*mm*]	41	35	40	41	49	37	27	20	31	51	56	55	483

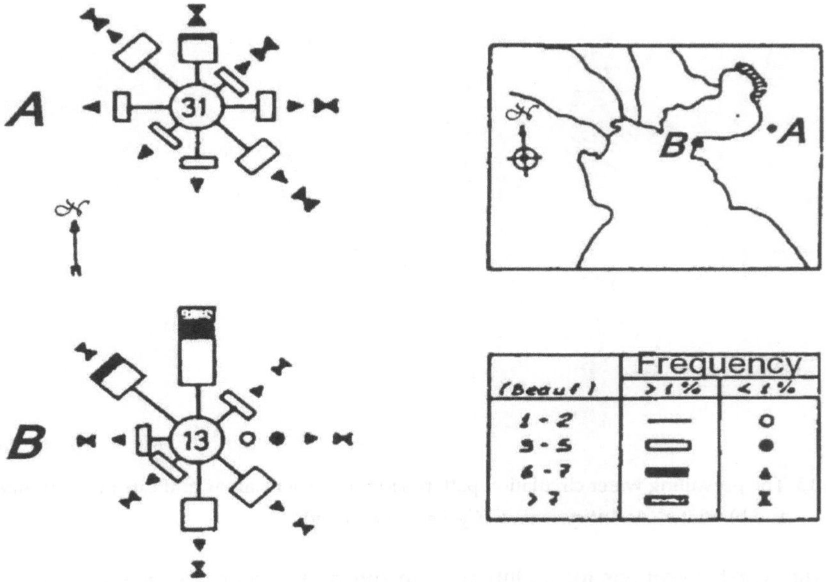

Figure 15-14. Meteorological characteristics in the region of Thermaikos Gulf: average yearly wind roses at two locations (A=Thessaloniki Mikra airport, B=Epanomi) during 1950-68 (Ganoulis, 1988a)

The main climatological data in the region are shown in Tables 15-5 and 15-6. The prevailing winds are S-SE, during summer, and N-NW during winter (Fig. 15-14). Strong winds (>15 m/s) are infrequent and of short duration, lasting one or two days and arising usually during winter.

Water Circulation Patterns

On the basis of hydrodynamic simulations and in situ measurements the average water circulation patterns during winter and summer have been determined and are shown in Figure 15-15 (Ganoulis, 1988a and 1991). Tidal currents are insignificant, with total tidal elevation in the inner gulf not exceeding 30 cm. The measurements show that, during summer, strong stratification of the water occurs with the warmer surface layer remaining stable over the colder depth layer; this leads to relatively anoxic conditions at the bottom. In contrast, during winter, the colder and denser surface layers destroy the stratification resulting in a satisfactory vertical mixing in the water column. Conse-

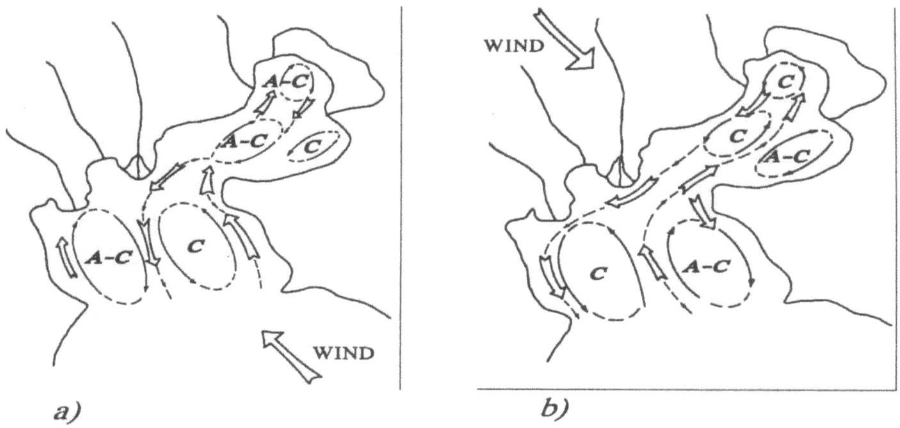

Figure 15-15. The prevailing water circulation patterns in inner Thermaikos gulf during a) summer and b) winter (*A: Anticyclonic, C: Cyclonic circulation*)

quently, the worst conditions for pollution occur during summer. This is the period of the year chosen to show the influence of climatic change on water quality.

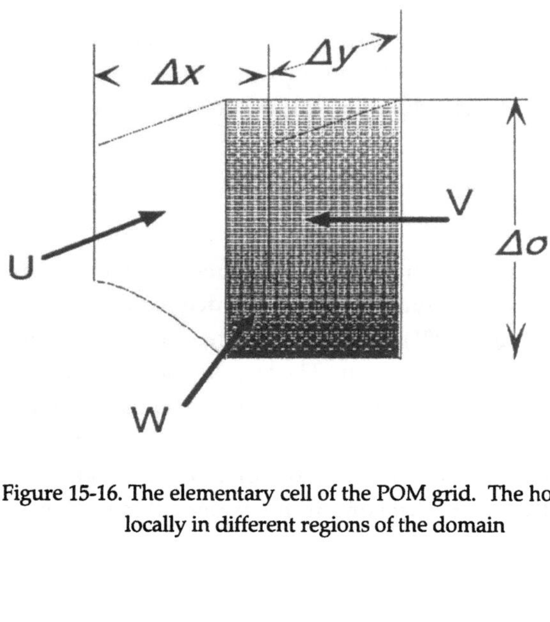

Figure 15-16. The elementary cell of the POM grid. The horizontal step sizes Δx and Δy may vary locally in different regions of the domain

Methodology

Water quality in the gulf is determined as follows. First, on the basis of prevailing summer winds, the water circulation in the gulf is found using the three-dimensional Princeton Ocean Model (POM) (Blumberg and Mellor, 1987). The gulf is divided horizontally in a rectangular 2x2 km grid. The POM model solves the Navier-Stokes equations in the finite-difference grid shown in Figure 15-16, using Cartesian coordinates along the horizontal directions -x,-y and the transformed coordinate $\sigma=(z-\zeta)/H$ along the vertical -z; H is the water depth and ζ the surface elevation. In using the POM model, the water velocities can be found everywhere in the gulf and are fed into a suitable ecological model. As such the Water Analysis Simulation Program (WASP) (di Toro *et al.*, 1981) has been chosen. The model is based on a link-node representation of the flow field, divided into interconnected segments. Using the flow exchanges between neighbouring segments, it solves the transport and diffusion equations for each pollutant. Different program modules deal with toxic or non-toxic pollutants, separately.

The spatial relationship between the WASP segments and the POM grid is shown in Figure 15-17, horizontally. Along the vertical direction, WASP follows POM's layering with each WASP layer following a user-specified number of POM layers in transformed depth coordinate σ. The linkage between POM and WASP is automatic and is accomplished, by means of software whose main features are shown in Figure 15-18.

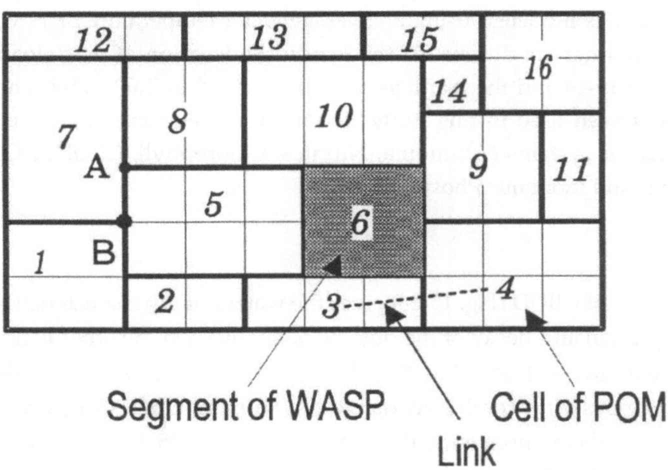

Figure 15-17. Relationship between the POM (fine lines) and WASP grids (thick lines) on the horizontal plane

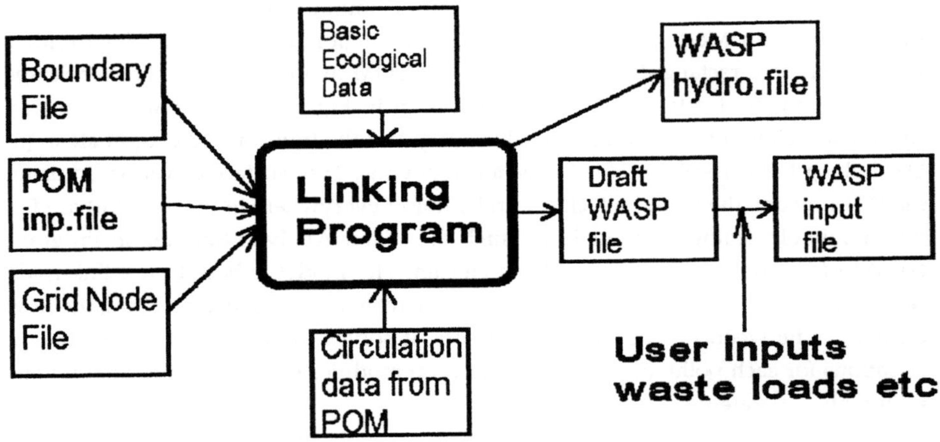

Figure 15-18. Functional description of the program linking the circulation (POM) and ecological (WASP) models. The rectangles indicate program Input/Output files

More details on the technical aspects of this linkage are available from the authors.

Though water temperature influences water circulation through density and salinity changes, this influence is not considered during the application of POM, at this stage. Instead, the water circulation is modelled using an average water temperature of 19°C, representative of mean summer conditions. In this study the location of wasteload inflows is shown in Figure 15-19 and the loadings are summarized in Table 15-7. The POM/WASP software has been used in this study for the case of eutrophication, i.e., using all 8 available physical systems (Ammonia, Nitrates, Chlorophyll, CBOD, DO, Organic Nitrogen, Organic and Inorganic Phosphorus).

Results

The spatial concentration of CBOD (Fig. 15-20) gives the water quality characteristics due to the advective dispersion and decay of the organic pollutants. On the other hand, the DO distribution near the seafloor (Fig. 15-21) is useful to indicate possible development of anoxic regions with conditions dangerous to marine life. Comparing these results with experimental measurements, at the same depth (Fig. 15-22), shows that an accurate prediction of pollution dispersion in the gulf is achieved using the two-model combination.

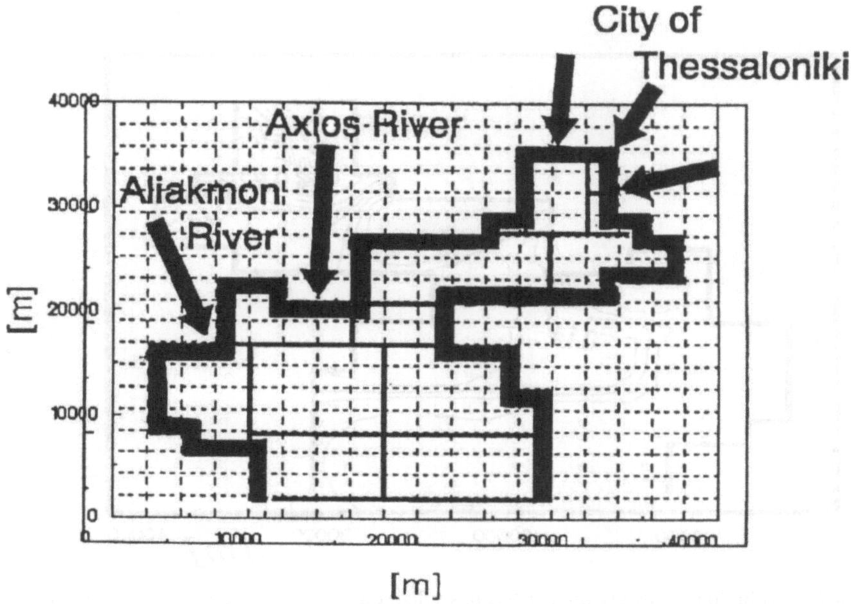

Figure 15-19. Outline of Thermaikos gulf showing the position of entry of the waste inflows used in the study

Table 15-7. Waste loads used in the study

	Loads [kg/day]	
Locations	CBOD	DO
Thessaloniki (total)	160,000	160
Axios	43,200	34,600
Aliakmon	23,800	7,000

The modelling approach described above was used to predict the effect of climate change on water quality of the gulf. Using the same overall circulation pattern (for summer), the ecological model was used with different daily temperature time series. Each consecutive run of the model resumed from the concentrations in each segment in the gulf as at the end of the last time step. Thus, a long run of many years duration could be done, whilst safeguarding the long-term data consistency of the model.

Two sets of ambient temperatures for Thessaloniki were used as inputs, both supplied from the methodology described previously (Bardossy and Plate, 1992; Ganoulis *et al.*, 1993a; Matyasovszky *et al.*, 1993b). One time series reflected the "historic

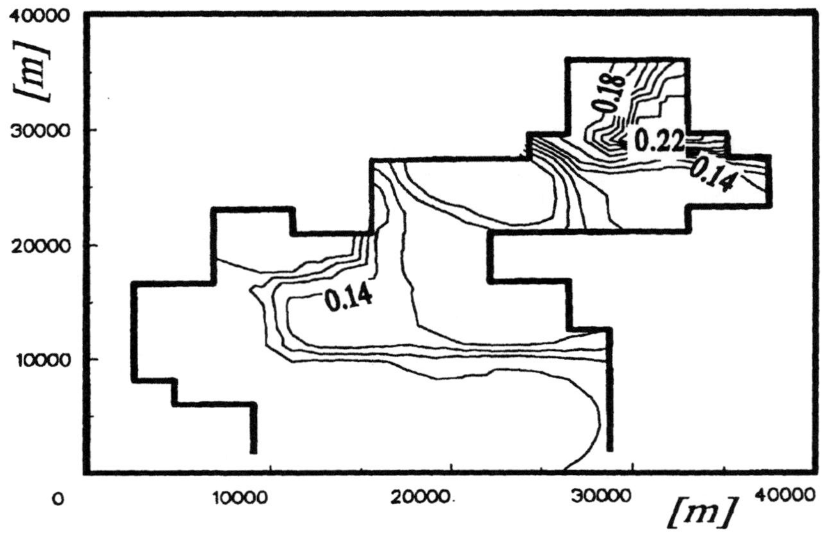

Figure 15-20. Distribution of CBOD close to the Gulf seafloor during the S-SE summer wind circulation

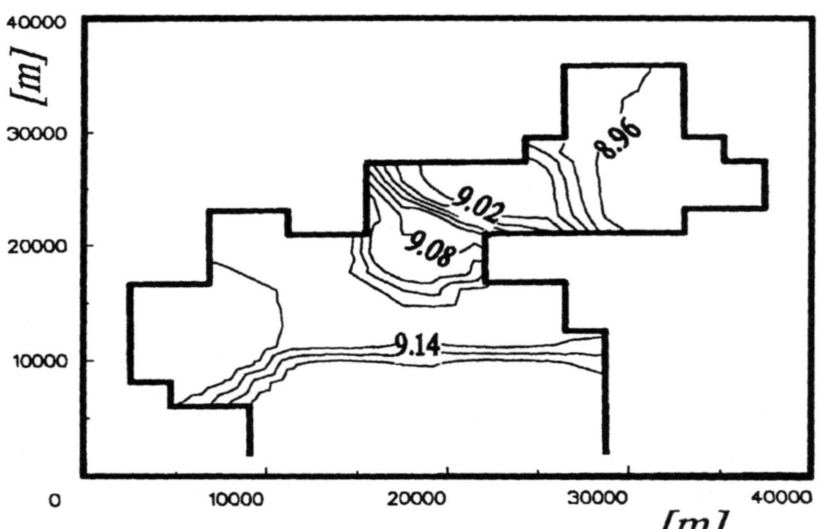

Figure 20-21. Distribution of DO close to the seafloor of the Gulf for summer circulation

case," i.e., under the $1xCO_2$ scenario. The second time series was based on a speculative $2xCO_2$ scenario ("modelled case"). From each time series, covering a total of 30 years, the daily air temperatures in May have been extracted to reflect typical summer conditions. A further reason for choosing this month was that then, as shown in Figure 15-7, a significant average global warming is predicted so that the effects on water quality, if any, would be easier to see.

Presently, in a first approximation, the temperature of water is assumed to be equal that of the air. Use of different temperatures in each segment of WASP, both horizontally and along the depth, is possible and may be used to refine further the study. Using the two temperature time series and with summer water circulation conditions, the trends in Figures 15-23 and 15-24 are obtained, for CBOD and DO, respectively. These trends reflect the minima and maxima of the respective water quality parameters throughout the body of the gulf, i.e., at all horizontal sites or depths. Thus, they represent the best- and worst-quality water environments which marine life may encounter within the water body, due to pollution.

From the simulation it emerges that:

a) Climate change will affect dramatically the overall dispersion of pollution in the gulf, with a significant increase in CBOD occurring.
b) Climate warming will lower the DO in the water body, especially its absolutely minimum values; although this will deteriorate the living conditions for fish and marine life, from this work it does not appear that the minimum DO threshold critical for sustaining life is to be reached soon.

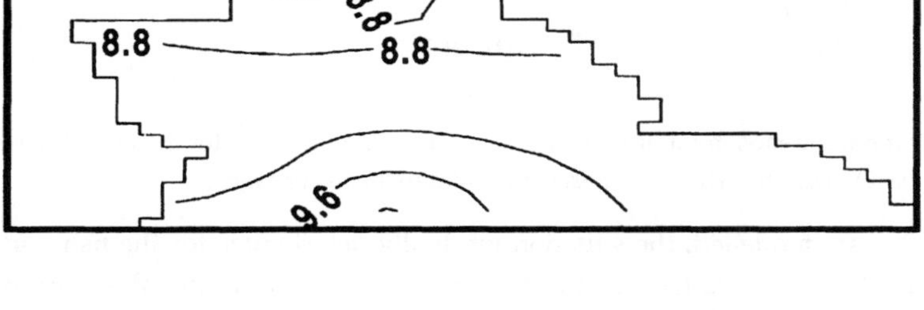

Figure 15-22. Experimental results for the distribution of DO near the bottom of the Gulf (averaged annual 1992 measurements) (Ganoulis, 1988a, 1991)

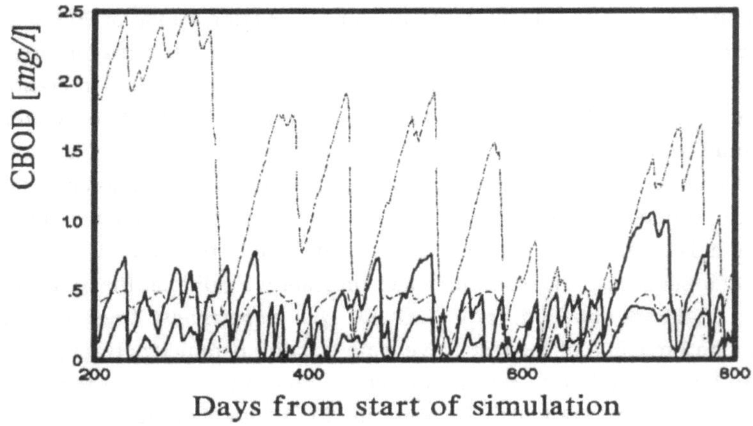

Figure 15-23. The expected trends in minimum and maximum CBOD throughout the gulf for the historic(1xCO_2) and 2xCO_2 (modelled) scenarios. The historic case is shown by the curves in solid lines

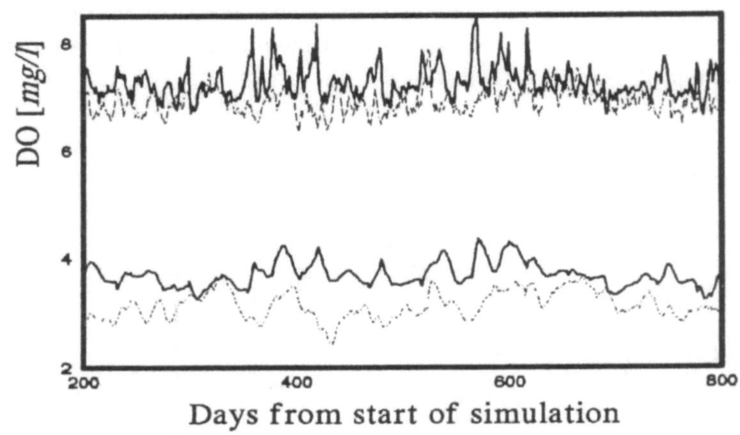

Figure 15-24. The expected trends in minimum and maximum DO in the gulf for the historic case and 2xCO_2 scenario. The solid lines correspond to the historical case

During the 30 years modelled, the situation gradually deteriorates for the fish and the aquatic life. Further work is needed to answer when (and if) a critical situation would be encountered, or whether some other compensating mechanism might come into play.

15-6. CONCLUDING REMARKS

This discussion has shown that:

a) An intimate link exists between climate and water quality, especially in coastal waters.

b) On the basis of climate models it is possible to determine with satisfactory accuracy local climatic conditions caused by global changes.

c) Using water circulation and ecological models, possibly linked directly with atmospheric circulation models too (Mironov, 1993), enables a comprehensive analysis to be made on the effect of climate change on water quality.

The work reported here shows that, in Mediterranean coastal waters, the $2xCO_2$ condition will cause a gradual decrease in the DO levels, whereas a more substantial increase in the overall pollution levels is likely to occur. This, in the long run, has important repercussions for marine ecosystems, reducing their capacity to sustain marine life. The significant ramifications for biodiversity from further DO reductions, combined with the added physiological stress on biological organisms from an actual water temperature rise, make urgent the need for further work in this area. The research should be aimed at refining existing models, improving our understanding of the very complex underlying physical, chemical and biological mechanisms, and ensuring that reliable measurements become available for input and/or model verification.

Last, but not least, attention must be paid in improving the computational efficiency of these models for use with the computer resources currently available. Benefit should be made from emerging advances in computer technology, such as parallel and distributed processing, networking of workstations, etc. (Furtney and Taylor, 1993). This will render even the more CPU intensive models usable, enabling multi-model combinations in large computation domains combining information at micro-, meso- and/or macro-scales.

REFERENCES

Anderberg M.R. (1973). Cluster Analysis for Applications. Academic Press, New York.

Anderson D.R., Dracup J.A., Fogarty T.J., and Willis R. (1976). Water quality modeling of deep reservoirs. J. Water Pollution Control Fed. 48 (1):136-46.

American Public Health Association. (1985). APHA Standard Methods for the Examination of Water and Wastewater, 16th Edn. APHA, Washington D.C..

Association of State and Interstate Water Pollution Control Administrators (ASIWPCA), America's Clean Water. The States Non-Point Source Assessment Appendix, 1985.

Baca R.G. and Arnett R.C. (1976). A Limnological Model for Eutrophic Lakes and Impoundments. Batelle Pacific Northwest Laboratories, Richland, Washington, D.C.

Bardossy A. (1991). Conditional Local Hydrological Variables on Circulation Patterns in West Germany. Working Paper, Institute for Hydrology and Water Resources, University of Karlsruhe, Karlsruhe, Germany, (in German)

Bardossy A. and Caspary H. (1990). Detection of climate change in Europe by analysing European circulation patterns from 1881 to 1989. Theor . Appl. Climatol., 42:155-67.

Bardossy A., Plate E. (1992). Space-time model for daily rainfall using atmospheric circulation patterns. Water Resour. Res., 28:1247-59.

Bartholy J. (1992). Methodological Study on Clustering Local and Regional Data Series of Precipitation. 6th Int Meet on Stoch Clim, June 22-26. Toronto, Canada.

Baur F., Hess P., and Nagel H. (1994). Kalender der Grosswetterlagen Europas 1881-1939. Bad Homburg, (in German)

Blumberg A.F. and Mellor G.L. (1987). A Description of a 3-D coastal ocean circulation model. *In*: Heaps NS (Ed.) Three-Dimensional Coastal Ocean Models, 4. Coastal Estu. Sci., American Geophysical Union, pp. 1-16.

Boer G.J., McFarlane N.A., Laprise R., Henderson J.D., and Blanchet J.P. (1984). The Canadian climate centre Spectral atmospheric general circulation model. Atmos-Ocean, 22:397-429.

Bogardi I. and Bardossy A. (1984). A Concentration Model of Phosphorus Stemming from Agricultural Watersheds. Research Report. Tiszadata, Budapest, (in Hungarian)

Bogardi I. and Duckstein L. (1978). A Stochastic Model of Phosphorus Loading from Non-point Sources. Res. Memor (RM-78-33). Int. Inst. for Appl. Syst. Anal., 2361 Laxenburg, Austria.

Bogardi I., Matyasovszky I., Bardossy A., and Duckstein L. (1993a). Application of a space-time stochastic model for daily precipitation using atmospheric circulation patterns. J. Geophys. Res., 98 (9): 16653-16667.

Bogardi I., Ganoulis J., Duckstein L., and Matyasovszky I. (1993b). Sediment yield and pollution from small watersheds in changing climatic conditions. *In*: Duckstein L. and Parent E. (Eds.) Engineering Risk and Reliability in a Changing Physical Environment, NATO ASI Series. Kluwer, Amsterdam.

Bowie G.L., Mills W.B., Porcella D.B., and Campbell C.L. (1985). Rates Constants and Kinetics Formulations in Surface Water Quality Modelling, 2nd edn. U.S. Environmental Protection Agency, Athens, Georgia 30613, Contract #68-03-3131.

Canale R.P. and Vogel A.H. (1975). Effects of Temperature on Phytoplankton Growth. ASCE J. Env. Eng. Div., 100(EE1):231.

Chen C.W., Smith D.J., and Lee S.S.(1975). A Comprehensive Water Quality Ecological Model for Lake Ontario (final Report). Prepared for National Oceanic and Atmospheric Administration, Tetra Tech Inc, Lafayette, California.

Copeland B.J. (1966). Effects of Industrial Waste on the Marine Environment. J Water Poll Contr Fed., 38:1000.

Copeland B.J. and Bechtel T.J. (1971). Species Diversity and Water Quality in Galveston Bay, Texas. Water Air Soil Poll. 1:89.

Craddock J.M. and Flood C.R. (1969). Eigenvectors for Representing the 500 mb Geopotential Surface over the Northern Hemisphere. Q J. Roy Met. Soc. 95:576-93.

Crane R.G. and Hewitson B.C. (1991). Regional Climates in the GISS GCM. Preprints of the 2nd Symp. on Global Change Studies, Am Meteorol Soc.

De Groot W.T. (1983). Modelling the multiple nutrient limitation of algal growth. Ecol. Model. 18:99-119.

Di Toro D.M., Fitzpatrick J.J., and Thomann R.V. (1983). Water Analysis Simulation Program (WASP) and Model Verification Program (MVP)-Documentation. Hydroscience Inc, Westwood, NY, for U.S. EPA, Duluth, MN, Contract #68-01-3872.

Duckstein L., Bardossy A., and Bogardi I. (1993). Linkage Between the Occurrence of Daily Atmospheric Circulation Patterns and Floods: An Arizona Case Study. J. Hydrol., Elsevier Science Publishers BV, Amsterdam, 143:413-28.

Duckstein L., Bardossy A., and Fogel M. (1978). An event-based stochastic model of phosphorus loading into a lake. Adv. in Water Res., 1:321-9.

Dzerdzeevskii B.L. (1970). Circulation Mechanisms in the Atmosphere of the Northern Hemisphere in the Twentieth Century. Inst for Geog. Sov. Acad. of Sci, Moscow. Goedecke R (transl), Berryman B.F. (Ed.), University of Wisconsin, 1968.

Elliott R.D. (1943). Studies of persistent regularities in weather phenomena. *In*: Synoptic Weather Types of North America, Calif. Inst. of Techn., Pasadena, California.

Elliott R.D. (1949). The Weather Types of North America. Weatherwise, 2; 15; 40; 64; 86; 110; 136, 1949.

Frexes P., Jirka G.H., and Brutsaert W. (1984). Examination of recent field data on stream reaeration. ASCE J. Env. Eng. Div., 110 (6):1179-83.

Furtney M. and Taylor G. (1993). Of Workstations and Supercomputers. IEEE Spectrum, 30 (5):64-8.

Ganoulis J. (1988a). Oceanographic Conditions and Environmental Impact from the Sewage Works of Thessaloniki. Report to the Greek Ministry of the Environment, Town Planning and Public Works, Thessaloniki, (in Greek)

Ganoulis J. (1988b). Modelling the Sewage Assimilative Capacity of Coastal Waters in the Mediterranean. *In*: Rapp. Comm. Int. Mer. Medit., 31(2):321.

Ganoulis J. (1989). Marine Pollution Assessment from Wastewater Discharges. Proc. IAHR Congress, Ottawa D:123-9.

Ganoulis J. (1991). Water quality assessment and protection measures of a semi-enclosed coastal area: the Bay of Thermaikos (NE Mediterranean Sea). Mar. Poll. Bull., 23:83-7.

Ganoulis J. (1992). Dispersion et disparition des bactèries coliformes dans la Baie de Thessaloniki. Rev. des Sci. de l Eau, 5:541-54.

Ganoulis J., Duckstein L., Bogardi I., and Matyasovszky I. Water Quality in Coastal Zones under Variable Climatic Conditions. loc cit Bogardi *et al.* 1993b, 1993a.

Ganoulis J, Rodolakis N., and Argyropoulos D. (1993b). Protection and development of sensitive coastal regions in the Mediterranean: the Example of the Axios Delta. *In*: HELECO '93 1st Intl Conf. on Env. Techn. April 1-4 Technical Chamber of Greece I:451-60 (in Greek)

Golfi P., Dagli K., Kavadias D., Krantonellis K., Paschali P. (1993). Coastal Regions and Tourist Development. loc cit Ganoulis *et al.* (1993b), I:480-95 (in Greek).

Gong X. and Richman M.B. (1992). An Examination on Methodological Issues in Clustering North American Precipitation. ibid Bartholy.

Gulliver J.S. and Stefan H.G. (1981). Air-Water Surface Exchange of Oxygen: Theory and Application to the USEPA Monticello Experimental Field Channels. St. Anthony Falls Hydraulic Laboratory External Memorandum No.173, University of Minnesota.

Hartigan J . (1975). Clustering Algorithms. Wiley, New York, 1975.

Hess P. and Brezowsky H. (1969). Katalog der Groswetterlagen Europas. Bericht des Deutsches Wetterdien Nr 113, Bd. 15, 2. Neu bearbeitete und ergänzte Aufl. Selbstverlag des Deutschen Wetterdienstes. Offenbach am Main (in German).

Higgins B.F. (1990). Stochastic Phosphorus Model for Agricultural Watersheds. Master Thesis, University of Nebraska, Lincoln, NE 68588.

Key J. and Crane R.G. (1986). A comparison of synoptic classification schemes based on "Objective Procedures." J. Climatol., 6:375-86.

Kirchofer W. (1973). Classification of European 500 mb patterns. Arbeit der Schweiz Meteorol Zentral 43. Geneva.

Krick I.P. (1943). Synoptic Weather Types of North America. Calif Inst of Techn. Pasadena, California .

Kutzbach J.E. (1970). Large-scale features of monthly mean northern hemisphere maps of sea-level pressure. Mon. Weath. Rev. 98:708-21.

Lamb H.H. (1972). British Isles Weather Types and a Register of the Daily Sequence of Circulation Patterns, 1861-1971. Geoph Memoran. London, 16, no. 116.

Leidy G.R. and Ploskey G.R. (1980). Simulation Modeling of Zooplankton an Benthos in Reservoirs: Documentation and Development of Model Constructs. U.S. Fish and Wildlife Service, Fayetteville, Arkansas. For U.S. Army Corps of Engineers, Waterways Experiment Station (WES), Vicksburg, Mississippi.

Lund I.A. (1963). Map-pattern classification by statistical methods. J. Appl. Meteorol., 2:56-65.

MacQueen J. (1967). Some methods for classification and analysis of multivariate observations. Proc. 5th Berkeley Symp. on Math. Stat .and Probab., 1:281-97.

Matyasovszky I., Bogardi I., Bardossy A., and Duckstein L. (1993a). Estimation of Local Precipitation Statistics Reflecting Climate Change. Water Resour Res., 29 (12): 3955-3968.

Matyasovszky I., Bogardi I., Ganoulis J., and Duckstein L. (1993b). Local Climatic Factors under Global Climatic Change in a Greek Watershed. Working Paper, Department of Civil Engineering, University of Nebraska-Lincoln.

Mironov D.V. (1993). Personal Communication. ICSC-World Laboratory, CCSEM, Erice, Italy.

Nelsen T.A. (1995). Coastal Water Resources: Examples from the Northern Gulf of Mexico with application to the Mediterranean basin. *In*: Angelakis A. and A. S. Isaar (Eds.) NATO ARW on Diachronic Climatic Impacts on Water Resources. Springer-Verlag, Berlin-Heidelberg-New York, this volume.

Odum E.P. (1971). Fundamentals of Ecology. W.B. Saunders, Philadelphia.

Raytheon Company New England River Basins Modeling Project. (1974). Vol. III-Documentation Report, Part 1-RECEIV II Water Quality and Quality Model. For Office of Water Programs, U.S. Environmental Protection Agency, Washington D.C.

Scavia D. (1980). An ecological model of Lake Ontario. Ecol. Model., 8:49-78.

Sobsey M. and Olson B. (1983). Microbial Agents of Waterborne disease, assessment of microbiology and turbidity standards for drinking water. *In*: Berger P.S. and Argaman Y. (Eds.), U.S. Environmental Protection Agency, EPA, 570-9-83-001.

Soil Conservation Service (SCS). (1972). National Engineering Handbook, Section 4. Hydrology.

Stakhiv E.Z., Ratick S.J., and Wei Du Risk. (1991). Cost aspects of sea level rise and climate change in the evaluation of shore protection projects. *In*: Ganoulis J. (Ed.). Water Resources Engineering Risk Assessment. NATO ASI Series, Vol. G29, Spinger-Verlag, Berlin-Heidelberg-New York, pp. 337-55.

Steele J.H. (1965). Notes on some theoretical problems in production ecology. *In*: Goldman C.R. (Ed.) Primary Production in Aquatic Environments. University of California Press, Berkeley, California, pp. 393-8.

Tetra Tech (1980). Inc. Methodology for Evaluation of Multiple Power Plant Cooling System Effects, Vol V, Methodology Application to Prototype-Cayuga Lake. Tetra Tech Inc, Lafayette, California, Report EPRI EA-1111 for Electric Power Research Institute.

U.S. Environmental Protection Agency (US.EPA) (1984). Report to Congress: Non-Point Source Pollution in the United States. Water Planning Division, Washington D.C.

Williams J.R. (1975). Sediment-yield prediction with universal equation using runoff energy factor. *In*: Present and Prospective Technology for Predicting Sediment Yields and Sources, ARS-40:244-51.

Wilson L.L., Lettenmaier D.P., and Skyllingstad E. (1992). A hierarchical stochastic model of large scale atmospheric circulation patterns and multiple station daily precipitation. J. Geoph. Res., 97(D3):2791-809.

Woodwell G.M., C.F. Wurster, and P.A. Isaacson. (1967). DDT Residues in an East Coast Estuary: A Case of biological concentration of a persistent insecticide. Science, 156:821-4.

CHAPTER 16

MANAGEMENT OF WATER RESOURCES AND WASTEWATER REUSE IN PALMA DE MALLORCA, BALEARIC ISLANDS, SPAIN

PEDRO SAMPOL[1] AND MIQUEL SALGOT[2]

[1] *EMAYA, Palma de Mallorca, Baleares.*
[2] *Lab. Edafología, Universidad de Barcelona, Barcelona, SPAIN.*

16.1. INTRODUCTION

The Mediterranean islands are especially prone to suffer droughts because of the natural climate variability (Fontanals, 1984). The relatively scarce surface and the impossibility to catch resources from distant basins, make this environment sensible to water deficits. With a surface of 3,623 km^2, the island of Majorca is one of the biggest in the Mediterranean sea. It has some 300 km of coastline and a mild climate, with sunshine quite every day. Majorca is a main tourism destination for all Europeans. Unfortunately, the island has no surface water resources, with irregular rains (Figs. 16.1 and 16.2), and a highly pervious soil, which implies the necessity to obtain resources from groundwater.

Since colonization of the island, there has been the necessity of creating an important water network, which has been developed all around the centuries. Rainwater and gully water were their other resources. With the Arab conquest the water infrastructure reached its maximum. Catalan people took advantage of all those works when Arabs were defeated in the XIII century. There is an important lack of knowledge about this period of time and the following three centuries, but some research has been done (Cañellas and Tortella, 1992, and Fuster, 1973; EMAYA, 1993). Some important characteristic of the period from the agricultural and hydraulic point of view was the building of terraces and drainages (Cañellas and Tortella, 1992, and Fuster, 1973).

When searching in the history of the wells and springs serving the city we find some interesting things; for example, the discovery that one of the most known springs of the city, the "Font de'n Baster" is not a natural spring but a typical "qanat" built during the arab domination. Another spring, the "Font de la Vila" is the prolongation of an ancient open canal, a "sequia", which used to irrigate fields located near the city and moved several mills (Cañellas and Tortella, 1992). For a full description of the hydraulic characteristics and names during the arab domination, and on the later first Catalonian period, refer to Fontanals' and Cañellas' books. In these books, one finds also some

NATO ASI Series, Vol. I 36
Diachronic Climatic Impacts on Water Resources
Edited by Andreas N. Angelakis and Arie S. Issar

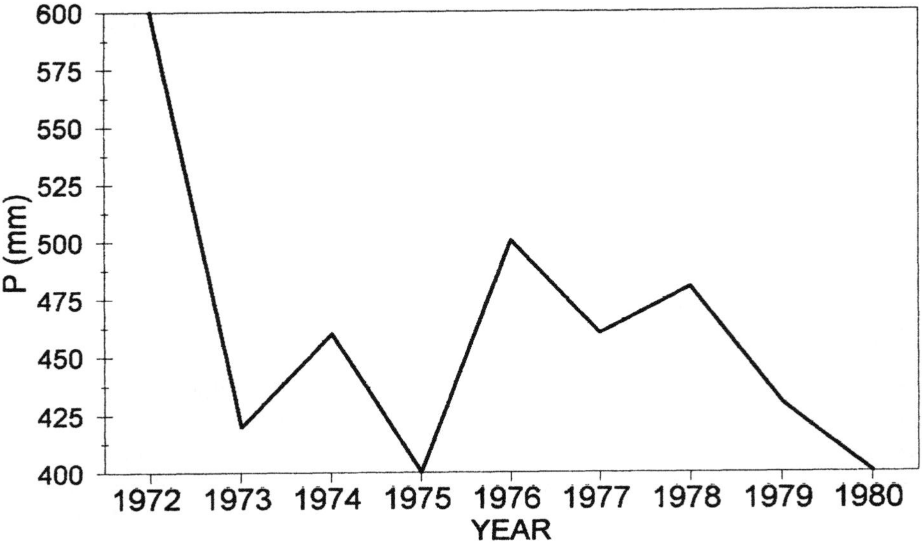

Figure 16-1. Palma's pluviometry from 1972 to 1980

references to drought incidents. For example, in 1381, a drought forced the Authorities of the island to close all the water points issuing from the "sequia". A similar drought is forcing the Authorities to decree special measures while writing this paper (October 1993).

An important characteristic of the Majorca landscape, since 1840 and until mid-century has been the windmill which pumps water up. At present, there are still some of them working, but the majority of them have disappeared during the last 30 years. Now groundwater is pumped using electrical devices. Apart from tourism, the main resources of the island used to be agriculture and cattle breeding. Nowadays both are disappearing quickly because of the lack of water resources, designated to supply water for cities and villages, and the use of good lands to build touristic resorts.

In an attempt to improve the quality of tourism and to extend the season, several golf courses have been built. There is an estimation that every course will need an average of 2,500 m^3 of water in summertime to keep the grass in good condition. Because of problems associated with low-spending tourism during the last 5 years, some studies have been undertaken to reorganize the sector, especially trying to attract high income tourists and to extend the season. This organization requires the availability of good quality and sufficient water resources, and the capability to cope with the usual seasons and years of drought in the island.

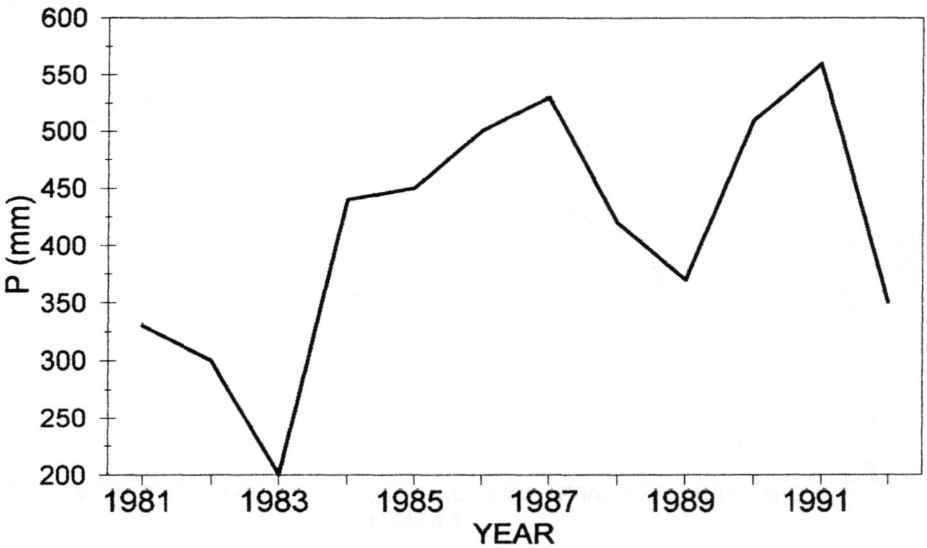

Figure 16-2. Palma's pluviometry from 1981 to 1992

16-2. WATER RESOURCES

Water resources are obtained from groundwater and surface waters. Studies have been made on desalination of sea water, but this solution has not been implemented in Mallorca yet.

Groundwater

Since the ancient times, the unique water resource of the island has been groundwater. In 1973 nearly 11,000 wells were known (EMAYA, 1993), and it was estimated that 500 applications to build new ones are made every year. Those great extractions caused, in a lot of cases, salinization of groundwater, either by the existence of saline water in the interior of the island or by marine intrusion. For example, due to this overexploitation, one of the biggest resources of the island, Pont d'Inca aquifer, was salinized nearly 25 years ago (Figs. 16-3 and 16-4).

Perhaps due to the lack of quality of the water supplied by the official networks, the establishment of wells experienced such a great increase. There are some indications of the growth of private exploitation by particulars of groundwaters, described in the book by Fontanals.

It is known that the majority of surface soil materials in the island are of pervious

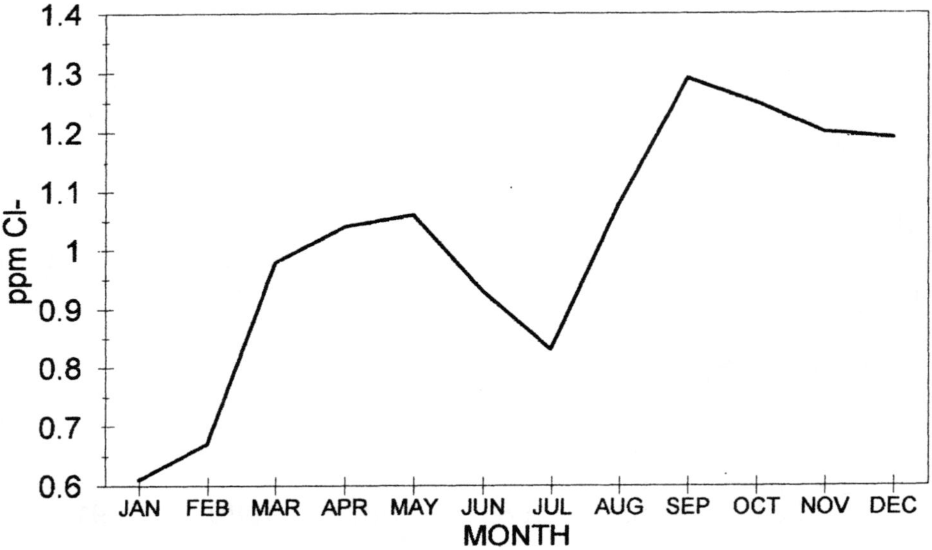

Figure 16-3. Palma's salinity during 1992

nature and that the main quantity of rain falls during autumn, with reduced evapotranspiration. This permits an important recharge of groundwaters. Unfortunately, this same characteristic implies a high possibility of aquifer contamination. In the northern range, it is difficult to use groundwater due to its extreme geological complexity, and only a few sources are used, especially for bottled water.

Nevertheless, some aquifers are used to supply the city of Palma, mainly the "S'Estremera", "Na Burguesa" and the "Pont d'Inca" ones, despite the excessive salinity of the last two. Other less important groundwater resources are used too (see below existing resources). Notice that the same water is used to supply some other neighbouring municipalities apart from Palma city (i.e.: Calvià).

Surface Waters

The northernmost part of the island has a really steep mountain range, with heights of more than 1,400 m, that occupies more than 25% of the total surface of the island. Other 15% is occupied by the "Llevant" (East) range, with mild relief and heights of a maximum of 500 m. The central part, 60% of the surface, is quite flat, with maximum heights of 120 m.

The rainwater is absolutely unevenly distributed, with some places in the North with 1,400 mm of rain per year and other places in the South with less than 400 mm. The fluctuations among the different years are quite important: 1.6 or 0.6 compared with the

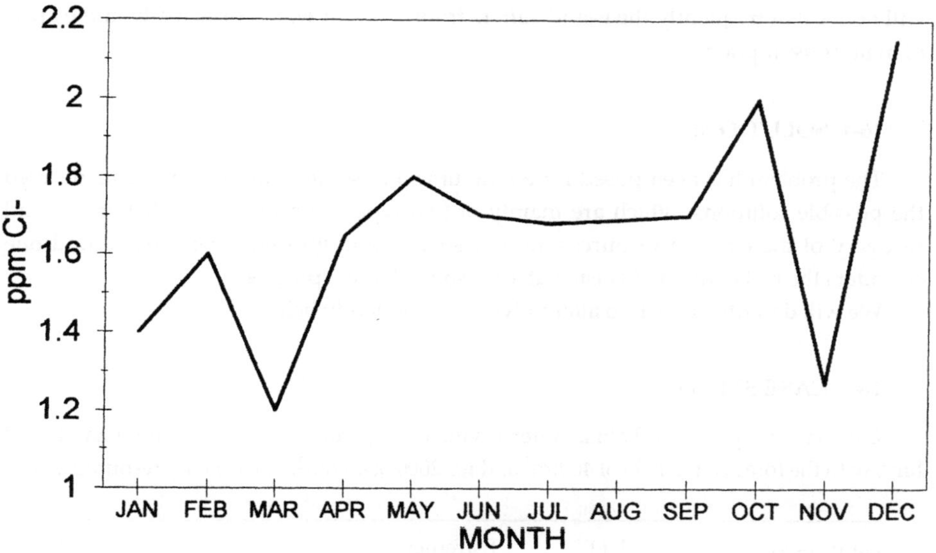

Figure 16-4. Palma's water salinity during 1993

mean figures and some years of extreme drought (EMAYA, 1993). The island has a Mediterranean type climate, with two rain maximum (50% between 1st November and 1st April, 25% between 1st of September and 1st of November) and a period of no rain in June-JulyAugust. This implies that a certain quantity of water in the North can be saved in a dam instead of being lost into the sea. This dam (named Gorg Blau) can store 11 hm^3.

The problem of surface resources is that they are mainly due to heavy rains and the majority of them are lost directly into the sea, with quite neither physical nor economic possibility to be used or captured. The drainage basins have quick concentration, usually less than one day.

16-3. GLOBAL RESOURCES

In 1973 a study of the resources of the island was published (EMAYA, 1993), and there was made an estimation of the possibilities of their use: a) The surface resources were estimated to be 26-43 hm^3/year. b) The groundwater resources were estimated as 260-343 hm^3/year

So the total resources per year are between 286-386 hm^3/year. Some of the stated resources are still not in use or are really difficult to obtain. If we compare the total amount of obtainable resources with the yearly demand of 290 hm^3 divided among urban (70 hm^3), agricultural (200 hm^3), and industrial (20 hm^3) uses (EMAYA, 1993), we will notice that frequently the island suffers from a lack of water, principally in the areas of main consumption.

16-4. SOLUTIONS

The problem has been posed for a long time, but some political constraints delayed the possible solutions, which are mainly of two types. The first one is better use and recovery of the existing resources, mostly surface and groundwater. The second one considers the reclamation of wastewater for some definite purposes.

We will describe these two alternatives in a case study below.

16-5. CASE STUDY

Cadenas in a paper on Palma water resources, explains that the overall demand is 44 hm^3, with the forecast for 1995 of 46 hm^3 and for 2000 of 48 hm^3. The existing resources are:

Dams (sw)	3-11	Estremera (gw)	7-12
Font de la Vila (gw)	2-6	Son Perot (gw)	4
Can Negret (gw)	2	Can Arabi (gw)	2
Son Vida (gw)	0.5	Son Veri (gw)	2

where:

figures are in hm^3, sw: surface water and gw: groundwater.

This implies that in the good years the shortage of water is roughly 6.5 hm^3 and in the bad years 23.5 hm^3. The only one solution is to use the brackish waters of "Na Burguesa" and "Pont d'Inca" aquifers to supply the deficient resources. The result is that in some places of the network the amount of chloride reaches 3.0 g/l, which makes water undrinkable.

A project was prepared to solve the problem with the following objectives:

a) Intensification of the programme for network leakages reduction.

b) Hydrogeological studies to identify aquifers not in use.

c) New forages ready to be used to obtain 5.5 hm^3/year from 1993 on.

d) Captation of "Sa Costera" spring, whose water is nowadays lost into the sea (6 hm^3/year).

All these actions will allow to obtain at least 40 hm^3/year in an average rain year. For security reasons, and not for the average rain years, the following actions will be performed:

- Modification of the brackish water extraction in "Pont d'Inca" aquifer.

- Building of a desalination plant near the city to treat brackish waters of "Pont d'Inca" aquifer.

In any case, the water distributed for consumption is microbiologically safe.

16-6. WASTEWATER REUSE

The Balearic Authorities have a strong will to develop wastewater reclamation and reuse in the islands. For this reason, some legislative work has been done in the Balearic Islands Autonomous Community, especially dealing with wastewater disposal and reuse. Notice the Decree about reclamation of treated wastewater (BOCAIB, 1988), the Law on golf courses (BOCAIB, 1992), or the Decree about disposal of treated wastewater (Cadenas, 1992). These three issues are:

a) Declare a priority on the use of reclaimed wastewater for agricultural purposes.
b) Reinforce the suggestion to irrigate golf courses with reclaimed wastewater.
c) Establish the legal and quality bases for reclamation and reuse of treated wastewater.

Palma de Mallorca city has two wastewater treatment plants, which are located near the airport, with a capacity of treatment of 20,000 m^3/day (Palma I) and 100,000 m^3/day (Palma II) (Vargas, 1992). The quantity per year is 25.7 hm^3, with an average BOD reduction of 82% and 90% respectively for plants I and II.

Treated wastewater of Palma I plant started to be reused in 1973, irrigating some 200 ha in the "Pla de Sant Jordi" area, with two main objectives: obtaining irrigation water and fighting against seawater intrusion. Given the good results of this first area, a complementary plan for wastewater reuse using reclaimed water from Palma II plant started in December 1992. The treated wastewater is disposed of initially in a 80,000 m^3 lagoon and after a retention time of 2.5 days, it is pumped into a 400,000 m^3 reservoir, which has nowadays a retention time of 10 days, and provides an additional treatment. Only a reduced amount of the water, between 30,000 and 40,000 m^3/day, is reused at the moment (October 1993), but in a few months the total quantity will be reclaimed and used for the previously defined purposes.

The reason the retention time is so long at the moment in the two lagoons is that the whole plan started with the building of the lagoons. When fully operative, the retention time will be shorter. Both reclamation systems are thought to operate jointly and in the near future a more ambitious plan is envisaged, using wastewater for:

a) Agricultural purposes. Irrigation of 700 ha. Possibility to reach 1,500 ha.
b) Fighting against salty water intrusion.
c) Groundwater recharge.
d) Implementing and recovering forest areas.

The plan is intended to be operative in full in 1994, and will allow to change water uses in all the area of "Pla de Palma" (Palma valley) and conserve better quality water to be used in the city. In addition, it is planned to catch a new resource of groundwater which is now lost into the sea and use it for city consumption (Cadenas, 1993, and Vargas, 1992).

Other developments for wastewater reclamation and reuse are being considered in the rest of Majorca island and in the other islands of the archipelago, e.g. Sant Lluís, Minorca (Govern Balear and SEARSA, 1989). A complete study on the possibility to reclame and reuse wastewater in the all Balearic Islands was done in 1989 (Govern Balear and SEARSA, 1989).

16-7. DISCUSSION

The drought situations in the Mediterranean are quite common, due to the special climatological circumstances that define the Mediterranean climate. This is a particularly hard problem in the islands, where the scarce surface in relation to the number of inhabitants or visitors leads to a chronic scarcity of water, which becomes more serious when drought conditions appear.

It is almost impossible to establish any water resource management on a yearly basis, but planning should be made for a minimum number of years. That is why the Mediterranean islands do need good planning and management of all water resources and that a maximum amount of people can live there and visit the places. This can be done by limiting the establishments (hotels, seaside resorts) or communication systems with the mainland (planes and ships).

If this limit is not established, recourse should be made to unconventional ways of water supply either for municipal, industrial or agricultural purposes. Those unconventional ways to obtain water resources could be mainly the following:

a) Desalination of seawater, which is a really expensive way to obtain water resources and has to be implemented as a last option. On the oher hand, the impact on prices could be important (if this solution is envisaged).
b) Reclamation of brackish water, less expensive than seawater desalination.
c) Wastewater reclamation and reuse for specific purposes, mainly landscaping, green areas and parks in the cities, agricultural purposes.

On Majorca island and especially in Palma de Mallorca it is necessary to implement those solutions, but also to coordinate a management plan for the resources of the whole island, combining all the possibilities. The diachronicity of resources, especially rainwater and consequently surface and groundwater will lead to control of the resorts existing on the island, when at a given moment in the future a maximum of water use will be reached.

The development of the economy is linked to the situation taking into account that cheap water is necessary to maintain a minimum of agriculture and industry. It is very well known that the dependence on only one economic resource (tourism, in this case) is not a good solution in modern times for a more or less isolated community. An equilibrium should be reached between the products-especially the agricultural ones-that arrive to the island from mainland and the ones produced in the island. Otherwise the economic development of the island and the capital city will be menaced.

16-8. CONCLUSIONS

Although it is quite difficult, a good global planning of Mallorca's water resources will help to maintain a good development of the island and in the capital city, even taking into account the diachronicity of rain. A good use of first hand water resources, with the establishment of priorities is absolutely necessary. Wastewater reclamation and reuse is an essential tool for the future development of the economy of Mallorca. In the past, the lack of water resources and the uneven distribution of rain in a yearly and more than yearly pattern has limited the population and development of the island, but nowadays, good management of water resources will not stop this development.

REFERENCES

Cadenas A. (1993). Personal communications.

Cadenas A. (1992). Plan de mejora de calidad del agua en el abastecimiento a la ciudad de Palma de Mallorca. Seminario Franco-Español: Nouvelles technologies en potabilisation des eaux. Reutilisation et gestion des eaux usées urbaines. Décembre 1992, Palma de Mallorca, Espagne.

Cañellas N. S., Tortella J. (1992). La gestió tradicional de l'aigua a Valldemossa: Son Ferradell, Son Oleza, Vistamar. Col.lecció Gorg . Blau. núm. 1. Editat per EMAYA / Ajuntament de Palma.

Conselleria D'Obres Públiques i Ordenació del Territori. Direcció General del Medi Ambient. Govern Balear / SEARSA. (1989). Estudio y seguimiento de los impactos ambientales en las Islas Baleares ocasionados por el reuso en regadíos de aguas y fangos de origen residual. Ciutat de Mallorca.

Decreto 13/1992 de 13 de Febrero. BOCAIB 07.03.92. "Por el que se regula la evacuación de vertidos líquidos procedentes de plantas de tratamiento de aguas residuales urbanas".

Decreto 33/1987 de 21 de Mayo. BOCAIB 02.06.87. "Utilidad pública de la utilización de agua residual depurada".

Duplessy J. C., Moyes J., Petit-Maire N., and Guiot N. (1989). Variabilité naturelle: De la fôret au désert. Le courrier du CNRS, no. 72. pp. 10-11.

EMAYA / Ajuntament de Palma. (1993). Ejercicio 1992. Informe Anual 1992.CNRS, no. 72. pp. 10-11.

Fontanals R. (1984). Un plànol de la sèquia de la vila del segle XIV (Ciutat de Mallorca). Ajuntament de Palma.

Fuster J. (1973). Estudio de los recursos Hidráulicos Totales de Baleares. Comité de Coordinación: Ministerio de Obras Públicas, Ministerio de Industria, Ministerio de Agricultura. Madrid.

Ley 12/1988, de 17 de Noviembre. BOCAIB 03.12.1988. "De campos de golf".

Vargas E. (1992). Aprovechamiento de aguas residuales para la agricultura en las Baleares. Seminario Franco-Español: Nouvelles technologies en potabilisation des eaux. Reutilisation et gestion des eaux usées urbaines. Décembre 1992, Palma de Mallorca, Espagne.

AUTHOR INDEX

SUBJECT INDEX

The ASI Series Books Published as a Result of Activities of the Special Programme on Global Environmental Change

---

This book contains the proceedings of a NATO Advanced Research Workshop held within the activities of the NATO Special Programme on Global Environmental Change, which started in 1991 under the auspices of the NATO Science Committee.

The volumes published as a result of the activities of the Special Programme are:

Vol. 1: **Global Environmental Change.**
Edited by R. W. Corell and P. A. Anderson. 1991.

Vol. 2: **The Last Deglaciation: Absolute and Radiocarbon Chronologies.**
Edited by E. Bard and W. S. Broecker. 1992.

Vol. 3: **Start of a Glacial.** Edited by G. J. Kukla and E. Went. 1992.

Vol. 4: **Interactions of C, N, P and S Biogeochemical Cycles and Global Change.**
Edited by R. Wollast, F. T. Mackenzie and L. Chou. 1993.

Vol. 5: **Energy and Water Cycles in the Climate System.**
Edited by E. Raschke and D. Jacob. 1993.

Vol. 6: **Prediction of Interannual Climate Variations.**
Edited by J. Shukla. 1993.

Vol. 7: **The Tropospheric Chemistry of Ozone in the Polar Regions.**
Edited by H. Niki and K. H. Becker. 1993.

Vol. 8: **The Role of the Stratosphere in Global Change.**
Edited by M.-L. Chanin. 1993.

Vol. 9: **High Spectral Resolution Infrared Remote Sensing for Earth's Weather and Climate Studies.**
Edited by A. Chedin, M.T. Chahine and N.A. Scott. 1993.

Vol. 10: **Towards a Model of Ocean Biogeochemical Processes.**
Edited by G. T. Evans and M. J. R. Fasham. 1993.

Vol. 11: **Modelling Oceanic Climate Interactions.**
Edited by J. Willebrand and D.L.T. Anderson. 1993.

Vol. 12: **Ice in the Climate System.** Edited by W. Richard Peltier. 1993.

Vol. 13: **Atmospheric Methane: Sources, Sinks, and Role in Global Change.**
Edited by M. A. K. Khalil. 1993.

Vol. 14: **The Role of Regional Organizations in the Context of Climate Change.**
Edited by M. H. Glantz. 1993.

Vol. 15: **The Global Carbon Cycle.**
Edited by M. Heimann. 1993.

Vol. 16: **Interacting Stresses on Plants in a Changing Climate.**
Edited by M. B. Jackson and C. R. Black. 1993.

Vol. 17: **Carbon Cycling in the Glacial Ocean: Constraints on the Ocean's Role in Global Change.**
Edited by R. Zahn, T. F. Pedersen, M. A. Kaminski and L. Labeyrie. 1994.

Vol. 18: **Stratospheric Ozone Depletion/UV-B Radiation in the Biosphere.**
Edited by R. H. Biggs and M. E. B. Joyner. 1994.

Vol. 19: **Data Assimilation: Tools for Modelling the Ocean in a Global Change Perspective.**
Edited by P. O. Brasseur and J. Nihoul. 1994.

Vol. 20: **Biodiversity, Temperate Ecosystems, and Global Change.**
Edited by T. J. B. Boyle and C. E. B. Boyle. 1994.

Vol. 21: **Low-Temperature Chemistry of the Atmosphere.**
Edited by G. K. Moortgat, A. J. Barnes, G. Le Bras and J. R. Sodeau. 1994.

Vol. 22: **Long-Term Climatic Variations – Data and Modelling.**
Edited by J.-C. Duplessy and M.-T. Spyridakis. 1994.

Vol. 23: **Soil Responses to Climate Change.**
Edited by M. D. A. Rounsevell and P. J. Loveland. 1994.

Vol. 24: **Remote Sensing and Global Climate Change.**
Edited by R. A. Vaughan and A. P. Cracknell. 1994.

Vol. 25: **The Solar Engine and Its Influence on Terrestrial Atmosphere and Climate.**
Edited by E. Nesme-Ribes. 1994.

Vol. 26: **Global Precipitations and Climate Change.**
Edited by M. Desbois and F. Désalmand. 1994.

Vol. 27: **Cenozoic Plants and Climates of the Arctic.**
Edited by M. C. Boulter and H. C. Fisher. 1994.

Vol. 28: **Evaluating and Monitoring the Health of Large-Scale Ecosystems.**
Edited by D. J. Rapport, C. L. Gaudet and P. Calow. 1995.

Vol. 29: **Global Environmental Change Science: Education and Training.**
Edited by D. J. Waddington. 1995.

Vol. 30: **Ice Core Studies of Global Biogeochemical Cycles.**
Edited by R. J. Delmas. 1995.

Vol. 31: **The Role of Water and the Hydrological Cycle in Global Change.**
Edited by H. R. Oliver and S. A. Oliver. 1995.

Vol. 32: **Atmospheric Ozone as a Climate Gas.**
Edited by W.-C. Wang and I. S. A. Isaksen. 1995.

Vol. 33: **Carbon Sequestration in the Biosphere.**
Edited by M. A. Beran. 1995.

Vol. 34: **Climate Sensitivity to Radiative Perturbations: Physical Mechanisms and Their Validation.**
Edited by H. Le Treut. 1996

Vol. 35: **Clouds, Chemistry and Climate.**
Edited by P. J. Crutzen and V. Ramanathan. 1996.

Vol. 36: **Diachronic Climatic Impacts on Water Resources.**
Edited by A. N. Angelakis and A. S. Issar. 1996.

Vol. 37: **Climate Change and World Food Security.**
Edited by T. E. Downing. 1996.

Vol. 38: **Evaluation of Soil Organic Matter Models.**
Edited by D. S. Powlson, P. Smith and J. U. Smith. 1996.

Vol. 39: **Microbiology of Atmospheric Trace Gases.**
Edited by J. C. Murrell and D. P. Kelly. 1996.

Vol. 40: **Forest Ecosystems, Forest Management and the Global Carbon Cycle.**
Edited by M. J. Apps and D. T. Price. 1996.

Springer-Verlag and the Environment

We at Springer-Verlag firmly believe that an international science publisher has a special obligation to the environment, and our corporate policies consistently reflect this conviction.

We also expect our business partners – paper mills, printers, packaging manufacturers, etc. – to commit themselves to using environmentally friendly materials and production processes.

The paper in this book is made from low- or no-chlorine pulp and is acid free, in conformance with international standards for paper permanency.

FSC
www.fsc.org
MIX
Papier aus verantwortungsvollen Quellen
Paper from responsible sources
FSC® C105338

If you have any concerns about our products,
you can contact us on
ProductSafety@springernature.com

In case Publisher is established outside the EU,
the EU authorized representative is:
Springer Nature Customer Service Center GmbH
Europaplatz 3, 69115 Heidelberg, Germany

Printed by Libri Plureos GmbH
in Hamburg, Germany